Intermediate Optical Design

Intermediate Optical Design

Michael J. Kidger

SPIE PRESS
A Publication of SPIE—The International Society for Optical Engineering
Bellingham, Washington USA

Library of Congress Cataloging-in-Publication Data

Kidger, Michael J.
Intermediate optical design / by Michael J. Kidger.
p. cm. – (SPIE Press monograph ; PM134)
Includes bibliographical references and index.
ISBN 0-8194-5217-3
1. Geometrical optics. I. Title. II. Series.

QC381.K54 2004
535'.32—dc22 2003065920

Published by

SPIE—The International Society for Optical Engineering
P.O. Box 10
Bellingham, Washington 98227-0010 USA
Phone: +1 360 676 3290
Fax: +1 360 647 1445
Email: spie@spie.org
Web: http://spie.org

Printed in the United States of America.

CONTENTS

Foreword xi

Preface xv

List of Symbols xxi

Chapter 1 Optimization **1**

1.1 Special characteristics of lens design as an optimization problem 2
1.2 The nature of the merit function 2
  1.2.1 The Strehl ratio 2
  1.2.2 MTF optimization 2
  1.2.3 General comments 3
  1.2.4 Comparison with the optical thin-film design problem 3
  1.2.5 Nonlinearity of the aberrations 4
  1.2.6 Changes needed to reduce high-order aberrations 5
  1.2.7 A method of visualizing the problem of optimization in lens design 5
1.3 Theory of damped least squares (Levenberg-Marquardt) 6
1.4 Some details of damped least squares as used in lens design 8
  1.4.1 Paraxial (first-order) properties 8
  1.4.2 Seidel and Buchdahl coefficients 9
  1.4.3 Transverse ray or wavefront aberrations 9
  1.4.4 Aberration balancing and choice of weighting factors 9
  1.4.5 Damping 11
  1.4.6 Control of physical constraints 12
  1.4.7 Control of glass boundary conditions 20
  1.4.8 Solves 23
  1.4.9 Lagrange multipliers 25
1.5 Some reasons for the success of the DLS method 25
1.6 Experiments with optimization programs 26
  1.6.1 Effect of changing the damping factor 26
  1.6.2 Effect of scaling the parameter changes 27
1.7 An optimization example 29
References 34

Chapter 2 Buchdahl Aberrations 37

2.1 Third-order coefficients 37
2.2 Fifth-order coefficients 38
2.3 Comparison with H.H. Hopkins notation 39
2.4 Examples 39
2.4.1 Double Gauss 39
2.4.2 Shafer lens with zero third- and fifth-order aberrations 45
References 49

Chapter 3 Synthesis of New Lens Designs 51

3.1 Choice of a starting point 51
3.1.1 Modification of an existing design 51
3.1.2 Purchase of a competing lens 52
3.1.3 Analytic solutions 52
3.1.4 Nonanalytic synthesis of new design forms 52
3.2 Examples 53
3.2.1 A unit magnification telecentric doublet pair 53
3.2.2 A simple zoom lens 59
3.3 The use of catalog components 67
3.3.1 Singlets 68
3.3.2 Doublets and triplets 68
3.3.3 Meniscus singlets 69
3.3.4 Field flatteners 70
3.3.5 Cemented triplets 71
References 72

Chapter 4 Lenses for 35-mm Cameras 73

4.1 The triplet 74
4.2 The tessar 75
4.3 The double-Gauss (planar-type) 78
4.4 The Sonnar 82
4.5 Wide-angle lenses for rangefinder cameras (Zeiss Biogon) 84
4.6 Wide-angle lens for rangefinder camera (Schneider Super-Angulon) 86
4.7 Wide-angle lenses for SLR cameras 89
4.8 Telephoto lens 91
4.9 Long-focus telephoto lens 93
4.10 Lens for compact point-and-shoot camera 95
4.11 Single lens for disposable cameras 97
References 99

Chapter 5 Secondary Spectrum and Apochromats 101

5.1 Apochromatic doublets 101
5.2 Apochromatic triplets 104
5.3 Petzval lenses 105
5.4 Double-Gauss lenses 106
5.5 Telephoto lenses 108
5.6 Zoom lenses 108
5.7 Microscope objectives 108
5.8 Secondary spectrum correction with normal glasses 108
 5.8.1 Liquids 108
 5.8.2 Diffractive optics 109
 5.8.3 McCarthy-Wynne principle 109
 5.8.4 Schupmann principle 112
5.9 Transverse secondary spectrum 115
References 115

Chapter 6 Lenses for Laser Applications 117

6.1 Gaussian beams 117
6.2 Laser beam expanders 118
 6.2.1 Two-lens beam expanders 118
 6.2.2 Three-lens beam expanders 122
6.3 F-Theta lenses 124
6.4 Lenses for optical disks 126
6.5 Laser diode collimators 129
References 130

Chapter 7 Microscope Objectives 133

7.1 Classical microscope objectives 133
7.2 Flat-field microscope objectives 135
7.3 Oil-immersion objectives 141
References 144

Chapter 8 Microlithographic Projection Optics 145

8.1 Unit-magnification zero-power monocentric systems 145
 8.1.1 Dyson 1× relay 146
 8.1.2 Offner 1× relay 148
8.2 Wynne-Dyson 1× relay 149
8.3 Wynne-Offner 1× relay 152
8.4 Reduction lenses 153
8.5 Catadioptric reduction systems 163
8.6 Catoptric reduction systems 167
References 170

Chapter 9 Zoom Lenses 173

9.1 General principles 173
9.1.1 Control of chromatic aberration 173
9.1.2 Field curvature 173
9.1.3 Minimization of movements 174
9.2 Two-component zooms 174
9.2.1 Minus-plus plastic disposable zoom 174
9.2.2 Plus-minus plastic disposable zoom 175
9.2.3 A typical minus-plus zoom 177
9.2.4 A typical plus-minus zoom 178
9.3 Three-component zooms 179
9.4 Four-component zooms 181
9.5 Zoom relays 188
9.6 Zoom telescopes 189
9.7 Zoom modules 190
References 190

Chapter 10 Decentered and Asymmetric Systems 193

10.1 General properties of decentered systems 193
10.2 Coordinate systems 194
10.3 Interpretation of results 196
10.4 New-axis surface 197
10.5 Toroids 197
10.6 Offset surfaces (or off-axis surfaces) 198
10.7 Convention for mirrors 198
10.8 Kutter system 199
10.9 Single parabolic mirror 202
10.9.1 Alpha rotations 204
10.9.2 Beta rotations 205
10.9.3 Alpha and beta rotations 206
10.10 Scanning systems 207
10.11 The "active" side of a surface 209
10.12 X-ray telescopes 210
10.12.1 WOLTER2 example 210
10.12.2 WOLTER1 example 211

Chapter 11 Design for Manufacturability 215

11.1 Tolerancing 215
11.2 Simplicity of design 216
11.3 Air spaces 217
11.4 Glass components 218
11.5 Glass choice 220
11.6 Mirror surfaces 220

11.7 Redesign for actual "melt" data 220
11.8 Use of existing tools and test plates 221
11.9 Selective assembly and adjustment after assembly 221
11.10 General points 221
References 222

Index 223

FOREWORD

In following the work for *Fundamental Optical Design*, I think it is helpful for the reader to understand the background of how the two books were made possible, and why we ask all readers of this volume to read the earlier volume. It is rather simple: whatever we read, fact or fiction, we have a better understanding if we start from the beginning.

At the time of the initial publishing discussions, we all assumed that Michael's book would fit the style and category of SPIE's Tutorial Text Series, which is based on their short courses. The idea was to capture Michael's very successful style of teaching and be reflective of his association with many of the internationally famous optical designers at Imperial College London, where he trained and taught and with whom he was a consultant, an innovator in optical design teaching, and of the optical design program later known as SIGMA. This group included Prof. H.H. Hopkins, Prof. Charles Wynne, Prof. W.D. Wright, and Prof. Walter Welford.

Michael Kidger died four days after he signed the original publishing contract with SPIE Press. Following his death, hundreds of colleagues and former students wrote to me from all over the world expressing their admiration for Michael's warmth, caring expertise, and style of teaching. It was clear to me from this unbounded and overwhelming expression that I should find a way to maintain, not only with practicing optical designers, but with future students of optical design, a continuity of this relationship between student and teacher. Posthumous publication of his book was the obvious choice.

To this end, I reviewed each and every set of course notes that Michael had given, sifting through the material to avoid duplication. Eventually, I had collected and sorted all the relevant source material. However, this material could not be Michael's book per the original agreement.

This compiled source material included special courses aimed at specific optical design requirements for whichever group he was teaching. For example, the material included courses given at the Institute Galileo, Florence. It also contained a hands-on course in practical lens design given at the Winter School in Optics, held at The International Centre for Theoretical Physics in Trieste, Italy. Michael gave courses in Australia, Singapore, and Europe, including many in Germany and the UK. Also, a course was given at Imperial College London for 30 attendees from Samsung, Korea. The optical designers from Samsung came with their specific problems, and these were worked through individually with Michael's assistance.

The material also included information from SPIE courses, for which Michael's course attendees always awarded him top scores in clarity, understanding, and presentation. This is the material we are talking about, part of which was used in *Fundamental Optical Design*, and the remainder in this volume. Hundreds, probably thousands, of students attended Michael's courses and have benefited from, and indeed still benefit from, this material.

It had always been a discussion within Kidger Optics (the company which Michael and I owned) about whether we called the "Intermediate Course" an "Advanced Course," but we decided to stay with "Intermediate." Some of those attending the Intermediate Course were extremely experienced optical designers, but they came to have the help of Michael's expertise. Note, however, that they would have attended the first course, as indeed we expect readers of the present volume to have previously read *Fundamental Optical Design*.

I worked with Rick Hermann, who was then with SPIE Press, and over a period of time it was decided that there was enough information for two books. This is how the two volumes were born. However, the challenge was to put this information into a form that enabled readers to feel that they were in a comfortable learning situation.

In the initial stages of discussion concerned with publishing the material, it was agreed that Prof. Donald O'Shea (Georgia Institute of Technology) would undertake the initial work on the first volume. Don gave generously and unselfishly of his time, for which we remain tremendously grateful. Following Don's work, David Williamson added further valuable content and made modifications to reflect Michael's own writing style. David has been irreplaceable in knowing Michael's teaching style and audience, especially having been a student of Michael's during his own days at Imperial College. In completing *Intermediate Optical Design*, David has made every attempt to absolutely preserve Michael's original intent and material, whilst also including some of his own experience and expertise. He also captures the very appropriate British usage of the English language, which best reflects Michael.

Optical designers will know that the style of teaching, and even some methods of optical design expression, were different at Imperial College in the UK, compared with, say, The University of Rochester or The Optical Sciences Center, University of Arizona, in the U.S. None are incorrect; they are simply different ways of reaching the same result.

Many famous optical designers, also students, wrote to me after the publication of *Fundamental Optical Design*. Here are some examples:

"It is a big world with billions of people, but Michael showed by example that a single man can make a broad and long lasting impression."

"Michael was an inspired and inspiring teacher...."

"I have a copy of Mike's book and I've been using it to put together a basic training course of optical design. The book has been invaluable."

"Michael always had a knack for teaching and simplifying to the essentials. I'm glad all this is being preserved."

These and many others stated that they were looking forward to reading the second volume.

As I have described, in looking for a way to make sure that all of his teachings and profound gift in optical design could reach as many old and new students of optical design as possible, it was decided that publication of Michael's lecture notes in book form was an excellent way to accomplish this goal. This would carry on some of Michael's own thoughts of documenting his teaching experiences. His style was pure empathy with, and concern for, the student.

David Williamson has kept true to Michael's work. In doing so, he has preserved the original genius of this work, and has shown his own genius in allowing you and future optical designers to have the benefit of the clarity, style and perfection of understanding of optical design as given by Michael J. Kidger.

Tina Emily Kidger
December 2003
Crowborough, East Sussex
United Kingdom

PREFACE

The title of this book originates from Michael Kidger's short courses for SPIE titled "Intermediate Optical Design." It is a compilation of material from these courses and a number of similar ones given for Kidger Optics in Australia, Germany, Italy, Sweden, Singapore, and the UK during the 1990s. It forms the second of two volumes, the first being *Fundamental Optical Design*, published by SPIE Press in 2002.

These intermediate-level courses were aimed at students and practicing optical designers who already had a thorough knowledge of geometrical optics and third-order (Seidel) aberration theory. The courses continued to use the same "Imperial College" nomenclature and the Sigma optical design program as their basis. They were workshops, or master classes, rather than academic discourses or promotional material for Sigma. In the interests of authenticity and continuity with the first volume, the same Sigma output format is used here for tables of lens prescriptions, aberration data, and graphics. This second volume does not review the material in the first volume; the assumption is that the reader is already familiar with it, preferably having already worked through the examples in that volume, and is able to refer back to it when needed. The connections between the two volumes are not often explicit, but rather left for the perceptive reader to discover as he or she works through these design examples—I think this is in keeping with the intent of the courses.

While the first volume carefully avoided the subject of optimization, this volume starts with a general but wide-ranging discussion of it. As any optical designer quickly discovers, the key to this art and science (as Professor Shannon has referred to our field) is optimization. In fact, a desire to learn more about optimization probably brought many people to these courses, even some who had been designing lenses for many years. More specifically, the courses taught "local" optimization, where the choice of starting design is crucial to the success of the optimized design. Michael was also interested in the more general problem of "global" optimization, which he somewhat wryly described as a user inputting a series of "flat plates" and the program finding the best possible arrangement of lens elements—ideally, the program would even decide the optimum number of plates, mirrors, diffractive elements, and so on.

The first chapter in this volume has been compiled from Michael's writings on local optimization, from several published papers as well as his course notes. This material is inevitably very closely related to the Imperial College and Sigma software that he developed and used. Unfortunately, readers of this book will not have access to Sigma, or Michael's unique ability to answer questions as they arise. They will instead be using another commercial optimization program, perhaps without supervision. Recognizing this inevitable limitation, I have attempted to rise above the trees of the Sigma program and give my interpretation of Michael's overview of the forest that is any optical design optimization program, at least one that is based on the damped least squares (DLS) method.

Included here are some design examples to illustrate certain points, such as the effects of lens diameter and thickness constraints on aberration performance. In the courses, these examples were in a separate section under the heading of "unusual or novel optical designs." One of these is the "monochromatic quartet," which I was fortunate enough to find with Sigma for the International Lens Design Conference problem in 1990. Several other designers also found the same form with other local optimization programs, and the design has subsequently been rediscovered by global optimization—from flat plates—but I have been gratified to hear that such programs have not yet found a better design. Although totally impractical (which was the intention), it is a good illustration of the importance of size constraints, which is one of the key factors in the control of higher-order aberrations.

The chapter concludes with the Kingslake double-Gauss example that became a tradition of Michael's courses. This was the start of the practical work with Sigma, but in this book readers are left to apply it to whichever optical design program they are using. Naturally, the detailed lens and ray setup will be quite different from one program to another. However, the example is retained with Sigma output because this is the way it was in the courses, and it also serves to illustrate the importance of damping factor and step size on the ultimate performance of the optimized design. It is unfortunate that in most other commercial programs these numbers remain invisible to the designer—Sigma was a model of transparency.

Perhaps the most important message from the first chapter, and indeed the whole book, is the role of higher-order aberration correction in most optical designs. Accordingly, the second chapter is a relatively brief discussion of Buchdahl fifth-order aberrations, with references provided for those mathematically inclined readers who want all the gory details. Over the years, after "graduating" as a formal student of Michael's, I had several discussions with him about Buchdahl aberrations while I went through a phase in my career when I was enthusiastic about them. I think he remained rather skeptical, even though Sigma became one of the few commercial lens design codes that included Buchdahl's monochromatic aberration coefficients in the main program (other programs have them as reluctant extensions). This is not only because they cannot simply be attributed to individual optical surfaces, but also because they are not usually sufficiently accurate, since they are based on paraxial rays. Michael made the point that it is usually quicker—and always more accurate—to trace a limited number of finite rays, and this is indeed the more common approach in this age of very fast personal computers.

It remains useful to think about fifth- and higher-order aberrations such as oblique spherical aberration, coma, astigmatism, and field curvature, because these often dominate the lens design problem. Usually I look for them in the transverse ray aberration plots, of which there are many examples in these two books. Someone (I forget who) has defined an optical designer as one who understands transverse ray aberration curves, and so this is considered a prerequisite for the reader of this intermediate-level text. Later in my career, thanks to Abe Offner and Juan Rayces, I became a convert to Zernike coefficients, which more accurately describe and optimize higher-order aberrations. While the last version of Sigma did contain these coefficients, the original course material contains little discussion of them, so I have decided not to add that here.

The remaining chapters contain a wide range of design examples that are dominated by higher-order monochromatic and chromatic aberrations. The choice of examples partly reflects Michael's background and interest in certain design types, but it also evolved from student requests. It does not attempt to cover all design types, but rather it uses these examples to illustrate specific or generally applicable design approaches.

Many readers may find in the third chapter the most useful insights into an optical designer's thought processes, such as they are. The synthesis of starting points for optimization is the aspect of optical design that is the least written about but most important—at least until the truly global optimization program arrives. While there have been numerous attempts to apply databases and artificial intelligence, here the approach to the challenge of the blank computer screen is to use human memory, common sense, experience, and knowledge. It involves using the (local) optimization program as a field in which to play with new ideas, discarding those that show less promise, and then refining the more promising ones to include greater accuracy and all of the required practical aspects prior to finalizing a design for manufacture. To the outsider this may seem like trial and error, but readers of this book are likely to appreciate that there is more to it than that!

Michael had a particular interest in the history and practice of photography, so the fourth chapter covers many of the classical photographic lenses. Although these will already be well known to most readers, this is a brief review, with some valuable insights and historical observations.

The fifth chapter attempts to cover a wide range of approaches to the problem of secondary spectrum correction, which is a higher-order chromatic aberration that the practicing designer will often encounter, and which can be the most stubborn imaging defect to reduce or eliminate. The most common approach has been to use anomalous glass types, and this is briefly reviewed. I have also added references to many of the published systematic methods of glass selection. Some less well known approaches using normal glasses are also described in more detail, and I have added a brief reference to the use of diffractive optics, since they are starting to be used in photographic lenses. I have also added the Schupmann medial telescope, from a prescription kindly provided by Richard Bingham. This is not often thought of as a means of secondary spectrum correction, since it avoids the problem rather than corrects it. An implicit lesson of these two volumes is that avoidance is usually better than correction! Many other catadioptric designs also use the

Schupmann principle, including some described in the last chapter of *Fundamental Optical Design*.

Chapter 6 illustrates the wide variety of situations that an optical designer may encounter, where in this case chromatic aberrations are unimportant, but other considerations specific to lasers need to be taken into account. Even for the simplest designs the opportunity is taken to remind us of some basic aberration theory, such as the sine condition, which was covered in more detail in the first volume.

The material in Chapters 7 and 8 were in the courses combined as a section on high-numerical-aperture designs. I have separated them into two chapters here, on the basis that microscope objectives operate over a relatively small field size and broad spectral bandwidth, whereas microlithographic objectives cover a relatively large field size and narrow spectral range. No doubt this also reflects my own interest in microlithography, from where I have added a few more designs—some my own—with the justification that the courses did include one of my earlier designs, and I wanted to include the substantial progress that has taken place since then.

Microlithographic objectives are the best illustration of the remarkably high and uniform performance across the image format that can be obtained by the combination of large size and complexity to "relax" the optical design, a term first used by Glatzel in 1980. Essentially, this strategy simply minimizes ray incidence angles on optical surfaces. It is one of the more remarkable stories in the history of optics that such large and complex designs are now routinely produced with aberrations measured in a small number of milliwaves rms in the ultraviolet spectrum.

Another class of systems that one might more reasonably expect to find in a book on advanced optical design is zoom lenses. As a designer with little experience in this field, and one who skips over the numerous complicated papers on such lenses, I found Michael's course notes on the subject remarkably easy to understand—this is the genius of Michael's teaching style. I have not changed the material significantly from the original, other than making the tables and graphics compatible with the rest of the book. After working with this chapter, I felt that I had sufficient understanding of zoom lenses to at least have the confidence to begin to design one.

Chapter 10 discusses some of the more basic issues involved in the design of tilted or decentered optical systems. Originally, I was not going to include this chapter, since it is specific to the particular way that Sigma treated such systems. But, there are some interesting illustrations here of apparently simple decentered systems with unusual ambiguities and challenges, which I felt made the chapter of general interest to users of all optical design programs.

The final chapter is a brief discussion of some of the practical issues involved in designing a lens that has to be manufactured within specified price and performance requirements. The original courses also included a more detailed discussion of tolerancing, using Sigma's methodology. However, I decided not to include that here, since it is rather too specific to Sigma. The user of another optical design program will generally find appropriate documentation provided with it, or may have to develop tolerancing methods for specific applications, but I considered such a discussion to be beyond the scope of this book.

I have not, for similar reasons, reproduced here other features of Sigma that were in the original course notes or program documentation. These include nonsequential raytracing, illumination system design, Gaussian beam propagation, fiber coupling, and so on.

I have also not provided a list of recommended reading, beyond the references given in each chapter, which I have expanded considerably from those given in the courses. However, mention should be made of Don O'Shea's extensive compilation of optical design papers on CD-ROM (published by SPIE). This includes, among others, all of Michael Kidger's published papers from the *Proceedings of SPIE*.

Many of the design examples in these two volumes were included in Sigma's optical design database. These have been converted and now form a part of the Zebase© lens database. There may be subtle differences in some of these lens prescriptions, such as glass refractive index data, apertures, and fields, which the reader is encouraged to explore; perhaps the prescriptions can be improved, or starting points for new designs can be formed. This is one reason why the paraxial ray data and Seidel aberration tables have been included for many of the designs: they can be used to check the accuracy of translation of the lens prescription into other optical design programs. Study of these tables is also a good way to understand how a design is working; it is one of the more important skills of the optical designer. Of particular interest are the marginal and chief ray incidence angles (A and ABAR), showing how "relaxed" the design is, and the magnitude of surface aberrations and how they are cancelled within the system. This will affect the magnitude of residual higher-order aberrations and sensitivity to manufacturing errors, and is the implicit theme of these two volumes—the simple secret, if there is one, of optical design.

I am grateful to a number of colleagues, including Brian Blandford, Tom Matsuyama, and Juan Rayces, who have read the manuscript and offered comments and constructive criticism. I am also very grateful for Tina Kidger's constant encouragement to "get it out there!" My hope is that this book retains as much as possible of Michael Kidger's original brevity and clarity, and that it will be a useful and practical resource for students of classical optical design for many years to come.

David M. Williamson

List of Symbols

A	ni (refraction invariant)
B	thin-lens conjugate variable (or magnification variable)
c	curvature of a surface (= $1/R$)
C	thin-lens shape variable (or bending variable)
C_1	Seidel coefficient of longitudinal chromatic aberration or "axial color"
C_2	Seidel coefficient of transverse chromatic aberration or "lateral color"
d	axial distance between two surfaces
D	distance between two surfaces, measured along an exact ray
e	eccentricity of a conicoid
ε	$(1 - e^2)$ for a conicoid
E	defined by $E \cdot H = \bar{h}/h$ ($\bar{h}$ is the paraxial chief ray height)
f, f'	focal length
h	paraxial ray height
H	Lagrange invariant (= $n\,u\,\eta$)
i	paraxial angle of incidence
I	exact angle of incidence
k	conic constant of a conicoid = $- e^2 = \varepsilon - 1$
K	power of a surface or system
l	Object distance, measured from surface (or lens) to the object
L, M, N	direction cosines of exact ray
n	refractive index
q	ratio of n / n'
R	radius of a surface (= $1 / c$)
S_1	Seidel spherical aberration coefficient
S_2	Seidel coma coefficient
S_3	Seidel astigmatism coefficient
S_4	Seidel field curvature coefficient (Petzval sum = S_4/H^2)
S_5	Seidel distortion coefficient
u	paraxial ray angle
V	Abbe V-value = $(n_d - 1) / (n_F - n_C)$
W	wavefront aberration
x, y, z	ray coordinates at a surface
x, y, z	ray coordinates at the vertex plane of a surface
η, ξ	coordinates in the object space

Note that a primed quantity refers to the image space.

Any quantity associated with a chief ray is denoted by a bar, e.g., $\bar{h}, \bar{u}$

CHAPTER 1
OPTIMIZATION

For the first 40 years of the twentieth century, optical design was done using a mixture of Seidel theory, a little ray tracing, and a great deal of experimental work. All of the computations were done using log tables. The usual method was to design a prototype lens using Seidel theory, then to manufacture this prototype and measure its aberrations. The designer would then modify the target values for the Seidel aberrations and produce a new design. Ray tracing was relatively little used; in most companies skew rays were almost never traced until about 1940. This seems strange to us nowadays, but it must have been sensible at the time; ray tracing, after all, is only a simulation technique, and if it is cheaper and quicker to do the real experiment, there is little point in carrying out a simulation.

Mechanical calculators were little used until about 1940. Again, this is now difficult to comprehend, but the reason appears to be that techniques for the use of log tables were developed to the point that an experienced "computer," usually a young lady, could actually trace rays faster than with calculators.

Computers were first used in the UK for optical design in 1949, when C.G. Wynne had some ray tracing done at Manchester University. Unfortunately, this pioneering work took so long that the lens was made and delivered to the customer before the ray tracing results were ready! The use of computers for ray tracing did not really become practical until about 1957, when Taylor Hobson acquired an Elliott computer for this purpose, but for several years it was the only British company with a computer dedicated to optical design.

However, rapid ray tracing, in itself, was not in most cases sufficient to enable the design of more advanced lenses, except where lens manufacturers were building very complex systems. For example, the author was fortunate to be employed in the early 1960s by Rank Taylor Hobson, where G.H. Cook and P.A. Merigold were designing complex zoom lenses.[1] For these systems, the use of a computer for ray tracing did enable the design of lenses that would have been impractical otherwise, but this was an exceptional situation.

For most lens designers, the major impact of computers came when optimization techniques began to be used in lens design. While early workers in this field suggested various methods, the dominant method has become that of damped least squares (DLS), proposed by Wynne[2] in the UK and Girard[3] in France, and now known to computer scientists as the Levenberg-Marquardt method.[4,5] Many optical designers have developed variations on this basic technique.[6–12] Some designers have also used other methods with success, such as the adaptive method of Glatzel[13] and Rayces.[14]

1.1 Special characteristics of lens design as an optimization problem

In lens design, we seek to optimize the parameters that determine performance while at the same time producing a design that is manufacturable and, preferably, can be made at a reasonable cost. The optical performance can be most conveniently described, for optimization purposes, in terms of the sum of the squares of some appropriate aberrations, and optimizing these aberrations is indeed the main problem.

1.2 The nature of the merit function

From the beginning, it has generally been implicitly assumed that in lens design a useful merit function can be defined by the sum of squares of the aberrations of a lens. This assumption follows from recognizing that the effect of a positive aberration is the same as that of a negative aberration, at least in the sense that both are equally harmful to image quality. However, there are good theoretical reasons for using the sum of the squares of the aberrations as a merit function, as we will now show.

1.2.1 The Strehl ratio

One criterion of image quality is the Strehl ratio, defined as the ratio of the intensity at the maximum of the actual point spread function (the image of a point object, such as a star or pinhole) to the maximum of the aberration-free point spread function. It is well known that for small aberrations, the Strehl ratio depends on the variance of the wavefront aberration.[15–17] Although this is not the same as the sum of the squares of the wavefront aberrations, as they are normally defined, it is still essentially dependent on the squares of the aberrations and not on the absolute values of the aberrations.

1.2.2 MTF optimization

A more realistic criterion for many lenses that image extended objects with larger residual aberrations is the modulation transfer function (MTF). It has been shown that the MTF depends on the squares of a set of terms that are closely related to the aberrations. To be specific, Gostick[18] and Kidger[19] showed that the following could approximate the geometrical optics approximation to the MTF:

$$L_r(f) \propto \sum [1 - 2\sin^2(\pi \cdot f \cdot \delta\xi')], \tag{1.1}$$

where $L_r(f)$ is the real part of the sagittal geometric MTF at a spatial frequency of f and $\delta\xi'$ is the x-component of transverse aberration. Naturally, a similar expression applies for the tangential component of the MTF.

It follows that an optimization program that minimizes $\sum \sin^2(\pi \cdot f \cdot \delta\xi')$ will maximize the MTF, according to the geometrical optics approximation. Naturally a similar but more complex expression can be derived for the accurate diffraction-based MTF value, but this will not be followed up here, as we are only concerned with showing that the sum of the squares of a set of terms is a logical form—and perhaps the only logical form—for a merit function for lens design optimization problems.

1.2.3 General comments

There are, of course, exceptions to the rule that the sign of an aberration does not have to be considered in a lens design optimization program, but they are few. One class of systems that does introduce some exceptions is almost any lens that is designed for visual use. In most visual lenses, such as eyepieces, it is effectively impossible to correct field curvature. However, in visual lenses the importance of the defocus introduced by field curvature is dependent on the sign of the aberration, because the normal observer is capable of accommodating to an image that is formed in front of, but not behind, the eye. Most lens design programs deal with this by setting a nonzero target for the appropriate aberration; the merit function is still considered to be the sum of squares of these modified "aberrations." It should be pointed out that, apart from lens design, not all optimization problems have this characteristic that the merit function is naturally formed by the sum of squares of a set of terms.

1.2.4 Comparison with the optical thin-film design problem

To take one example, but still in an optical context, optimization programs are used in the design of thin-film filters (e.g., antireflection coatings) where a given reflectance or transmittance is required. Of course this reflectance or transmittance is dependent on wavelength, angle of incidence, and polarization, but this is not relevant to the present discussion.

In the case of an antireflection coating, reflectances should be zero, so in that sense they can be likened to aberrations in lens design; but it is not logical to define a thin-film merit function as the sum of the squares of the reflectances. It is much more logical to define the merit function as the sum of the reflectances to the first power, because the mean reflectance, integrated over the whole of the relevant waveband, is proportional to the weighted sum of the reflectances and not at all on the sum of the squares.

It can be argued that the intensity is itself the square of the amplitude, and indeed a thin-film optimization program could use a merit function consisting of the sum of the squares of the real and imaginary components of the amplitude; but a general-purpose optical thin-film optimization program is often required to optimize for nonzero reflectance, as in the case of beamsplitters, where a reflectance of 50% may be desired. In this case, it is not appropriate to define a merit function in terms of the squares of its components. However, we will not consider any further the optical thin-film design problem; we merely wish to emphasize that many optimization problems, apart from lens design, do not lend themselves automatically to a merit function that is built up as a sum of squares.

1.2.5 Nonlinearity of the aberrations

If the aberrations were linear functions of the design parameters (curvatures, separations, refractive indices, etc.), lens design would be relatively simple. It would be possible to easily construct a set of linear equations that could be solved to find a set of parameters that minimize the merit function. Apart from relatively minor problems such as the control of boundary conditions, and the detailed definition of the merit function, an optimization program would locate the minimum of the merit function very rapidly.

However, apart from a few special and nontypical cases, in lens design the aberrations are nonlinear functions of the design parameters. This has a great impact on the methods needed in lens design optimization, and on the effectiveness of lens design optimization programs.

Consider one very simple example. The Seidel spherical aberration of a single surface, S_1, is given by

$$S_1 = -A^2 h\left(\frac{u'}{n'} - \frac{u}{n}\right), \tag{1.2}$$

where $A = ni$,

- i is the angle of incidence of the paraxial marginal ray,
- h is the height of the paraxial marginal ray,
- u is the angle of the paraxial marginal ray,
- n is the refractive index,

and primes indicate quantities after refraction or reflection at a surface.

Since i is a linear function of the curvature, c, and $(u'/n' - u/n)$ is also a linear function of c, clearly the spherical aberration is a cubic function of c. The expressions for several other Seidel aberrations also demonstrate that they must be cubic functions of c. As we shall see in Chapter 2, the relationship between fifth-order spherical aberration and curvature is much more complex, but it certainly depends on at least the third power of c. In addition, all high-order aberrations are affected by induced aberrations; in other words, the aberration introduced by a surface is affected by the aberrations of previous surfaces.

It should be clear from the above discussion that in general there are no simple relationships between the aberrations of a lens and its design parameters. Admittedly, a few aberrations are simple. The Petzval sum of a surface is exactly linearly proportional to the curvature, as are first-order axial color and lateral color, but the effectiveness of an optimization procedure is determined by the most difficult cases, not by the most simple ones, and the most difficult aberrations to optimize are the highly nonlinear aberrations.

1.2.6 Changes needed to reduce high-order aberrations

Another factor also determines the strategy of lens design, even before we start to think about using computers and optimization programs. It has been well understood for many decades, as a generalization, that high-order aberrations change relatively slowly, while third-order terms change quickly with changes in the lens parameters. In the precomputer period of lens design, this fact was used quite consciously. As already mentioned, it was a common procedure for designers to set up a prototype lens design with given Seidel aberrations, and then either to measure the aberrations of an actual prototype lens, or to compute the aberrations by ray tracing. In both cases, it was possible to determine a new set of Seidel aberrations, which would improve the lens performance, on the assumption that the high-order aberrations would not change much.

All of these factors, when applied to the lens design optimization problem, mean that in interesting cases we expect that an optimization program should be able to reduce the high-order aberrations. However, to do this, large parameter changes are often required, and the nonlinearity of the Seidel aberrations therefore implies that Seidel aberrations will usually change rapidly, and nonlinearly, while the high-order aberrations are gradually being reduced.

1.2.7 A method of visualizing the problem of optimization in lens design

It is convenient to think of the simplest case, a hypothetical two-dimensional problem. Given that large parameter changes are normally required to reduce high-order aberrations, we can predict that the distance from the starting point to the minimum is, in general, large, and therefore contours of the merit function will be elongated ellipses. This method of visualization is in good agreement with the general experience that in a complex system it is easy to correct the Seidel aberrations, but that small changes from the optimum design parameters will degrade the performance significantly.

If we visualize the contours of the merit function as representing a valley, the bottom of the valley represents the region where the Seidel aberrations are corrected. As we travel downhill along the valley floor, the high-order aberrations are slowly reducing, and of course the valley is not straight, because the Seidel aberrations are nonlinear functions of the design parameters.

The basic problem in quickly finding the minimum is therefore to find the lowest point at the bottom of this curved valley; of course, in important cases the number of variables is quite large, typically between about 20 and 50, and the two-

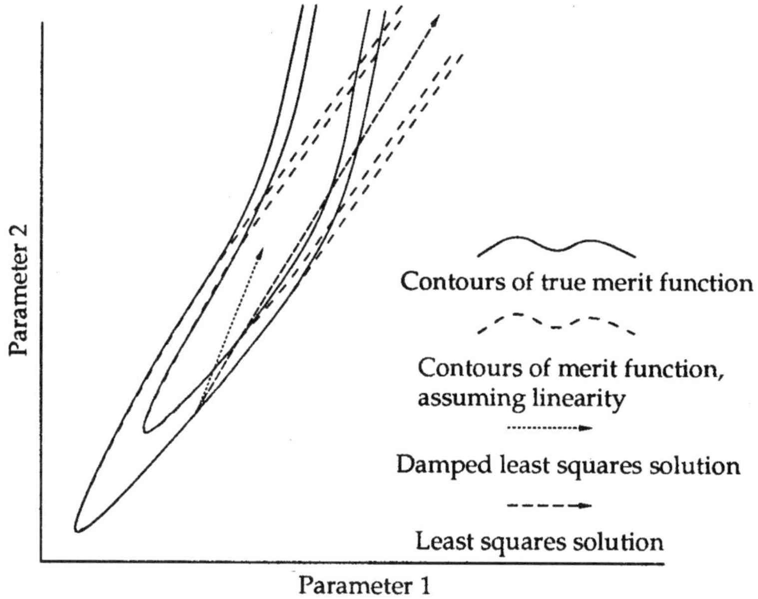

Figure 1.1 An illustration of the merit function.

dimensional picture in Fig. 1.1 is highly simplified, but it nevertheless gives a useful model for visualizing the optimization process.

1.3 Theory of damped least squares (Levenberg-Marquardt)

Suppose that the aberrations of an optical system can be described by a vector **g**. The components of this vector, g_i, are the aberrations of the lens, and typically there might be about 100 of these aberrations, although in extreme cases there can be many hundreds of aberrations. In practice these aberrations are often multiplied by a weighting factor, and they are also sometimes related to a nonzero target value, but this does not affect the following discussion.

The derivatives of the aberrations with respect to the variable design parameters are written as $a_{ij} = \partial g_i / \partial x_j$, and together they are written as a matrix **A**. The individual parameter changes (i.e., changes to curves and separations, typically) are written as x_j, and the set of parameter changes is written as a vector **x**. Typically, the number of variable parameters is a few tens.

We define a merit function as

$$\Psi = \mathbf{g}^2. \tag{1.3}$$

After a set of parameter changes has been made, the aberrations can be described by a new vector, $\mathbf{g}'$, where

$$\mathbf{g}' = \mathbf{g} + \mathbf{A} \cdot \mathbf{x} + \ldots \tag{1.4}$$

assuming that the parameter changes are linear. It may be shown[5] that the new merit function is minimized if the parameter changes **x** are the solution of

$$\mathbf{A}^t\mathbf{A} \cdot \mathbf{x} = -\mathbf{A}^t\mathbf{g}. \tag{1.5}$$

However, as explained above, the parameter changes **x** given by Eq. (1.5) are normally very large, as the high-order aberrations change slowly. In addition, the nonlinearities discussed above limit the range of validity of Eq. (1.4), and therefore we normally find that a lens with parameter changes given by Eq. (1.5) will not have a smaller merit function. For this reason, Levenberg, Wynne, and Girard suggested that the equations should be "damped," as follows:

Instead of minimizing Ψ, we minimize

$$\Psi' = (\mathbf{g} + \mathbf{A} \cdot \mathbf{x})^2 + p^2\mathbf{x}^2, \tag{1.6}$$

where we have added a term p^2 to the merit function. p is known as the *damping factor*. It is squared in Eq. (1.6) merely to illustrate that it must be positive. Again it may be shown[5] that the parameter changes that minimize this new merit function are given by

$$(\mathbf{A}^t\mathbf{A} + p^2\mathbf{I}) \cdot \mathbf{x} = -\mathbf{A}^t\mathbf{g}. \tag{1.7}$$

The effect of this is that the size of the parameter changes is reduced, compared with the simple least squares method, and by a suitable choice of p the parameter changes can be chosen so that they are within the region of validity of Eq. (1.4). In practice, however, there is no requirement to do exactly this; it is more useful simply to choose the value of the damping factor that minimizes the merit function, and this implies that some departure from Eq. (1.4) is accepted. This is the basis of the Levenberg-Marquardt optimization method, known to lens designers as damped least squares (DLS).

It is important to point out that one of the properties of DLS is that the changes computed by this method are the smallest that can give a specific reduction in the merit function. This assumes that we can define the size of the parameter changes, but this point is not important in practice, except that it leads us to one variation of the basic method, the use of multiplicative damping, which is described below.

One variant of the basic DLS method, proposed by Spencer[6, 7], is to add the condition that some equations must be exactly solved, while at the same time the merit function is to be minimized. If we add the condition that the following equation must be exactly satisfied:

$$\mathbf{B} \cdot \mathbf{x} = \mathbf{e}, \tag{1.8}$$

where **B** is a matrix representing a set of derivatives, b_{ij}, and **e** is a vector representing a a set of quantities, e_j, the equations to be solved are the following:

$$
\begin{aligned}
(\mathbf{A}^{t}\mathbf{A} + p^2\mathbf{I}) + \mathbf{B}^{t} \cdot \lambda &= -\mathbf{A}^{t} \cdot \mathbf{g}, \\
\mathbf{B} \cdot \mathbf{x} &= \mathbf{e},
\end{aligned}
\tag{1.9}
$$

where λ is a vector representing a scalar set of multipliers. The solution of Eq. (1.9) consists of a vector combining λ and **x**, but only **x** is actually required. Several lens optimization programs have used equations such as this for the "exact" solution of boundary conditions, but this topic is discussed in more detail in section 1.4.9 below.

1.4 Some details of damped least squares as used in lens design

The merit function may comprise many different types of aberration, depending on the designer's preferences, experience, and the application for which the lens is intended.

1.4.1 Paraxial (first-order) properties

The most important paraxial properties are focal length, image position, and magnification. These are controlled with two "aberrations" and their targets and weights. The first aberration is the focal length or the magnification. In the case of lenses with an object at infinity, the first "aberration" is taken to be the focal length. For finite-conjugate lenses, the first aberration is the magnification. In the case of afocal systems, the first aberration is the angular magnification. If we wish to control the back-focus, we must use a second aberration, which is the back focus error.

Other first-order properties that may be controlled include

- Object-to-image distance, in the finite conjugate case. Often this distance can be quite large (several times the focal length), so a very small weight might be appropriate.

- Total thickness of "glass" elements in the system, that is, for spaces in which the refractive index is not 1.0. The effect of controlling this quantity is that the weight of the lens and the cost of the glass are more or less controlled.

- Distance from the first surface of the lens to the image surface.

- Distance from the first surface of the lens to the last surface of the lens.

- Defocus. Normally we assume that the image surface is at the paraxial image position, but we can specify a defocus from the paraxial position. If the variable label for the image surface is set, the program will use the defocus as a variable parameter in optimization. However, we do not want an excessive amount of defocusing, so a weight on defocus may be used to prevent this.

1.4.2 Seidel and Buchdahl coefficients

As we shall see in Chapter 2, the aberration function can be expanded as a power series in aperture and field, and analytic expressions exist for the calculation of the coefficients. The third-order (Seidel) aberrations can be simply and quickly computed. Computation of the fifth-order aberration coefficients is also possible. Expressions for seventh-order aberrations also exist, but are rarely used.

Even the seventh-order expressions (which are taken to include the third- and fifth-order terms) are not sufficient to infallibly describe the aberrations of a typical photographic lens, for example, because their calculation is based on paraxial rays. It is therefore almost universal for optimization programs to use real ray-tracing results.

1.4.3 Transverse ray or wavefront aberrations

Exact ray tracing is, in principle, a simple and infallible process for calculating the aberrations of particular rays as they pass through a lens. In practice, great care is needed to ensure that the results are correct and accurate in all cases. A general-purpose lens design program must be capable of ray tracing through many different types of surface, but these details are of no concern to the present discussion.

Since no approximations are involved, ray tracing is almost universally used in optimization programs. The results of ray tracing are most often expressed as either transverse or wavefront aberrations. Transverse ray aberrations are the first derivatives of the wavefront aberrations, so that if the transverse aberrations are fully known, the wavefront aberrations can be computed, and vice versa.[20]

1.4.4 Aberration balancing and choice of weighting factors

During optimization, the program does not differentiate between primary and higher-order aberrations, but they are all included in the ray-trace aberrations. To a certain extent, the fact that a least-squares process is used will automatically produce some sort of aberration balance, but a sensible choice of weighting factors is essential if we are to achieve anything like the best possible performance. The program therefore needs to select default weighting factors, according to the following criteria:

1. In some cases it is better to have a sharp center to the spread function, with a large flare, than a smaller flare with a less sharp center. To attempt to achieve this, we normally give low-aperture rays a larger weighting factor than marginal rays.

2. In some cases, it is permissible for the performance at the edge of the field to be worse than at smaller field angles. Very often, of course, it is inevitable, but sometimes (as in photocopiers and microlithography) it is not at all desirable. It is therefore sensible, often, but not always, for weighting factors to be smaller for larger field positions.

3. When using the Conrady chromatic aberration formula (i.e., nearly always), one must remember that this aberration is a wavefront aberration, and is normally smaller than the transverse ray aberration. To see why, remember that

$$\delta\eta' = \frac{-1}{n'\sin\alpha'} \cdot \frac{\partial W}{\partial y} \tag{1.10}$$

and

$$\delta\xi' = \frac{-1}{n'\sin\alpha'} \cdot \frac{\partial W}{\partial x}. \tag{1.11}$$

Assuming that $W = W_{40}y^4$, $\partial W / \partial y = 4W_{40}y^3$, so at the edge of the aperture, when $y = 1$,

$$\partial W/\partial y = 4W_{40} = 4W. \tag{1.12}$$

Taking a numerical aperture of 0.1 as typical, we have $dh' = -10\partial W/\partial y = -10 \cdot 4W = -40W$. What this means is that for a given ray, if the system has third-order spherical aberration, the transverse ray aberration for the marginal ray will be 40 times the wavefront aberration. If, therefore, we have a merit function that includes transverse aberrations as well as wavefront aberrations, the wavefront aberrations should have a much larger weight.

4. If we are using the Coddington astigmatism formula, these are longitudinal aberrations and will normally be larger than transverse aberrations. They should therefore have a smaller weight.

5. It is often important that we should not attempt to control aberrations that are uncorrectable. Examples of this are distortion in eyepieces, chromatic aberration in Ramsden eyepieces, and astigmatism in doublets.

1.4.5 Damping

It should be clear that the effectiveness of the DLS method is entirely dependent on the damping, so that the length of the parameter changes is minimized. This assumes, however, that the "length" is well defined. In some cases this may be true. If the only variable parameters are, for example, the curvatures of the lens surfaces, it is reasonable to define the "length" of the parameter change as the RSS value of the changes to the curvatures. However, in most optimization problems there are several types of variable parameters and there is no consistent way in which curvatures, separations, refractive indices, and aspheric coefficients can be compared, and therefore no consistent way in which we can think of minimizing the "length" of the parameter changes. However, this point does not introduce any insuperable difficulties, and it does lead to a useful variation of the damping method.

1.4.5.1 Additive damping

In the most obvious application of the DLS method, the variable parameters are generally scaled in such a way that the changes they introduce are similar in magnitude, although they are not similar in direction. By this method, we can simply add the damping factor to Eq. (1.9). In practice, this is usually very effective, but there are cases when it becomes difficult to equate changes to very different types of parameters; then it is often more appropriate to use the method of multiplicative damping, as proposed by Levenberg.

1.4.5.2 Multiplicative damping

In this variation of the basic method, the diagonal terms in the product matrix are all multiplied by the same term $(1 + p^2)$. The effect of this is to introduce a damping term that depends on the sensitivity of each parameter. Since the diagonal term depends only on the derivatives with respect to one parameter, a parameter that has little effect will then receive small damping, and a parameter that is very sensitive will be heavily damped. The justification for adopting this method of damping is that we might assume that there is some correlation between the sizes of the first derivatives, which determine the parameter changes, and the sizes of the higher derivatives, which introduce the nonlinearity. Clearly we can think of cases where this is or is not true.

As an example of a case where it is not true, consider a situation where the chromatic aberration is computed using the Conrady formula:

$$\partial W_\lambda = \sum (\bar{D} - D)\partial n. \tag{1.13}$$

The chromatic aberration is exactly linearly proportional to the dispersion, ∂n, and therefore the higher derivatives are automatically zero. Therefore, there is no reason to use multiplicative damping if we consider this situation to be typical.

As a second example, consider a case where a surface has been bent so that the angle of incidence, i, of the paraxial marginal ray is zero. Since the Seidel spherical aberration is proportional to i^2, the first derivative will be zero, but since we are at the apex of a parabola, the second derivative is clearly nonzero. Again, there is no justification for using multiplicative damping, and indeed it would probably be actually harmful in this case.

But there are situations where the opposite is true. Suppose we have a surface that introduces a large amount of aberration, and which is strongly curved. In general we expect such a surface to be very sensitive, and we would also expect that the aberration change is a nonlinear function of the curvature. Here, heavy damping would be sensible, and multiplicative damping will achieve this.

Another situation where multiplicative damping is useful occurs with aspheric coefficients. The asphericity introduced by an aspheric coefficient is given by equations such as

$$z = a_i \cdot r^i, \tag{1.14}$$

where r is the clear radius of the surface, and i is usually an even number between 2 and 10. Therefore, in a lens where $r = 10$ mm, if $i = 10$, an a_{10} coefficient of 10^{-12} represents 10 μm of aspheric deformation, which is perhaps typical, and introduces a large amount of aberration, in most systems. Therefore, it can be seen that the numerical value of an aspheric coefficient can be very small, even though it will introduce a large amount of aberration. In this situation, the magnitude of the derivatives with respect to changes in aspheric coefficients can vary by many orders of magnitude, depending on details of the optimization program. Here, multiplicative damping can be useful.

Arguments such as those given above cannot prove or disprove whether additive or multiplicative damping is best; the only proof is the actual use of the two methods on a variety of practical lens design problems, and this will depend greatly on the details of the algorithms used in the program. In practice, there are certainly cases where one technique is better, and there are other cases where the other is better. The arguments given above do allow a user to make an intelligent choice, if the program permits this.

1.4.6 Control of physical constraints

It has been known since optimization programs were first used that they could produce designs that are impractical, unless this is somehow prevented. For example, when optimizing a typical tessar design (described in more detail in Chapter 4), an unconstrained optimization program will generate designs that are unreasonably long. Figure 1.2 shows a typical example that illustrates this point, originally

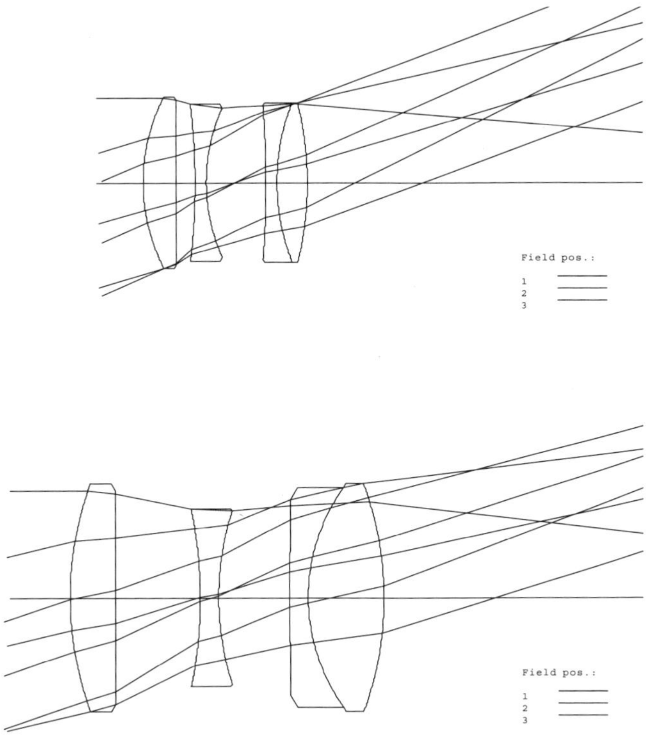

Figure 1.2 A tessar lens optimized with no limit on size, as from Wynne.

described by Wynne. Although the larger design is better, the difference is not very great, and in most cases another type of construction would be preferable, if significantly better performance is needed.

More recently, a problem posed at the 1990 International Lens Design Conference illustrated the importance of lens size and thickness. The problem was to optimize a "monochromatic quartet," using four BK7 components, subject to various conditions.[21] The best solutions were all similar, and all very long, as shown in Fig. 1.3, with aberrations at NA 0.1666 and over a 15-deg semifield, shown in Fig. 1.4. In comparison, a conventional four-element double-Gauss design is shown in Fig. 1.5, with aberrations at NA 0.125 and semifield of 10 deg, shown in Fig. 1.6 (note that the aberration scale of Fig. 1.6 is five times larger than that of Fig. 1.4).

```
EFL =    100.000
WAVELENGTHS  [nm]
      587.60
SURFACES
  # SURF SPACE    RADIUS        SEPN       INDEX1      V      CLR RAD  GLASS
  0                            0.00000   1.000000
  1 S          145.68750     2.00000   1.516798            63.416  S-BK7
  2 S           94.71160    10.80390   1.000000            60.314
  3 S          162.72740   231.49710   1.516798            60.233  S-BK7
  4 S         -143.65390     0.00000   1.000000            13.618
  5#S            Plane       0.10000   1.000000            13.567
```

6	S	68.40040	103.22940	1.516798	14.193	S-BK7
7	S	-1480.27920	6.72150	1.000000	23.204	
8	S	-43.03370	2.00000	1.516798	23.217	S-BK7
9	S	858.34200	2.00381	1.000000	25.890	
10	S	Plane			26.948	

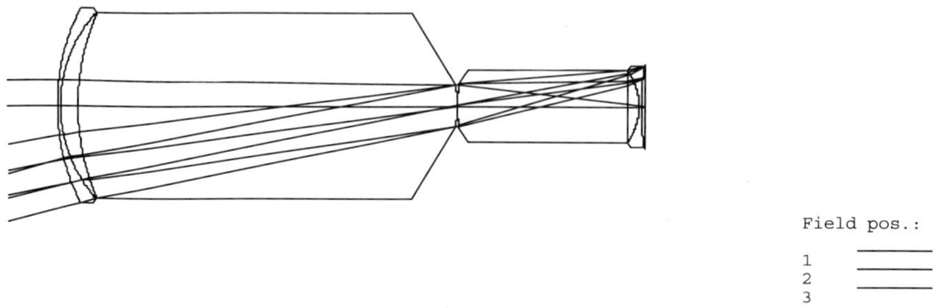

Figure 1.3 The optimum solution for the monochromatic quartet.

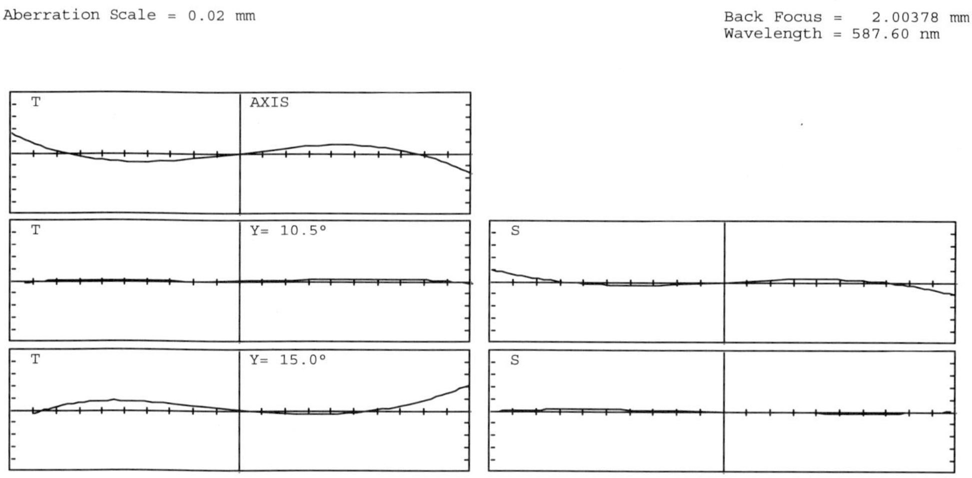

Figure 1.4 Transverse ray aberrations of the monochromatic quartet.

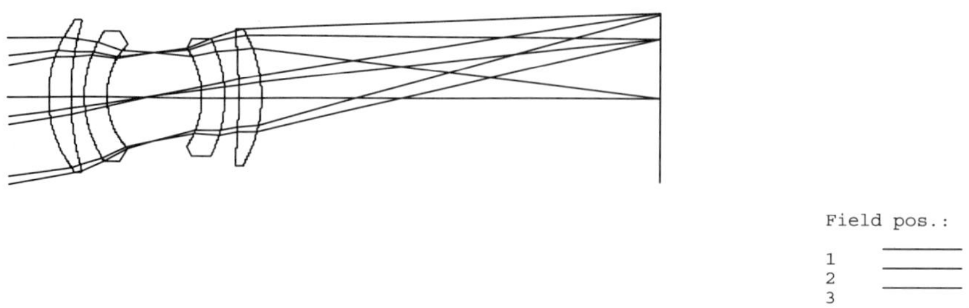

Figure 1.5 A typical four-element double Gauss.

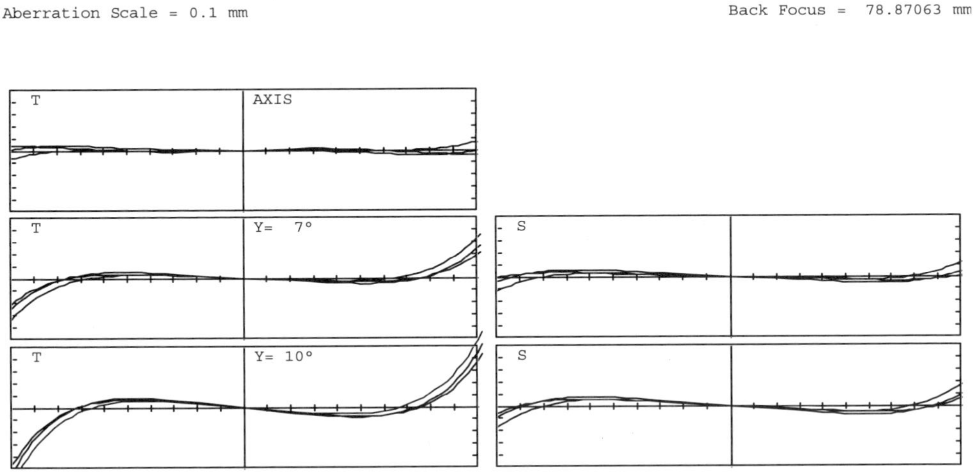

Figure 1.6 Transverse ray aberrations of four-element double Gauss.

EFL = 100.000

WAVELENGTHS [nm]

587.60 656.30 486.10

SURFACES

#	SURF SPACE	RADIUS	SEPN	INDEX1	V	CLR RAD	GLASS
0			-2.55949	1.000000			
1	S	26.70338	4.50000	1.640493	60.08	16.136	S-LAK21
2	S	54.64438	2.80000	1.000000		15.616	
3	S	24.11186	4.70000	1.805176	25.42	13.545	S-SF6
4	S	16.50433	9.40000	1.000000		10.984	
5	#S	Plane	9.40000	1.000000		9.362	
6	S	-19.13782	4.70000	1.805176	25.42	10.462	S-SF6
7	S	-28.42408	2.80000	1.000000		12.381	
8	S	-140.60829	4.50000	1.640493	60.08	13.918	S-LAK21
9	S	-30.43225	78.87027	1.000000		14.285	
10	S	Plane				17.692	

While it has to be said that excessive length and a short image clearance were the only ways to achieve the small residual aberrations of the winning monochromatic quartet, this was not intended to be anything more than a theoretical exercise, and there is no suggestion, as far as the author is aware, that it represents a sensible solution to any real lens design problem. However, less extreme practical lenses make use of the same negative-positive-negative power distribution and reasonably thick lenses. One famous example is the Zeiss Hologon, by Glatzel, shown in Fig. 1.7 at NA 0.05. The system is almost symmetric, and can cover a large field angle with little distortion. Flint glasses for the outer negative lenses achromatize the design, as may be seen in the transverse ray aberrations of Fig.1.8. The thickness of the central positive lens is a difficulty in that the aperture stop must be placed inside this lens, and it is clearly impossible for the stop diameter to be variable in the normal way.

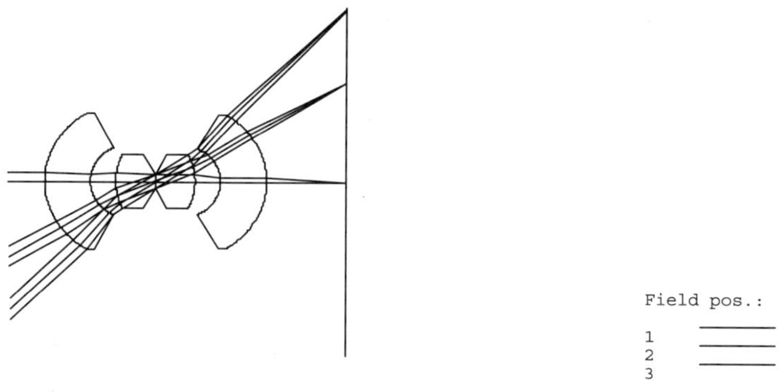

Figure 1.7 Zeiss Hologon.

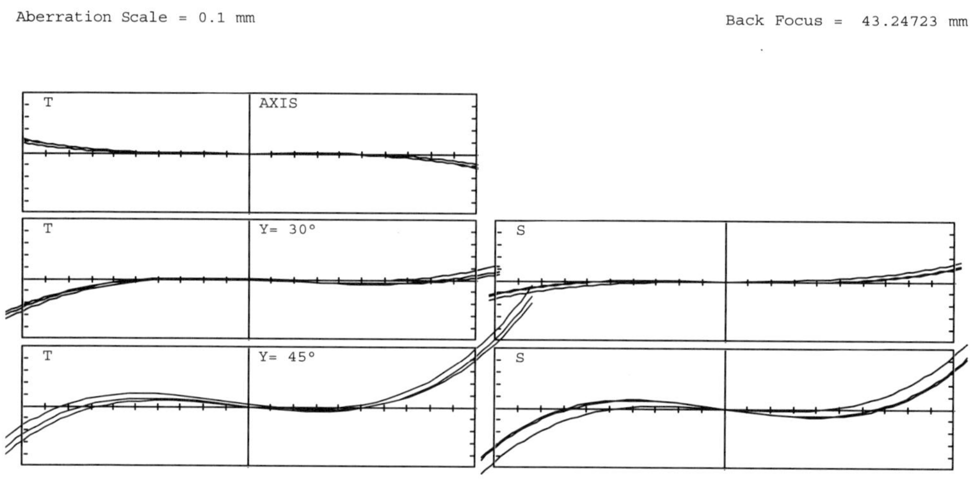

Figure 1.8 Transverse ray aberrations of Zeiss Hologon.

EFL = 100.000

WAVELENGTHS [nm]

587.60 656.30 486.10

SURFACES

# SURF SPACE	RADIUS	SEPN	INDEX1	V	CLR RAD	GLASS
0		13.48497	1.000000			
1 S	45.21491	25.25397	1.805176	25.42	39.512	S-SF6
2 S	20.92696	14.24974	1.000000		19.554	
3 S	34.69437	21.78921	1.640493	60.08	15.983	S-LAK21
4#S	Plane	21.78921	1.640493	60.08	4.209	S-LAK21
5 S	-32.72118	14.24974	1.000000		15.424	
6 S	-20.58436	25.25397	1.805176	25.42	19.071	S-SF6
7 S	-45.40425	43.24683	1.000000		38.736	
8 S	Plane				101.149	

Another invention making deliberate use of very thick lenses, by H.H. Hopkins, relates to endoscopes; these systems use relay lenses and field lenses to relay

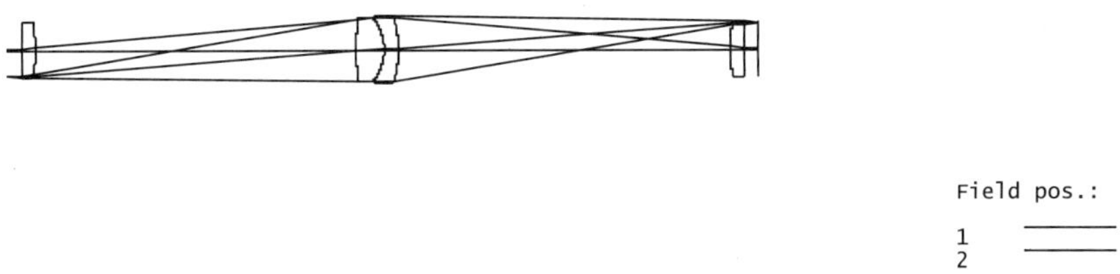

Figure 1.9 Telecentric unit magnification relay for an endoscope.

an image down a narrow tube. Classically, the relay lens is a simple doublet and the field lens is a single lens near the image. We can design a unit magnification system that is telecentric in the object space and the image space (i.e., the entrance and exit pupils are at infinity) using field lenses in both spaces, as shown in Fig. 1.9. We can then place a large number of these in the endoscope tube and relay the image as far as needed, but the aberrations shown in Fig. 1.10 will of course get worse and worse. The main aberration, as with simple doublets, is astigmatism, but field curvature is also significant.

EFL = 454.672
OBJECT DISTANCE = 1, CURVATURE = 0, TILT ANGLE = 0
WAVELENGTHS [nm]
587.60 656.30 486.10

SURFACES

# SURF SPACE	RADIUS	SEPN	INDEX1	V	CLR RAD	GLASS
0		0.00000	1.000000			
1 S	Plane	1.00000	1.516798	64.14	2.100	S-BK7
2 S	-12.80014	22.68210	1.000000		2.155	
3 S	14.13481	2.00000	1.516798	64.14	2.438	S-BK7
4#S	-3.88963	1.00000	1.620037	36.35	2.314	S-F2
5 S	-8.84150	23.30426	1.000000		2.500	
6 S	12.68945	1.00000	1.516798	64.14	2.110	S-BK7
7 S	Plane	1.00004	1.000000		2.054	
8 S	Plane				2.032	

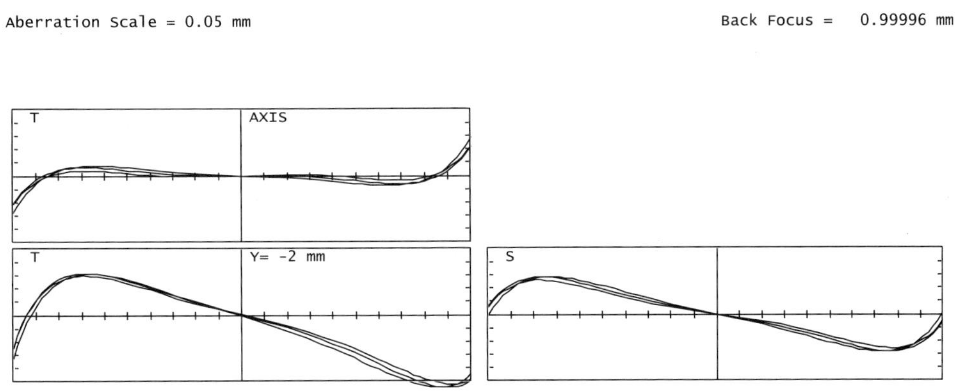

Figure 1.10 Transverse ray aberrations of unit magnification relay.

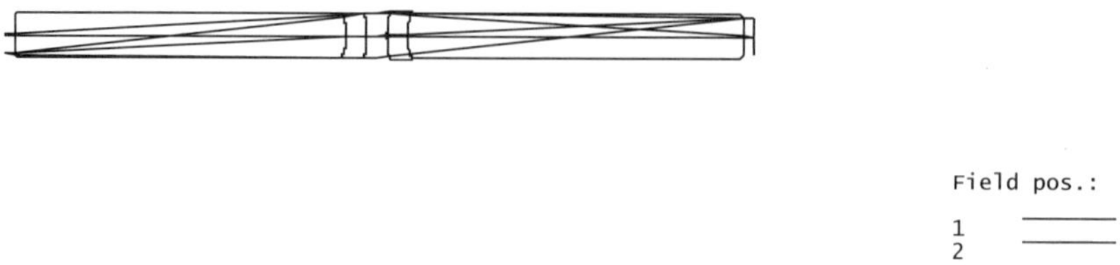

Figure 1.11 Hopkins solid endoscope.

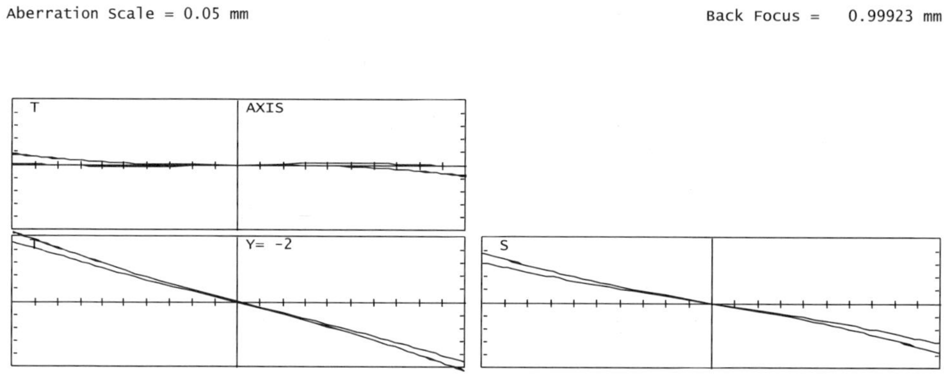

Figure 1.12 Transverse ray aberrations of Hopkins endoscope.

The vital feature of the Hopkins endoscope is that the air spaces are effectively filled with glass, increasing the path length of each unit so that a smaller number of lenses are required, and thus reducing the total aberration. In effect, the glass rod becomes the space between "air lenses" and we get a unit-magnification telecentric system as shown in Fig.1.11. Although the aberration of each unit is not significantly reduced, the total aberration will be smaller, as shown in Fig. 1.12.

```
EFL =   1245.233
OBJECT DISTANCE = 1, CURVATURE = 0, TILT ANGLE = 0
WAVELENGTHS  [nm]
    587.60       656.30       486.10
SURFACES
```

# SURF SPACE	RADIUS	SEPN	INDEX1	V	CLR RAD	GLASS
0		0.00000	1.000000			
1 S	12.83864	34.01516	1.516798	64.14	2.118	S-BK7
2 S	-6.14437	2.00000	1.647685	33.83	2.398	S-SF2
3 S	-10.91977	1.00000	1.000000		2.488	
4#S	Plane	1.00000	1.000000		2.420	
5 S	10.91977	2.00000	1.647685	33.83	2.523	S-SF2
6 S	6.14437	34.01516	1.516798	64.14	2.427	S-BK7
7 S	-12.83864	0.99925	1.000000		2.067	
8 S	Plane				2.074	

If we depart from the Hopkins patented design, letting the negative element become thicker, as shown in Fig.1.13, it becomes possible to correct astigmatism.

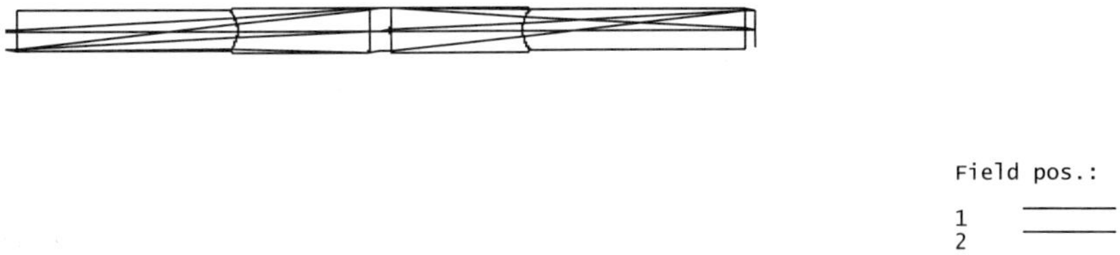

Figure 1.13 Solid endoscope with increased thickness negative lens.

The total aberration of the endoscope is very much reduced, as we can see in Fig. 1.14.

EFL = 4133.058
OBJECT DISTANCE = 1, CURVATURE = 0, TILT ANGLE = 0
WAVELENGTHS [nm]
587.60 656.30 486.10
SURFACES

# SURF SPACE	RADIUS	SEPN	INDEX1	V	CLR RAD	GLASS
0		0.00000	1.000000			
1 S	13.70409	22.59880	1.620408	60.30	2.111	S-SK16
2 S	-3.37717	13.40426	1.666796	32.99	2.078	S-SF19
3 S	-13.65274	1.00000	1.000000		2.360	
4#S	Plane	1.00000	1.000000		2.360	
5 S	13.65274	13.40426	1.666796	32.99	2.360	S-SF19
6 S	3.37717	22.59880	1.620408	60.30	2.078	S-SK16
7 S	-13.70409	0.99997	1.000000		2.111	
8 S	Plane				2.036	

These examples illustrate that lens thicknesses and diameters can be useful parameters, within the limits of practicality. Often there is a trade-off to be made between performance and size. A lens optimization program must therefore be able to control at least the following physical constraints, although the following list is not intended to be complete.

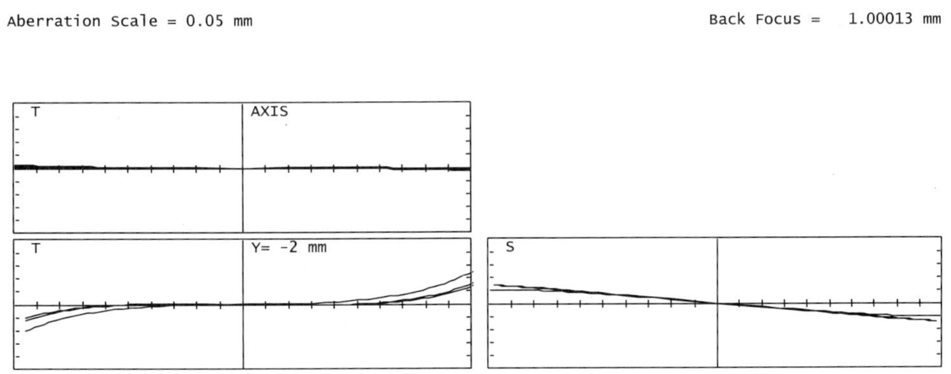

Figure 1.14 Transverse ray aberrations of endoscope with increased negative lens thickness.

1. Focal length or magnification is usually defined.
2. Back focus must often exceed some stated minimum.
3. Stop or pupil positions may be defined.
4. Object-image distance may be limited.
5. Overall size of the lens must be reasonable.
6. Glass components must not be too thin.
7. Clear radius must sometimes be limited.
8. Air spaces must have positive thicknesses.
9. Glass types, if variable, must be within defined cost limits.

Typically, at the beginning of each iteration, the program computes the axial and edge thicknesses of all spaces in the system, and compares them with limiting values. If a thickness is below the limit, the difference is added to the list of aberrations. The program sets default values of the limits, but these can easily be altered. Similarly, the program compares the maximum height of any ray with the limiting value. This is done at every surface. If the ray height is above the limit, the difference is added to the list of aberrations.

1.4.7 Control of glass boundary conditions

Figure 1.15 shows most of the optical glasses in the Schott glass catalog. When glasses are variable in optimization, we allow refractive index and dispersion to be continuously variable within a region that represents the glasses that are available for a given lens. Note that the materials on the extreme left of the plot are fluorite (CaF_2) and some FK glasses. The three materials on the right, below all of the others, are plastics (styrene, polycarbonate, and polystyrene).

For many design problems we do not need all of the available glasses.[22] Walker[23] has suggested a list of 49 glasses, which should normally be adequate. These glasses were chosen on the basis of optical characteristics and other properties such as workability, thermal properties, and cost. In 1978 the Schott designations for these glasses were PK50, BK7, PSK3, K7, K5, K10, BaK2, BaK1, BaK4, SK11, SK5, SK14, SK16, SK15, SK10, SK2, SK18A, LaK21, LaKN7, LaKN22, LaKN12, LaK8, LaK10, SSK2, SSKN5, SSKN8, BaLF5, BaLF4, KF3, KF9, LLF6, LLF1, BaF3, BaF52, BaF4, BaF51, BaFN10, BaF50, LF7, LF5, F2, F5, F1, F6, BaSF2, BaSF64A, BaSF51, LaFN23, LaF3, LaF2, LaF22A, LaFN7, SF2, SF5, SF8, SF15, SF1, SF53, SF4, SF14, and SF56A.

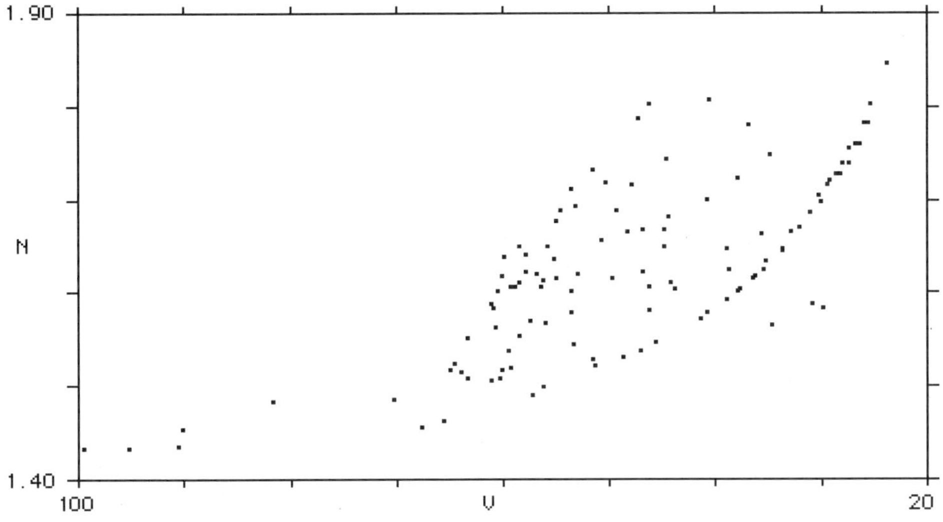

Figure 1.15 Glass map using V-values.

Recently, several other more restricted lists have been proposed. For example, Zhang and Shannon[24] suggested a list with only 10 Schott glasses: BaF52, BK7, FK5, LaF2, LaFN7, LLF6, SF2, SF4, SK16, and SSKN5. Tesar[25] has suggested several lists, including one of 20 glasses derived from the earlier lists, but with a more uniform distribution over the glass map, using Schott 2000 designations, and taking into account the discontinuation of certain glasses that are either uneconomical or contain environmentally unfriendly chemicals such as arsenic, lead, or thorium. These are N-FK5, N-BK7, N-K5, LLF6, BaLF4, BaF52, N-KZFS4, F2, N-SK16, SF2, BaF51, N-SSK5, SF5, N-LaK9, N-LaF3, SF1, BaSF51, LaF2, LaFN7, and SF4.

In general, the author would strongly recommend that inexperienced designers restrict themselves to some such list of readily available glasses. Naturally, the preferred list will tend to evolve with experience and special requirements, such as a high level of spectral or thermal aberration correction. Commercial considerations may require the list to include Hoya, Ohara, or other manufacturers' glasses. The designer may also at times need to use crystals, plastics, liquids, and ultraviolet or infrared materials outside the visible spectrum.

Rather than using the V-value directly, optimization programs use the dispersion, e.g., $n_F - n_C$ in the visible spectrum. A plot of refractive index against dispersion is shown in Fig. 1.16. It is clear that the region of low-index flint glasses (on the bottom edge of the diagram) becomes almost a straight line if we plot refractive index against dispersion, and optical design programs use a triangle or a quadrilateral to represent the region of available glasses.

The polygon is defined by three or four glasses, which may be preferred glass types, from a restricted list. In many cases the triangle of Fig. 1.17 can adequately

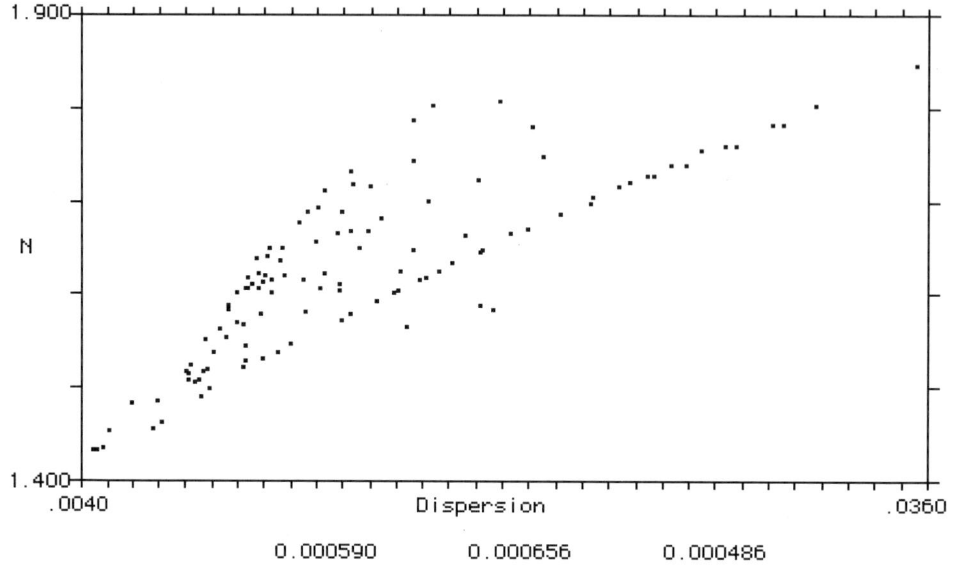

Figure 1.16 Glass map using dispersion.

represent the range of available glasses, but particularly when high-index crown glasses are being used (top left of Fig. 1.16), a quadrilateral is a little better. Note that in Fig. 1.17 we are plotting refractive index against dispersion, as discussed above.

At the beginning of each iteration, the program tests each variable glass to see if it is inside the polygon. If it is outside any line, the "distance" to the line is treated as an aberration, which will be minimized. This has the effect of moving the glass to the line in later iterations. Often a glass will move outside two lines, and the optimization will then hold the glass to the corner of the glass polygon. At the end of the optimization, the user should choose a real glass, and carry out another optimization using real glasses.

All lenses that are intended for use over an extended waveband must be corrected for chromatic aberration, and this implies the use of crown (low-dispersion)

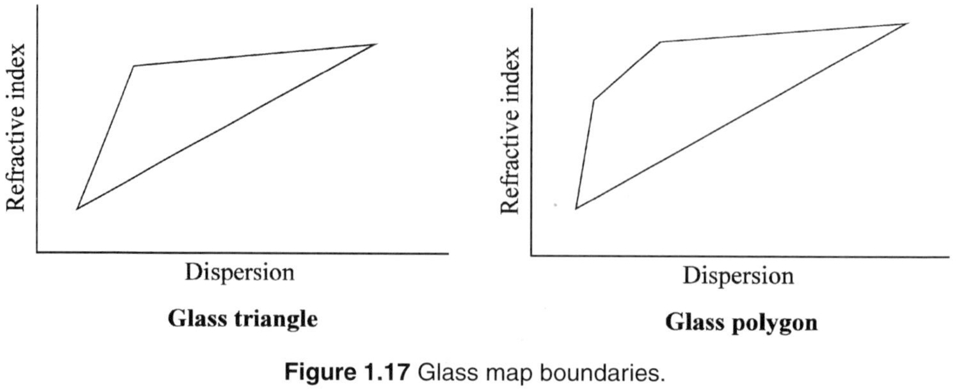

Figure 1.17 Glass map boundaries.

glasses and flint (high-dispersion) glasses in suitable combinations, unless we are using a reflecting or catadioptric system. For lens systems of positive power (the most common case), we therefore usually find positive elements made from crown glasses and negative elements made of flint glasses.

In most situations, thin lenses introduce spherical aberration that is of the same sign as the power of the lens (considering the wavefront aberration, or optical path difference). The minimum third-order spherical aberration is proportional to

$$\frac{n^2}{(n-1)^2} - \frac{n}{n+2}. \tag{1.15}$$

This expression is correct, providing that the lens is immersed in air and the object is at infinity. There is also a similar relationship for fifth-order spherical aberration. Consequently, over a range of refractive indices from 1.5 to 1.8 there is a variation in spherical aberration of approximately a 2.2:1 ratio.

It is well known that the astigmatism of a thin lens at the aperture stop does not depend on the refractive index of the lens. Although this result is only strictly true for thin lenses and not true at all for lenses away from the stop, the author would suggest that refractive index changes will very rarely have a significant effect on astigmatism correction. This is also true of coma and distortion. However, the field curvature of a thin lens is inversely proportional to the refractive index, so this gives a strong reason for using high-index glasses for positive elements. Note, however, that the variation of field curvature is much less than the variation of spherical aberration.

Since an increase in refractive index will improve both spherical aberration and field curvature, we find that many high-aperture lenses employ high-index glasses in their positive elements, even though they are more expensive. An obvious strategy to correct field curvature is to use high-index positive lenses and low-index negative lenses; in practice, this method is not frequently used, as we can see from Fig. 1.16 that the higher dispersion flint glasses preferred for minimizing the achromatizing powers of elements also have higher refractive index. Consequently, it is usually better to correct field curvature by using a design form with separated positive and negative elements.

1.4.8 Solves

In some cases, a simple analytic approach can be useful for controlling constraints. For example, if it is known that the focal length must be 100 mm, it is a simple matter to trace two paraxial rays through a lens, and then to calculate analytically the value of some chosen surface curvature required for a focal length of exactly 100 mm. Alternatively, all of the curves can be recalculated to obtain the given focal length. In principle it is possible to control several of the constraints this way, but again this is determined by the design of the program. It can be useful to adjust the

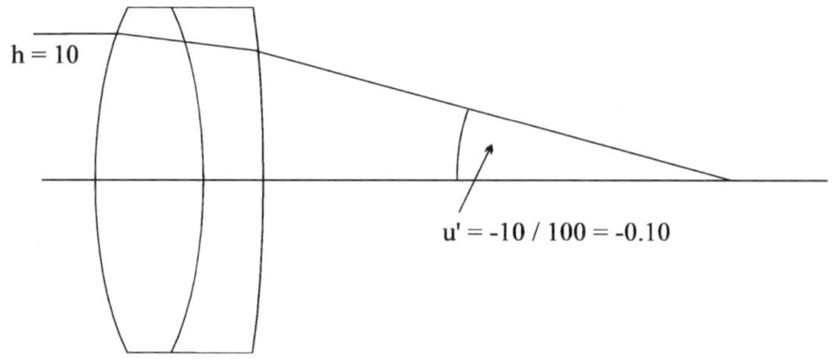

Figure 1.18 Angle solve.

curvature of one surface in a lens to achieve the correct value of magnification or focal length. To do this in optimization, it is only necessary to set a marker equal to the number of the required surface.

The use of angle solves in optimization has at least two advantages: (a) it reduces the number of variables; (b) in the case of afocal systems, it can often improve the convergence rate, because the system is automatically held afocal; and (c) in non-afocal systems, either the magnification or the focal length is automatically controlled exactly, and this effectively removes one aberration from the merit function. Sometimes this appears to speed the convergence of the optimization.

When we request an angle solve, either in optimization or manually, we adjust a specified surface so that a paraxial ray leaves that surface at the required angle. Most often we use the angle solve feature to get a given focal length. For example, suppose that we have a cemented doublet, shown in Fig. 1.18 with three surfaces, and we require the focal length to be exactly 100 mm. If we trace a paraxial ray, at an initial height of 10 mm, the final paraxial ray angle in the image space should be exactly – 10/100 = – 0.10, if the focal length is 100 mm. Suppose we trace the ray as far as surface 3, and find that its angle before surface 3 is u_3. The ray angle after surface 3 is given by

$$n_3'u_3' = n_3u_3 - h_3c_3(n_3' - n_3), \tag{1.16}$$

so the program computes the value of c_3 that gives the required value of u' from

$$c_3 = \frac{n_3u_3 - n_3'u_3'}{h_3(n_3' - n_3)}. \tag{1.17}$$

Note that if h is very small (the surface is close to the image or the object), a large value of c may be needed, so we should not use the angle solve feature if h is likely to be small. Some optimization programs actually use a more complex calculation than this, and the solve surface need not necessarily be the last in the system. It is therefore recommended that the angle solve is normally used at surface where h is large.

1.4.9 Lagrange multipliers

As stated in Sec. 1.3.1, it is possible to define a set of conditions that are solved exactly, while other conditions are solved in a least-squares sense. For example, if the object-image distance is to be exactly controlled, Eq. (1.9) can be set up to give exact solutions for this distance, assuming that it is a linear function of all of the variables. This will not, in practice, be true, but this is not a problem. After a few iterations, the distance will be accurately controlled, for all practical purposes. The author has used this technique for many years, but in the hands of an inexperienced user it can cause difficulties. If a user inadvertently defines incompatible conditions, for example, the resultant equations will become indeterminate and the optimization will not succeed.

A more simple method, and the author's preference, is to simply add extra terms to the list of aberrations, corresponding to the constraints that are, at any given time, violating the boundary conditions. This method has the advantage that constraints can be given large or small weight, depending on whether they are very important or only slightly important.

For example, it is clearly important that distances between surfaces must be positive; i.e., the surfaces must not cross over. Since this is of critical importance, these constraints may be given a large weight. On the other hand, as we have seen, there is almost always a demand to limit the size of a lens, even though there may be no specific constraint. In this situation, the author often applies a small weight to the overall length of a lens, without giving a specific target; in other words, the target is zero but the effect of the small weight is to bias the optimization toward compact designs. Clearly, by varying the appropriate weighting factor, a series of lenses can be designed with a range of sizes and with varying performance.

Another advantage of this method is that when an inexperienced user inadvertently defines incompatible conditions, the program will simply find the best compromise between the incompatible conditions, and it is then usually reasonably obvious what has happened.

1.5 Some reasons for the success of the DLS method

At first sight it is surprising that the DLS method, first used in 1944 and basically very simple, is still the most successful process for lens optimization more than 50 years after its invention. But after some thought one sees that there are good reasons for the success of this method.

First, it is a least-squares method, and we have already seen that it is natural to define the lens design merit function as a sum of squares of a set of aberrations. Second, as stated above, the DLS method automatically gives the best set of parameter changes for a given distance in parameter space. Although this statement presupposes that we know what we mean by "distance" in this context, this is not a difficulty in practice. This "best" set of parameter changes is very effective when a program has to make a series of large changes over many iterations in order to minimize high-order aberrations.

Third, although the aberrations are nonlinear functions of the parameters, near the minimum the aberrations are linear. In other words, aberrations can change from positive to negative on passing through the minimum of the merit function. This is not true in other optimization problems. One example is the case of optimizing thin-film antireflection coatings, where reflectance is necessarily positive and therefore cannot change from positive to negative, but must necessarily be a quadratic function in the region of the minimum. This property that the aberrations are approximately linear, while the merit function is nonlinear (because it is a sum of squares), seems to be one of the critical reasons for the success of DLS as an optimization method in lens design.

Fourth, a feature of the DLS method is that it does have at all times information about the first derivatives of the *components* of the merit function and not just the merit function itself. Thus, the amount of information available is much greater than in methods that only consider the merit function as a whole, such as the Simplex method.

1.6 Experiments with optimization programs

Many variations of the DLS method have been suggested since the method was first proposed for lens design. Some of these are concerned with the choice of the damping factor, and with possible scaling of the parameter changes.[26]

1.6.1 Effect of changing the damping factor

First, consider the way in which the merit function will depend on the damping factor in a single iteration. It is simple to set up an optimization program so that several values of the damping factor are tried. Figure 1.19 shows a typical situation.

When the damping factor is large, very small parameter changes will be computed and the merit function will not be changed very much. As the damping factor is reduced, the parameter changes increase and the merit function is reduced, as long as Eq. (1.4) is valid. At some stage, the damping factor is reduced to the point that the nonlinearities do become significant, and there is then a difference between the actual merit function, as computed by ray tracing, and the predicted merit function, computed from Eq. (1.4), assuming linearity. If the damping factor is further reduced, the nonlinearities will become dominant and the merit function will rapidly increase, often to a value that is much larger than the starting value. With a damping factor of zero, the result is often catastrophic.

Clearly, it is sensible to make some attempt to locate, at least approximately, the minimum of the curve in Fig. 1.19, but there are many possible ways of doing this, and it is the author's belief that the details of how this is done are not critical.

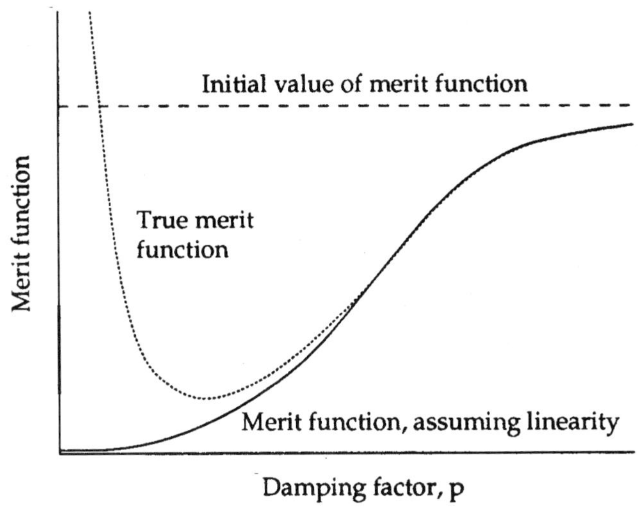

Figure 1.19 The relationship between the merit function and the damping factor.

One important point is that it is better to be on the left of the minimum than on the right, i.e., underdamping rather than overdamping, for the following reason. If one watches the progress of an optimization, what usually happens is that the changes at one iteration are similar but not identical to the changes at the previous iteration. Therefore, bearing in mind the model of Fig. 1.1, it is better to make large changes than small ones, as this will in principle allow the optimization to reach the minimum more quickly.

1.6.2 Effect of scaling the parameter changes

Another variation that has been proposed is that the parameter changes could be scaled. Originally this suggestion was made in the context of simple undamped least squares. As explained above, if the damping factor is zero, the optimization often diverges catastrophically; but if the parameter changes are scaled down so that they are within the range of linearity, the process must converge. In practice, it does converge, but very slowly. A simple explanation for the slowness can be seen from Fig. 1.1. The parameter changes computed with zero damping are very large, as the ellipses in Fig. 1.1 are very elongated. In general, unless we start at one "end" of the ellipse, the direction of the undamped parameter changes will be almost parallel to the contours, and therefore the rate of improvement will be very small.

Several experiments were carried out by Kidger and Wynne,[26] in which the damping factor and the size of the parameter changes were both varied. For each value of the damping factor, p, the parameter changes were computed from the damped least squares Eq. (1.9). The parameter changes computed this way were then scaled by varying amounts, and the true merit function of the resultant design was computed by ray tracing. In addition, the value of the merit function that would be predicted from Eq. (1.4) was computed for each of these designs. The RSS value of the parameter changes, s, was also computed. It is often helpful to think of s as the "distance" that the program has moved in parameter space.

Figure 1.20 shows contours of the merit function plotted as a function of p and s. Contours of the true merit function are drawn as solid lines. Contours of the predicted merit function are drawn as dotted lines. As would be expected, for small values of s, the two sets of contours are close together, but as the parameter changes increase, they diverge. The crosses show the solution computed by the DLS equations, for each value of p, without scaling.

These plots showed that, as expected from the above discussion, the DLS solution is close to an optimum, and that scaling the parameter changes does not improve the merit function except when the damping factor is much too large. This is to be expected, from an understanding of the topography in the merit function and of the DLS method. If it is working correctly, the DLS method should find points along the valley bottom, referring to Fig. 1.1, and if the parameter changes are scaled up, the new design is worse.

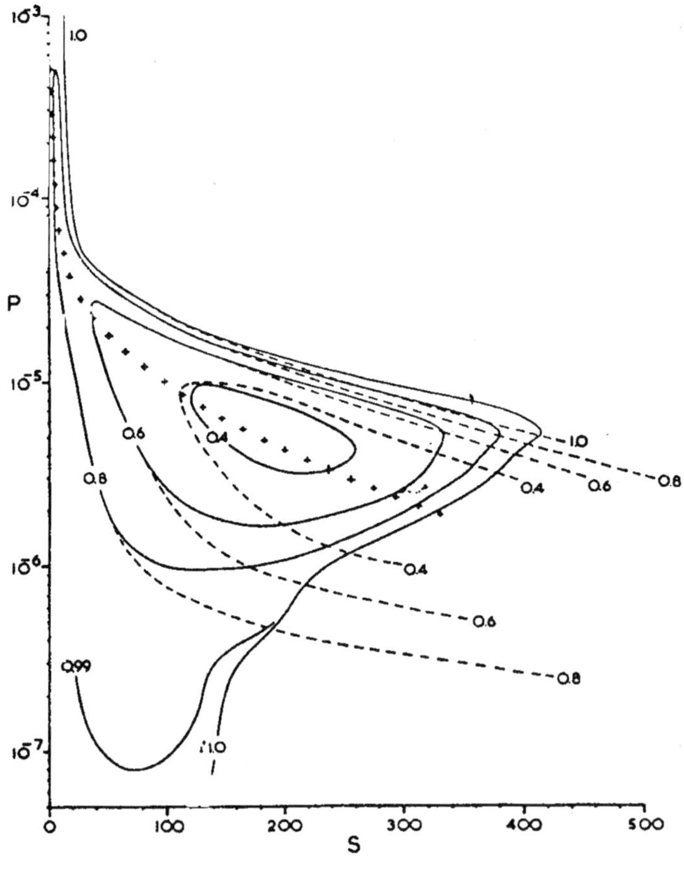

Figure 1.20 An experiment to illustrate the effect of scaling parameter changes.

1.7 An optimization example

In the rest of this chapter we discuss in detail the operation of DLS optimization, with reference to an example, the four-glass double-Gauss design described by Kingslake[27] and shown in Fig. 1.21.

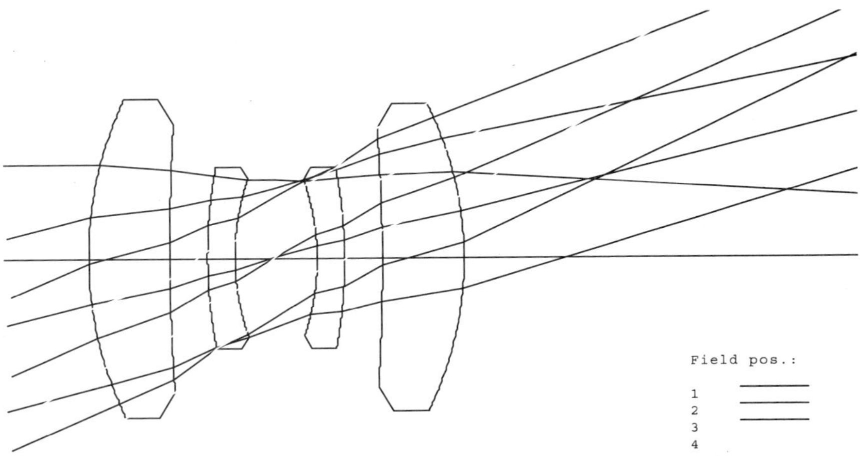

Figure 1.21 Four-glass double-Gauss derivative.

Before starting an optimization, we would always recommend that you should view the merit function. The results of this are shown below.

#	SF	SP	SFV	SPV	CURVE	RADIUS	SEPN	INDEX1	DISPN	CLR RAD	GLASS
0							0.000	1.00000	0.00000		
1	S		C	D	0.038600	25.907	5.083	1.61272	0.01045	10.60	S-SK4
2	S		C	D	0.006787	147.341	2.355	1.00000	0.00000	8.89	
3	S		C	D	0.028732	34.804	1.694	1.61659	0.01684	6.05	S-F4
4	S		C	D	0.057670	17.340	2.542	1.00000	0.00000	5.25	
5	S			D	0.000000	PLANE	2.542	1.00000	0.00000	5.15	
6	S		C	D	-0.057670	-17.34	1.694	1.61659	0.01684	5.20	S-F4
7	S		C	D	-0.028732	-34.80	2.355	1.00000	0.00000	6.00	
8	S		C	D	-0.007201	-138.87	5.083	1.61272	0.01045	8.60	S-SK4
9*	S		C	D	-0.040770	-24.53	89.832	1.00000	0.00000	10.24	
10	S				0.000000	PLANE				63.02	

Defocus = 0.000000 U' = -0.062500 EFL = 100.0000

L	M	X	Y	OPD	COLOR	S	T	DIST(%)
	-0.06249		0.00523	-0.00010	-0.00031			
	-0.04321		0.00249	-0.00003	-0.00017			
CHIEF RAY	0.25535		26.81549			-0.4741	0.1620	0.077%
	0.30434		-0.21880	-0.00265	-0.00048			
	0.29022		-0.07531	-0.00071	-0.00031			
	0.21924		0.04153	-0.00040	0.00008			
	0.20440		0.11877	-0.00152	0.00005			
-0.05979	0.25519	0.01813	-0.01976	0.00017	-0.00034			
-0.04199	0.25528	-0.00411	-0.00732	0.00026	-0.00019			

```
CHIEF RAY 0.41698           46.77193                        -0.5790 0.3422 0.303%
          0.45084           -0.37391 -0.00300 -0.00023
          0.44155           -0.12873 -0.00084 -0.00018
          0.39096            0.10351 -0.00071  0.00018
          0.38104            0.29784 -0.00256  0.00025
-0.05553  0.41681  0.11850  -0.04582 -0.00106 -0.00039
-0.03935  0.41691  0.02635  -0.01670 -0.00000 -0.00022

CHIEF RAY 0.52290           62.94093                        0.5276 -1.7497 0.727%
          0.54624           -0.22622 -0.00073  0.00026
          0.53988           -0.04593  0.00004  0.00012
          0.50506            0.05343 -0.00000  0.00014
          0.49838            0.23466 -0.00087  0.00029
-0.04922  0.52287  0.33138  -0.07178 -0.00417 -0.00042
-0.03535  0.52289  0.12113  -0.02526 -0.00123 -0.00022

FOCAL LENGTH          100.0000   TARGET =   100.0000     WEIGHT =    10.0000
```

NO		ABERN	TARGET	WEIGHT	CONTRIBN	MERIT FN
1	FOCAL LENGTH	100.00000	100.00000	10.000	0.00000	0.00000
2	BACK FOCUS	89.83151	0.00000	0.000	0.00000	0.00000
3	YBAR AT STOP	0.00002	0.00000	0.000	0.00000	0.00000
4	MBAR (IMAGE)	0.52290	0.00000	0.000	0.00000	0.00000
5		0.00000	0.00000	0.000	0.00000	0.00000
6	GLASS THICKNESS	13.55400	0.00000	0.000	0.00000	0.00000
7	LENGTH	113.17951	0.00000	0.000	0.00000	0.00000
8	THICKNESS	23.34800	0.00000	0.000	0.00000	0.00000
9	DEFOCUS	0.00000	0.00000	1.250	0.00000	0.00000
10	DY, RAY 2	0.00523	0.00000	40.000	0.04379	0.04379
11	COLOR	-0.00031	0.00000	2560.000	0.64247	0.68626
12	DY, RAY 3	0.00249	0.00000	83.623	0.04334	0.72960
13	COLOR	-0.00017	0.00000	5351.880	0.86984	1.59943
14	SAGITTAL 4	-0.47407	0.00000	1.600	0.57535	2.17478
15	TANGENTIAL 4	0.16196	0.00000	1.600	0.06715	2.24193
16	DISTORTION	0.00077	0.00000	100.000	0.00589	2.24782
17	DY, RAY 5	-0.21880	0.00000	12.293	7.23442	9.48224
18	COLOR	-0.00048	0.00000	786.748	0.14324	9.62548
19	DY, RAY 6	-0.07531	0.00000	25.088	3.56966	13.19513
20	COLOR	-0.00031	0.00000	1605.610	0.25204	13.44717
21	DY, RAY 7	0.04153	0.00000	23.656	0.96530	14.41247
22	COLOR	0.00008	0.00000	1513.960	0.01366	14.42613
23	DY, RAY 8	0.11877	0.00000	11.591	1.89544	16.32157
24	COLOR	0.00005	0.00000	741.841	0.00137	16.32294
25	DX, RAY 9	0.01813	0.00000	12.669	0.05278	16.37572
26	DY, RAY 9	-0.01976	0.00000	12.669	0.06264	16.43836
27	COLOR	-0.00034	0.00000	59.468	0.00041	16.43877
28	DX, RAY 10	-0.00411	0.00000	25.855	0.01130	16.45007
29	DY, RAY 10	-0.00732	0.00000	25.855	0.03579	16.48586
30	COLOR	-0.00019	0.00000	121.363	0.00053	16.48639
31	SAGITTAL 11	-0.57903	0.00000	1.000	0.33527	16.82167
32	TANGENTIAL 11	0.34224	0.00000	1.000	0.11713	16.93880
33	DISTORTION	0.00303	0.00000	100.000	0.09163	17.03043
34	DY, RAY 12	-0.37391	0.00000	10.662	15.89381	32.92424
35	COLOR	-0.00023	0.00000	682.389	0.02467	32.94890
36	DY, RAY 13	-0.12873	0.00000	21.760	7.84672	40.79562
37	COLOR	-0.00018	0.00000	1392.630	0.06079	40.85641
38	DY, RAY 14	0.10351	0.00000	19.625	4.12695	44.98336

39	COLOR	0.00018	0.00000	1256.010	0.05060	45.03396
40	DY, RAY 15	0.29784	0.00000	9.616	8.20315	53.23711
41	COLOR	0.00025	0.00000	615.445	0.02377	53.26088
42	DX, RAY 16	0.11850	0.00000	8.316	0.97100	54.23188
43	DY, RAY 16	-0.04582	0.00000	8.316	0.14518	54.37706
44	COLOR	-0.00039	0.00000	39.034	0.00024	54.37729
45	DX, RAY 17	0.02635	0.00000	16.971	0.19991	54.57720
46	DY, RAY 17	-0.01670	0.00000	16.971	0.08036	54.65756
47	COLOR	-0.00022	0.00000	79.661	0.00030	54.65786
48	SAGITTAL 18	0.52761	0.00000	0.800	0.17816	54.83602
49	TANGENTIAL 18	-1.74973	0.00000	0.800	1.95939	56.79540
50	DISTORTION	0.00727	0.00000	100.000	0.52786	57.32326
51	DY, RAY 19	-0.22622	0.00000	11.172	6.38765	63.71091
52	COLOR	0.00026	0.00000	715.021	0.03481	63.74572
53	DY, RAY 20	-0.04593	0.00000	22.800	1.09678	64.84250
54	COLOR	0.00012	0.00000	1459.230	0.03031	64.87281
55	DY, RAY 21	0.05343	0.00000	20.841	1.23968	66.11249
56	COLOR	0.00014	0.00000	1333.800	0.03568	66.14817
57	DY, RAY 22	0.23466	0.00000	10.212	5.74250	71.89067
58	COLOR	0.00029	0.00000	653.561	0.03516	71.92583
59	DX, RAY 23	0.33138	0.00000	6.420	4.52548	76.45131
60	DY, RAY 23	-0.07178	0.00000	6.420	0.21236	76.66367
61	COLOR	-0.00042	0.00000	30.133	0.00016	76.66383
62	DX, RAY 24	0.12113	0.00000	13.101	2.51850	79.18233
63	DY, RAY 24	-0.02526	0.00000	13.101	0.10953	79.29186
64	COLOR	-0.00022	0.00000	61.496	0.00019	79.29206

The first part of the above is the starting lens prescription and a normal ray trace, with L, M as direction cosines, and X,Y as coordinates of the rays at the image plane, as well as their optical path differences from the chief rays and Conrady chromatic aberrations. Note that an angle solve has been used on surface 9 (indicated by the *) to give exact control of the focal length.

The second part lists the aberrations that form the merit function, their targets, weights, the contribution of each aberration to the merit function (aberration times weight, squared), and the cumulative sum of the contributions. This merit function includes various paraxial and dimensional constraints, transverse ray aberrations, longitudinal sagittal and tangential field curvature, distortion, and Conrady chromatic aberrations. The final number, 79.29206, is therefore the value of this merit function for the initial design.

Note that some contributions to the merit function are large; aberration 17 (DY for ray 5) has a contribution of 7.23, and aberrations 34, 51, and 57 each contribute more than 5 to the merit function. This is quite reasonable, as these are the main factors limiting the performance of this lens.

```
ITERATION 1 ADDITIVE DAMPING
     P              FORECAST          MERIT FN       CHANGE
  0.010000          27.22025          33.12733       295.12
  0.005000          19.24940          48.05063       489.65
  0.020000          40.66273          41.11373       120.88
  0.005000          19.24940          48.05063       489.65
HT,SURF      1 RAY 22 TARGET  10.750 VALUE  11.025
HT,SURF      9 RAY 19 TARGET  10.500 VALUE  10.690
```

In the first iteration, the program chooses an arbitrary value of the damping factor, p, and forecasts that a new lens should have a merit function of 27.22; the parameter changes needed represent a "distance" in parameter space of 295.12 units (in units of the curvature and separation increments). The program sets up this new design, and finds, by ray tracing, that the merit function is actually 33.13. Since this is an improvement on the initial design, the program reduces p by a factor of 0.5 and tries again; this time the merit function is 48.05, so it increases p and gets a merit function of 41.1. It then returns to the second solution and uses this design as the starting point for the next iteration.

However, when the program tests this new design for edge thicknesses and clear radius violations, it finds that surfaces 1 and 9 (i.e., first and last) have clear radii greater than their respective maximum clear radius limits. (We have previously set clear radius limits of 10.75 mm and 10.5 mm in order to prevent the lens from becoming too large.) The errors in these two quantities are then included in the merit function for the next iteration, and the optimization continues.

```
ITERATION 2 ADDITIVE DAMPING
      P             FORECAST          MERIT FN       CHANGE
  0.010000          20.62711          20.41544        98.20
  0.005000          19.87371          19.88654       158.52
  0.002500          19.18912          19.63881       296.02
  0.001250          17.70930          25.44689       805.78
 EDGE,SPACE  2 RAY 22 TARGET    2.500 VALUE   -0.436
 EDGE,SPACE  9 RAY 19 TARGET    2.500 VALUE    1.062
```

In iteration 2, the merit function goes down to 25.44, but the program now finds that the edge thicknesses in spaces 2 and 9 are too small. (This can also be seen in the lens drawings.) So these are included in the merit function. From time to time, more violations occur, and these are included in the merit function:

```
ITERATION 3 ADDITIVE DAMPING
      P             FORECAST          MERIT FN       CHANGE
  0.010000          17.88976          18.82642       335.00
  0.005000          17.68926          18.58802       341.63
  0.002500          17.20306          18.09227       405.57
  0.001250          15.64286          26.36485       877.33
 EDGE,SPACE  8 RAY   7 TARGET   0.500 VALUE    0.100
 EDGE,SPACE  9 RAY  19 TARGET   2.500 VALUE    1.062

ITERATION 4 ADDITIVE DAMPING
      P             FORECAST          MERIT FN       CHANGE
  0.010000          15.86338          15.83307        64.14
  0.005000          15.74494          15.72034        82.37
  0.002500          15.45582          15.47744       186.84
  0.001250          14.75446          16.31408       542.39

ITERATION 5 ADDITIVE DAMPING
      P             FORECAST          MERIT FN       CHANGE
  0.001250          14.45281          14.48802       319.89
  0.000625          14.18773          15.86263       668.50
  0.002500          14.70007          14.68607       105.36
  0.000625          14.18773          15.86263       668.50
```

```
ITERATION 6 ADDITIVE DAMPING
    P              FORECAST          MERIT FN      CHANGE
 0.000625          13.72108          13.80075      442.98
 0.000313          13.67703          14.46337      631.38
 0.001250          13.83550          13.80000      212.17
 0.002500          13.94079          13.93326       82.82
 0.000625          13.72108          13.80075      442.98

ITERATION 7 ADDITIVE DAMPING
    P              FORECAST          MERIT FN      CHANGE
 0.000625          13.13830          13.40683      555.34
 0.000313          13.04831          15.72195      855.41
 0.001250          13.32327          13.27618      238.13
 0.002500          13.45788          13.45217       81.33
 0.000625          13.13830          13.40683      555.34
EDGE,SPACE  4 RAY 13 TARGET   1.250 VALUE   1.135
EDGE,SPACE  7 RAY  7 TARGET   1.250 VALUE   1.125

ITERATION 8 ADDITIVE DAMPING
    P              FORECAST          MERIT FN      CHANGE
 0.010000          13.23281          13.24784       34.35
 0.005000          13.05313          13.08599       71.36
 0.002500          12.90662          12.98322      130.50
 0.001250          12.84127          12.97553      191.89
 0.000625          12.82673          13.00431      230.04
EDGE,SPACE  3 RAY  6 TARGET   0.500 VALUE   0.464
EDGE,SPACE  7 RAY  7 TARGET   1.250 VALUE   1.125

ITERATION 9 ADDITIVE DAMPING
    P              FORECAST          MERIT FN      CHANGE
 0.010000          12.96800          12.96763       12.69
 0.005000          12.95437          12.95353       21.35
 0.002500          12.94563          12.94449       33.54
 0.001250          12.94375          12.94260       40.49
 0.000625          12.94356          12.94245       42.80
 0.000313          12.94355          12.94244       43.43
 0.000156          12.94355          12.94244       43.59

ITERATION 10 ADDITIVE DAMPING
    P              FORECAST          MERIT FN      CHANGE
 0.000156          12.94214          12.94217        3.34
 0.000078          12.94214          12.94217        3.34
 0.000313          12.94214          12.94217        3.32
 0.000625          12.94214          12.94217        3.25
 0.001250          12.94214          12.94217        3.00
 0.000313          12.94214          12.94217        3.32
EDGE,SPACE  3 RAY  6 TARGET   0.500 VALUE   0.500
EDGE,SPACE  7 RAY  7 TARGET   1.250 VALUE   1.243
EDGE,SPACE  8 RAY  7 TARGET   0.500 VALUE   0.491
EDGE,SPACE  9 RAY 19 TARGET   2.500 VALUE   2.495
HT,SURF     1 RAY 22 TARGET  10.750 VALUE  10.753
HT,SURF     9 RAY 19 TARGET  10.500 VALUE  10.504
```

The next few iterations proceed normally; the damping factor, p, gradually reduces over a series of iterations, which is usual; the merit function gradually reduces, and we can see that the parameter changes are quite large. If the program draws the lens at each iteration, we can also see that the inner lenses are moving toward the outer lenses and away from the stop. As the optimization approaches the minimum, the "changes" usually become quite small, and often P also becomes small, as we can see from iterations 11 and 12.

```
ITERATION 11 ADDITIVE DAMPING
    P             FORECAST        MERIT FN      CHANGE
 0.010000         12.90860        12.90844        9.10
 0.005000         12.88496        12.88364       25.21
 0.002500         12.83648        12.82840       71.48
 0.001250         12.78193        12.75843      151.67
 0.000625         12.76235        12.73281      212.86
 0.000313         12.75980        12.73144      236.86
 0.000156         12.75960        12.73210      243.73

ITERATION 12 ADDITIVE DAMPING
    P             FORECAST        MERIT FN      CHANGE
 0.000156         12.71247        12.71214       23.99
 0.000078         12.71247        12.71214       24.08
 0.000039         12.71247        12.71214       24.10
 0.000020         12.71247        12.71214       24.11
 0.000010         12.71247        12.71214       24.11
 0.000005         12.71247        12.71214       24.11
 0.000002         12.71247        12.71214       24.11
 0.000001         12.71247        12.71214       24.11

MERIT FUNCTION = 12.7121
OPTIMIZATION COMPLETED
```

The optimization does not show any sign at this stage of terminating, as the parameter changes are quite large (this is often more significant than the reduction in the merit function). For the purpose of this demonstration, however, we have selected the PAUSE button, which terminates the optimization after iteration 12. The student is encouraged to try optimizing this example to see if a similar or better reduction of the merit function can be achieved, as well as to explore the effects of specifying various merit functions, weights, variables, and constraints.

References

1. G.H. Cook and P.A. Merigold, "Optical Objectives of Variable Focal Length," British Patent 1,064,323 (1963).
2. C.G. Wynne, "Lens designing by electronic digital computer, I," *Proceedings of the Physical Society* **72**, 777 (1959).
3. A. Girard, *Revue d'Optique* **37**, 225, 397
4. K. Levenberg, "A method for the solution of certain non-linear problems in least squares, *Quart. Appl. Math.* **2**, 164 (1944).
5. R. Fletcher, *Practical Methods of Optimization,* John Wiley & Sons, 100 (1987).
6. G. Spencer, "A flexible automatic lens correction procedure," *Appl. Opt.* **2**, 1257 (1963).

7. J.L. Rayces and L. Lebich, "Experiments on constrained optimization with Spencer's method," *Opt. Eng.* **27**, 12 (1988).
8. J. Meiron, "Damped least-squares method for automatic lens design," *JOSA* **55**, 1105 (1965).
9. H. Brunner, *Optica Acta* **18**, 743 (1971).
10. D.C. Dilworth, "Pseudo-second-derivative matrix and its application to automatic lens design," *Appl. Opt.* **17**, 3372 (1978).
11. P.N. Robb, "Accelerating convergence in automatic lens design," *Appl. Opt.* **18**, 4191 (1979).
12. M.J. Hayford, "Optimization methodology," in *SPIE Proc.* Vol. 531, Bellingham, WA (1985).
13. E. Glatzel and R. Wilson, "Adaptive automatic correction in optical design," *Appl. Opt.* **7**, 265 (1968).
14. J.L. Rayces, "Ten years of lens design with Glatzel's adaptive method," in *SPIE Proc.* Vol. 237, 75, Bellingham, WA (1980).
15. W.T. Welford, *Aberrations of Optical Systems*, Adam Hilger, Bristol, UK (1986).
16. J. Meiron, "The use of merit functions on wavefront aberrations in automatic lens design," *Appl. Opt.* **7**, 667 (1968).
17. W.B. King, "Dependence of the Strehl ratio on the magnitude of the variance of the wave aberration," *JOSA* **59**, 1155 (1969).
18. R. Gostick, "Optimization Methods in Optical Design," Ph.D. thesis, London University (1974).
19. M. Kidger and P. Benham, "Optimization of the optical transfer function," in *SPIE Proc.* Vol. 1354, 92, Bellingham, WA (1990).
20. J.L. Rayces, "Exact relation retween wave aberration and ray aberration," *Optica Acta* **11**(2), 85 (1964.).
21. D.C. O'Shea, "The monochromatic quartet: A search for the global optimum," in *SPIE Proc*, Vol. 1354, 549, Bellingham, WA (1990).
22. M.J. Kidger, "Glass selection in optical design," *in SPIE Proc.* Vol. 531, Bellingham, WA (1985).
23. B.H. Walker, "The 20% rule: selecting optical glass," *Optical Spectra*, 42–45 (Dec. 1978).
24. S. Zhang and R. Shannon, "Lens design using a minimum number of glasses," in *SPIE Proc.* Vol. 2263, Bellingham, WA (1992).
25. J. Tesar, "Using small glass catalogs," *Optical Engineering* **39**(7), (2000).
26. M.J. Kidger and C.G. Wynne, "Experiments with lens optimization procedures," *Optica Acta* **14**, 279 (1967).
27. R. Kingslake, *Lens Design Fundamentals*, Academic Press, New York (1978).

CHAPTER 2
BUCHDAHL ABERRATIONS

The use of third-order (Seidel) aberration coefficients has been well established in lens design for very many years. The ability to determine which surface is responsible for the introduction of a certain aberration is almost essential if the designer is to intelligently correct that aberration, and Seidel coefficients form the basis of intelligent lens design. However, for lenses at higher numerical apertures and field sizes, third-order aberrations are insufficient for describing all of the aberrations. In fact, for the majority of lenses, higher-order aberrations need to be included in their design and analysis.[1]

Several authors have developed schemes for calculating the fifth-order and even the seventh-order coefficients. The equations developed by Buchdahl seem to be the most popular, partly due to Rimmer, who described the Buchdahl method in a master's thesis from the University of Rochester.[2,3] The equations discussed below follow those published by Rimmer, but they are exactly equivalent to Buchdahl's equations. These equations also include the seventh-order spherical aberration coefficients, and we will refer to the fifth-order aberration coefficients and seventh-order spherical aberration coefficients collectively as Buchdahl coefficients.

2.1 Third-order coefficients

Using a modified version of the notation of Rimmer, the third-order transverse ray aberration of a lens can be represented as a polynomial of degree 3, as follows:

$$\varepsilon_y' = \mu[B(\cos\theta)\rho^3 + F(2+\cos 2\theta)\rho^2\eta + (3C+\Pi)(\cos\theta)\rho\eta^2 + E\eta^3] \tag{2.1}$$

and

$$\varepsilon_x' = \mu[B(\sin\theta)\rho^3 + F(\sin 2\theta)\rho^2\eta + (C+\Pi)(\sin\theta)\rho\eta^2], \tag{2.2}$$

where

ε_y' is the y-component of transverse aberration,
ε_x' is the x-component of transverse aberration,
ρ is a normalized pupil coordinate,

θ is the azimuth angle of a ray in the pupil,
η is the normalized image radius,
μ is $-1/(2\, n'u')$ where $n'u'$ is the numerical aperture;

and

B is the third-order spherical aberration coefficient,
F is the third-order coma coefficient,
C is the third-order astigmatism coefficient,
Π is the third-order field curvature coefficient, and
E is the third-order distortion coefficient.

Note, then, that the coefficients B, F, C, Π, and E are effectively wavefront aberration terms, and they are in fact identical, apart from a sign reversal, to the S_1, S_2, S_3, S_4, and S_5 coefficients of Welford.[4] Therefore, a positive single lens that is uncorrected for spherical aberration will have a negative B coefficient; this tells us that the transverse aberration of a ray above the axis is negative. The S_1 coefficient is positive, which tells us that the wavefront aberration is positive.

2.2 Fifth-order coefficients

Following Rimmer's notation again, the fifth-order transverse ray aberration of a lens can be represented as a polynomial of degree 5, as follows:

$$\begin{aligned}\varepsilon_y'' = \mu[&B_5(\cos\theta)\rho^5 + (F_1 + F_2\cos 2\theta)\rho^4\eta \\ &+ (M_1 + M_2 + M_3\cos^2\theta)\cos\theta\rho^3\eta^2 \\ &+ (N_1 + N_2\cos^2\theta)\rho^2\eta^3 + (5C_5 + \Pi_5)(\cos\theta)\rho\eta^4 + E_5\eta^5] \end{aligned} \tag{2.3}$$

and

$$\begin{aligned}\varepsilon_x' = \mu[&B_5(\sin\theta)\rho^5 + F_2\sin 2\theta\rho^4\eta \\ &+ (M_2 + M_3\cos^2\theta)\sin\theta\rho^3\eta^2 \\ &+ (N_3\sin 2\theta)\rho^2\eta^3 + (C_5 + \Pi_5)(\sin\theta)\rho\eta^4]. \end{aligned} \tag{2.4}$$

It has become conventional in lens design to describe the fifth-order aberrations in the following way:

$\mu B_5 = \mathrm{SA5}$	Fifth-order spherical aberration
$\mu(F_1 + F_2) = \mathrm{COMA5}$	Fifth-order coma
$\mu(M_1 + M_2 + M_3) = \mathrm{TOBSA}$	Tangential oblique spherical aberration
$\mu M_2 = \mathrm{SOBSA}$	Sagittal oblique spherical aberration
$\mu(N_1 + N_2) = \mathrm{LCOMA}$	Elliptical coma

$\mu C_5 = \text{AST5}$ Fifth-order astigmatism
$\mu \Pi_5 = \text{PETZ5}$ Fifth-order Petzval
$\mu E_5 = \text{DIST5}$ Fifth-order distortion.

Note that, for systems in which third-order aberrations are small, the following are approximately true:

$$2F_1 = 3F_2, \quad M_1 = M_3, \quad N_1 = N_3. \tag{2.5}$$

2.3 Comparison with H.H. Hopkins notation

Following H.H. Hopkins,[5] we describe the wavefront aberration as a power series:

$W = {}_0w_{40}\,\rho^4$	Seidel spherical aberration
$+\ {}_1w_{31}\,\eta\rho^3 \cos\phi$	Seidel coma
$+\ {}_2w_{22}\,\eta^2\rho^2 \cos^2\phi$	Seidel astigmatism
$+\ {}_2w_{20}\,\eta^2\rho^2$	Seidel field curvature
$+\ {}_3w_{11}\,\eta^3\rho\cos\phi$	Seidel distortion

and the following nine "fifth-order" terms:

$+\ {}_0w_{60}\,\rho^6$	Spherical aberration
$+\ {}_1w_{51}\,\eta\rho^5 \cos\phi$	Coma
$+\ {}_2w_{42}\,\eta^2\rho^4 \cos^2\phi$	Oblique spherical aberration
$+\ {}_2w_{40}\,\eta^2\rho^4$	Oblique spherical aberration
$+\ {}_3w_{31}\,\eta^3\rho^3 \cos\phi$	Elliptical coma
$+\ {}_3w_{33}\,\eta^3\rho^3 \cos^3\phi$	Elliptical coma
$+\ {}_4w_{20}\,\eta^4\rho^2$	Fifth-order Petzval
$+\ {}_4w_{22}\,\eta^4\rho^2 \cos^2\phi$	Fifth-order astigmatism
$+\ {}_5w_{11}\,\eta^5\rho \cos\phi$	Fifth-order distortion, (2.6)

where W is the wavefront aberration, η is the object size, and ρ and ϕ are polar pupil coordinates.

2.4 Examples

2.4.1 Double Gauss

We discuss as an example of the use of Buchdahl coefficients the same four-glass double Gauss whose optimization with exact rays we looked at in Chapter 1, Fig. 1.21. The aberration curves show, in Figs. 2.1, that there is significant oblique

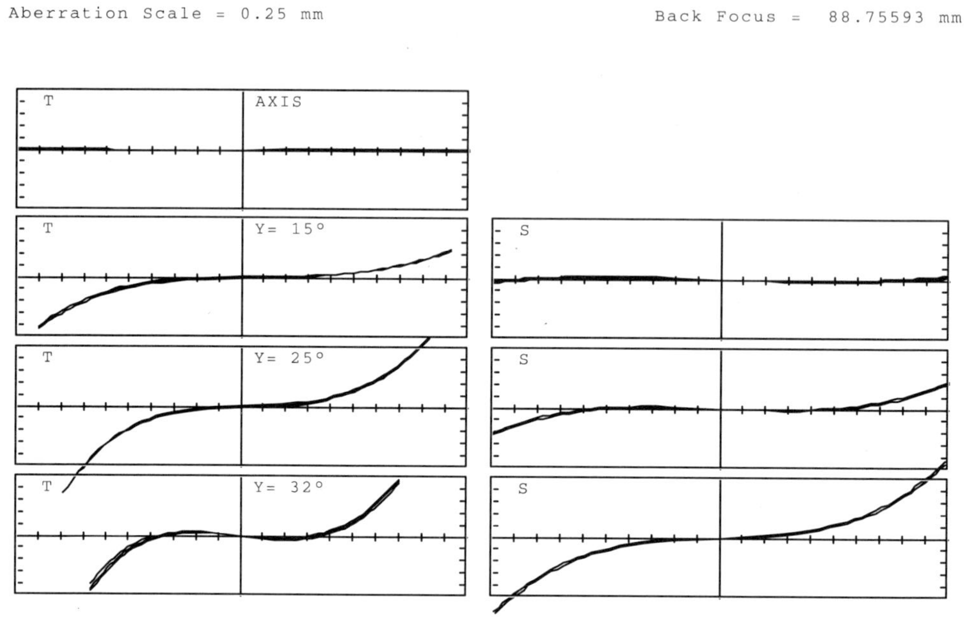

Figure 2.1 Transverse ray aberrations before optimization.

spherical aberration, both sagittal and tangential, i.e., SOBSA and TOBSA. They also show (although it is more clear in the astigmatism plots, Fig. 2.2), that AST5 and PETZ5 are quite significant.

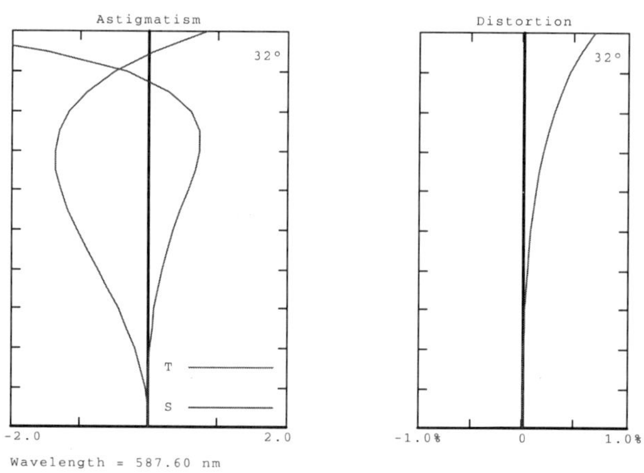

Figure 2.2 Field aberrations before optimization.

EFL = 98.803
WAVELENGTHS [nm]
587.60 656.30 486.10
SURFACES

#	SURF SPACE	RADIUS	SEPN	INDEX1	V	CLR RAD	GLASS
0			0.00000	1.000000			
1	S	25.90674	5.08300	1.612718	58.61	10.600	S-SK4

2 S	147.34050	2.35500	1.000000		8.890	
3 S	34.80440	1.69400	1.616589	36.61	6.050	S-F4
4 S	17.34004	2.54200	1.000000		5.250	
5#S	Plane	2.54200	1.000000		5.147	
6 S	-17.34004	1.69400	1.616589	36.61	5.200	S-F4
7 S	-34.80440	2.35500	1.000000		6.000	
8 S	-138.86960	5.08300	1.612718	58.61	8.600	S-SK4
9 S	-24.39619	88.75580	1.000000		10.240	
10 S	Plane				63.015	

In order to calculate the fifth-order aberrations, we first trace a paraxial marginal ray and a paraxial chief ray, as we do for third-order aberrations:

4-GLASS DOUBLE GAUSS DERIVATIVE LAGRANGE INVARIANT = -3.9054

H	U	I	HBAR	UBAR	IBAR
6.250000	0.000000	0.241250	-7.101831	0.624870	0.350739
5.784103	-0.091658	-0.052401	-4.602958	0.491614	0.460373
5.492637	-0.123765	0.034050	-2.780911	0.773693	0.693792
5.260979	-0.136752	0.166649	-1.918544	0.509071	0.398429
5.174556	-0.033998	-0.033998	-0.000000	0.754738	0.754738
5.088133	-0.033998	-0.327431	1.918544	0.754738	0.644096
5.242097	0.090888	-0.059728	2.780912	0.509072	0.429170
5.369411	0.054061	0.015396	4.602958	0.773693	0.740547
5.614470	0.048212	-0.181926	7.105511	0.492338	0.201083
	-0.063257			0.615545	

We then calculate a table of Seidel (third-order surface coefficients); these are essentially wavefront aberrations, but with simple scaling factors of 1/2, 1/4, and 1/8 to relate them to actual wavefront aberrations.

We also calculate the third-order pupil spherical aberration and pupil coma:

SA3	COMA3	AST3	PETZ3	DIST3	LCOLOR	TCOLOR	PUP-SA	PUP-CO
-S1	-S2	-S3	-S4	-S5	C1	C2	-S1bar	-S2bar
-0.02067	-0.03006	-0.04370	-0.22368	-0.38873	0.00978	0.01421	0.27960	0.19232
-0.00276	0.02429	-0.21340	0.03933	1.52932	0.00317	-0.02784	-1.18964	0.13541
0.00025	0.00508	0.10357	-0.16715	-1.29555	0.00195	0.03970	0.61413	0.03014
0.01932	0.04619	0.11043	0.33549	1.06612	-0.01476	-0.03530	-0.35007	-0.14642
0.00000	0.00000	0.00000	0.00000	0.00000	0.00000	0.00000	0.00000	0.00000
0.04922	-0.09681	0.19044	0.33549	-1.03458	-0.01736	0.03414	-0.35007	0.17796
-0.00011	0.00076	-0.00545	-0.16715	1.24022	0.00527	-0.03789	0.61413	-0.08547
-0.00003	-0.00148	-0.07116	0.04173	-1.41571	0.00054	0.02578	-1.18241	-0.02458
-0.04502	0.04976	-0.05500	-0.23753	0.32334	0.01068	-0.01180	0.23184	-0.20975
Total aberration								
0.00019	-0.00227	0.01572	-0.04346	0.02443	-0.00074	0.00100	-1.33250	0.06960

It is then possible to calculate the fifth-order terms. The most fundamental difference between the fifth-order aberrations and the third-order aberrations is that the fifth-order aberrations can all be induced by the presence of third-order aberrations in surfaces before the surface under consideration. In contrast, third-order aberrations are not affected by the presence of third-order aberrations in surfaces before the surface under consideration. They are, however, affected greatly by the first-order properties of the incoming beam, so the general principle is that aberrations of any order are affected by the lower-order properties of the beam. As a generalization, it is most common for fifth-order aberrations of a surface to be induced in this way, than for them to be intrinsic properties of the surface.

We see below a table of fifth-order coefficients and seventh-order spherical aberration. These are effectively wavefront aberrations, as they have not been multiplied by the term μ, referred to in the description given above.

```
   SA5      COMA5     LCOMA      AST5      PETZ5     SOBSA     TOBSA     DIST5      SA7
-0.00069 -0.00383  0.02014  0.00758  0.03999 -0.00617 -0.00460  0.06915 -0.00003
-0.00026  0.00383  0.04458 -0.06132 -0.01357 -0.00659 -0.01298  0.16578 -0.00002
-0.00004 -0.00071 -0.01142  0.05802  0.01905  0.00296  0.01104 -0.17261 -0.00000
 0.00046  0.00860  0.06231 -0.01040 -0.02210  0.01812  0.04509  0.13169 -0.00000
 0.00000  0.00000  0.00000  0.00000  0.00000  0.00000  0.00000  0.00000  0.00000
 0.00219 -0.02711 -0.05148 -0.01817 -0.02703  0.02462  0.08057 -0.00087  0.00014
 0.00001 -0.00210 -0.07798  0.04261  0.01166  0.00683  0.02584  0.16287 -0.00001
-0.00003 -0.00196  0.07167 -0.04392 -0.00849 -0.00810 -0.01896 -0.21459 -0.00001
-0.00215  0.01286 -0.06292  0.02206  0.04162 -0.00607 -0.00405 -0.11767 -0.00017
Total aberration
-0.00050 -0.01043 -0.00511 -0.00355  0.04112  0.02560  0.12196  0.02375 -0.00009
```

We then give a summary of the total aberrations, using the conversion factor of μ:

```
Total Aberration Coefficients converted to Transverse Aberrations
   SA3  0.00151  COMA3 -0.05446   AST3  0.12576  PETZ3 -0.34768  DIST3  0.19544
LCOLOR -0.01183 TCOLOR  0.01594

   SA5 -0.00403  COMA5 -0.08341  LCOMA -0.04086   AST5 -0.02837  PETZ5  0.32898
 SOBSA  0.20481  TOBSA  0.97567  DIST5  0.18999

   SA7 -0.00070
```

The fifth-order analysis confirms the results shown in the aberration curves, where we can see that the largest single fifth-order aberration is TOBSA (0.97567), but SOBSA and PETZ5 are also greater than 0.1 mm. However, the fifth-order analysis does not give us so effective diagnostic information concerning the reason for the existence of particular fifth-order aberrations at a given surface. This is in part because the equations are very complicated, but this complication is due to the complex way in which fifth-order aberrations can be induced.

After a DLS optimization, which includes a change of glass types to the system shown in Fig. 2.3, some of the high-order aberrations are significantly reduced, as shown in Fig. 2.4 and Fig. 2.5. Clearly, the aberration curves show that the astigmatism has been reduced by a large factor, and TOBSA and SOBSA are also much reduced.

```
EFL =    100.000
WAVELENGTHS  [nm]
      587.60       656.30       486.10
SURFACES
  # SURF SPACE     RADIUS        SEPN       INDEX1       V      CLR RAD  GLASS
  0                           -1.94982   1.000000
```

1 S	16.52123	3.78206	1.642496	57.94	10.750	S-LAKN6
2 S	35.19548	0.46939	1.000000		9.883	
3 S	23.49829	0.96657	1.612928	37.02	8.945	S-F3
4 S	13.52743	7.06558	1.000000		7.685	
5#S	Plane	7.07178	1.000000		5.185	
6 S	-11.67918	0.96657	1.612928	37.02	7.257	S-F3
7 S	-17.58258	0.46725	1.000000		8.356	
8 S	-32.22066	3.78206	1.642496	57.94	9.495	S-LAKN6
9 S	-16.77160	86.20271	1.000000		10.501	
10 S	Plane				62.737	

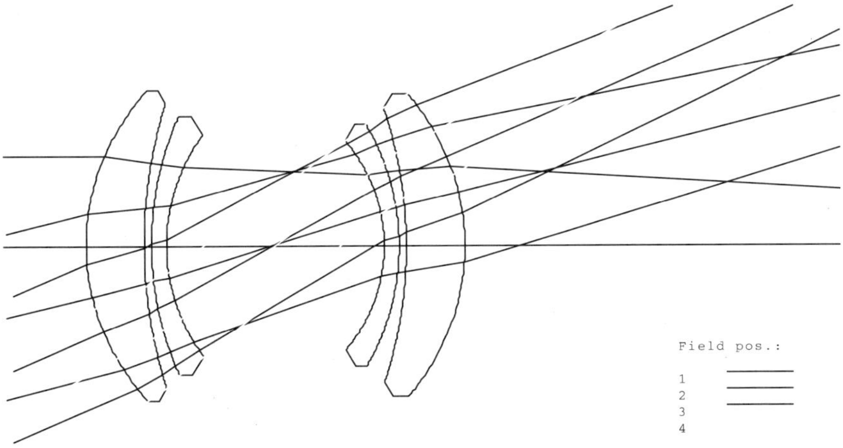

Figure 2.3 Four-glass double Gauss after optimization.

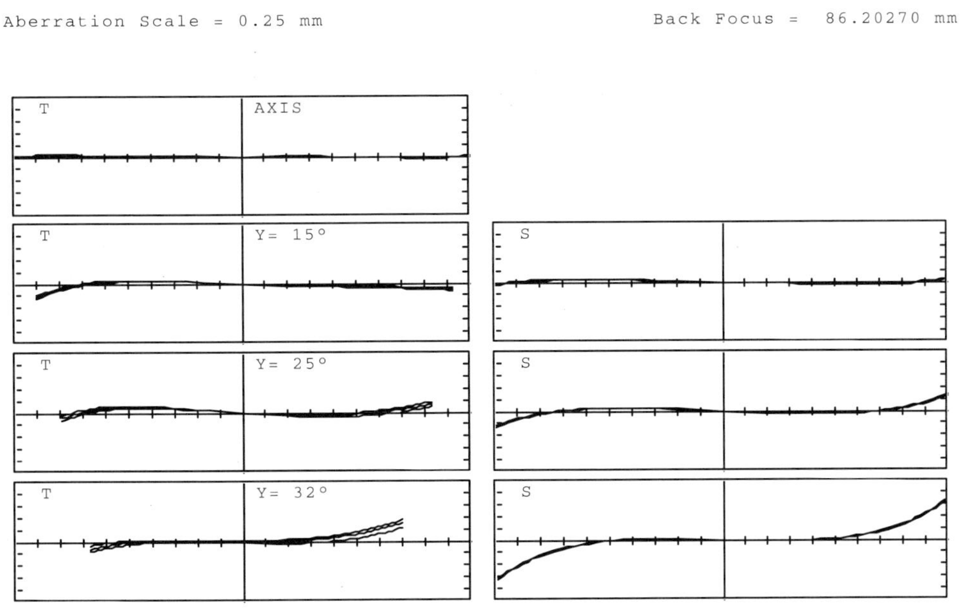

Figure 2.4 Transverse ray aberrations after optimization.

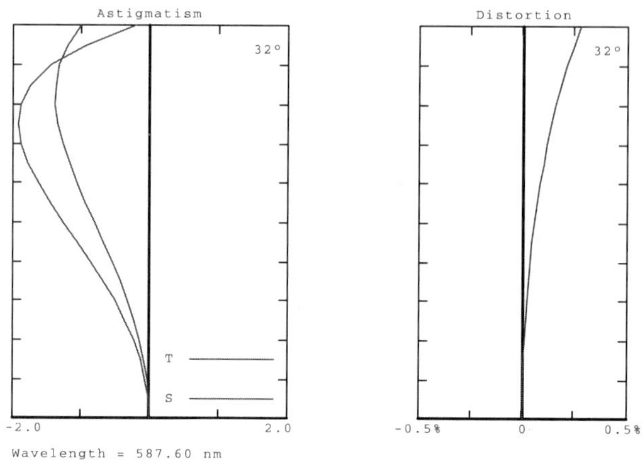

Figure 2.5 Field aberrations after optimization.

The third- and fifth-order contributions (below) show that this improvement has come from several factors. The fifth-order AST5 and PETZ5 are more than halved; this permits the third-order AST3 and PETZ3 to be similarly reduced, to maintain a good balance of aberrations. Why? For the case of AST5 and PETZ5, we can visualize a mechanism that causes this effect. Assuming that each component has little intrinsic aberration, induced aberrations arise because the third-order aberrations are nonzero, and lenses are operating in an aberrated beam. We can indeed see that the surface contributions to S3 are smaller in the optimized design.

Also, TOBSA is more than halved. We can also see that the large TOBSA contributions in the original design arise at the inner negative lenses; in the optimized design, the contributions are much smaller.

SA3 -S1	COMA3 -S2	AST3 -S3	PETZ3 -S4	DIST3 -S5	LCOLOR C1	TCOLOR C2	PUP-SA -S1bar	PUP-CO -S2bar
-0.08059	-0.02325	-0.00671	-0.36113	-0.1061	0.01596	0.00461	0.02744	0.09513
-0.00014	-0.00416	-0.12216	0.16952	1.3920	-0.00086	-0.02541	-1.34433	-0.04573
0.00165	0.00972	0.05710	-0.24666	-1.1139	0.00579	0.03399	0.94107	0.16015
0.05148	0.05089	0.05030	0.42847	0.4733	-0.02040	-0.02016	-0.25668	-0.25966
0.00000	0.00000	0.00000	0.00000	0.0000	0.00000	0.00000	0.00000	0.00000
0.12683	-0.08223	0.05331	0.49627	-0.3563	-0.02310	0.01498	-0.16825	0.25950
-0.01222	0.02450	-0.04911	-0.32965	0.7593	0.01258	-0.02522	0.60768	-0.30313
0.00067	-0.00372	0.02059	0.18517	-1.1379	-0.00389	0.02150	-1.18910	0.21502
-0.09190	0.02807	-0.00857	-0.35574	0.1113	0.01396	-0.00426	0.03197	-0.10467
Total aberration								
-0.00422	-0.00018	-0.00525	-0.01375	0.0217	0.00004	0.00002	-1.35019	0.01661

SA5	COMA5	LCOMA	AST5	PETZ5	SOBSA	TOBSA	DIST5	SA7
-0.00659	-0.00205	0.04573	0.00299	0.04301	-0.01863	0.01489	0.0133	-0.00064
0.00013	0.00401	-0.03069	-0.01244	-0.04807	-0.01382	-0.03866	0.1463	-0.00001
-0.00024	0.00194	0.03249	0.01653	0.03929	0.00746	0.03052	-0.1629	-0.00004
0.00299	0.00910	-0.02769	-0.00602	-0.02546	0.02988	0.02073	0.0456	0.00015
0.00000	0.00000	0.00000	0.00000	0.00000	0.00000	0.00000	0.0000	0.00000
0.01406	-0.04291	0.09370	-0.01661	-0.02431	0.04323	0.01504	0.0448	0.00191
0.00024	-0.00790	-0.04023	0.00469	0.02628	-0.00226	0.02217	0.0485	0.00001

```
-0.00011  0.00543  0.02997  0.00460 -0.03794 -0.01139 -0.03977 -0.0620  0.00002
-0.00741  0.01094 -0.07889  0.00851  0.04424 -0.01524  0.02023 -0.0786 -0.00084
Total aberration
 0.00306 -0.02145  0.02438  0.00225  0.01705  0.01921  0.04515 -0.0050  0.00057

Total Aberration Coefficients converted to Transverse Aberrations
   SA3 -0.03373  COMA3 -0.00429   AST3 -0.04203  PETZ3 -0.10996  DIST3  0.17363
LCOLOR  0.00065 TCOLOR  0.00029

   SA5  0.02444  COMA5 -0.17160  LCOMA  0.19504   AST5  0.01797  PETZ5  0.13639
 SOBSA  0.15371  TOBSA  0.36119  DIST5 -0.04012

   SA7  0.00454
```

2.4.2 Shafer lens with zero third- and fifth-order aberrations

Glatzel and Shafer have discussed at length the concept of designing a "relaxed" system, in which the aberrations introduced at each surface are kept as small as possible.[7,8] In this way, we might expect that the induced high-order aberrations will be smaller. Another benefit of this approach is that manufacturing tolerances of such designs will tend to be as large as possible, since surfaces with large aberration contributions are usually more sensitive to misplacement along or lateral to the optical axis.

Shafer has in fact gone further than this, and suggested that a profitable approach is to optimize for all monochromatic, third-, and fifth-order aberrations.[9] Figure 2.6 shows a design similar to Shafer's, in which it is found to be quite simple to obtain correction of all monochromatic Buchdahl coefficients. Figure 2.7 shows significant seventh-order spherical aberration. This is perhaps not surprising, as it was not included in the merit function.

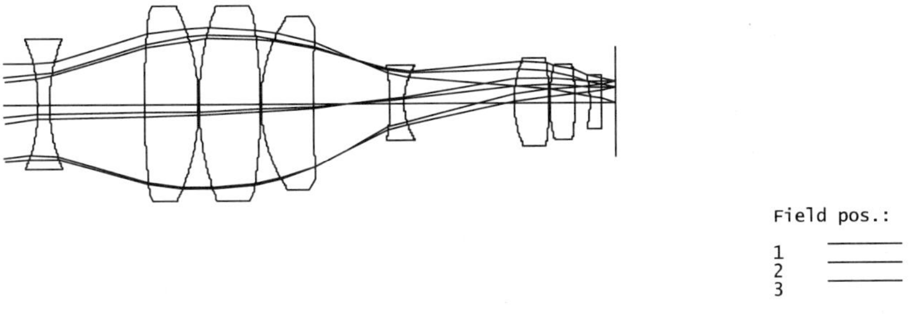

Figure 2.6 Shafer lens with zero third- and fifth-order aberrations.

```
EFL =     30.492
WAVELENGTHS  [nm]
      587.60
SURFACES
  # SURF SPACE    RADIUS        SEPN      INDEX1     V    CLR RAD  GLASS
  0                          0.00000   1.000000
  1 S            -32.94420   2.00000   1.600000          12.000  User-FICTIONL
  2 S             34.76590  16.53800   1.000000          12.000
  3 S            170.23020   9.40000   1.600000          17.000  User-FICTIONL
```

```
  4 S         -45.32520   0.10000   1.000000          18.000
  5 S          56.70250  10.70000   1.600000          18.000  User-FICTIONL
  6 S        -106.00180   0.10000   1.000000          17.000
  7 S          31.22220   9.00000   1.600000          16.000  User-FICTIONL
  8 S         241.22540   5.97400   1.000000          14.000
  9#S           Plane     7.47600   1.000000           8.636
 10 S         -36.97010   2.40000   1.600000           7.000  User-FICTIONL
 11 S          14.08060  19.16200   1.000000           7.000
 12 S          21.73820   6.10000   1.600000           8.000  User-FICTIONL
 13 S         -54.62330   0.10000   1.000000           8.000
 14 S          29.82150   4.45000   1.600000           7.000  User-FICTIONL
 15 S         -49.01340   2.60200   1.000000           7.000
 16 S         -22.75420   2.00000   1.600000           5.000  User-FICTIONL
 17 S           Plane     2.45530   1.000000           5.000
 18 S           Plane                                 10.000
```

```
LAGRANGE INVARIANT =     -0.9362
     H          U          I                HBAR       UBAR       IBAR
  7.625000  0.000000 -0.231452           -2.674157  0.122785  0.203957
  7.798589  0.086794  0.311112           -2.581555  0.046301 -0.027955
 12.321095  0.273461  0.345841           -2.093223  0.029528  0.017231
 13.672545  0.143771 -0.157883           -1.876402  0.023066  0.064465
 13.677450  0.049041  0.290256           -1.870227  0.061745  0.028762
 13.037542 -0.059804 -0.182798           -1.324963  0.050959  0.063459
 13.020594 -0.169483  0.247547           -1.316060  0.089034  0.046883
 10.659774 -0.262313 -0.218123           -0.672980  0.071453  0.068664
  8.310874 -0.393187 -0.393187            0.000000  0.112651  0.112651
  5.371407 -0.393187 -0.538478            0.842182  0.112651  0.089871
  4.912388 -0.191258  0.157618            1.031661  0.078950  0.152218
  3.059670 -0.096687  0.044064            4.294576  0.170280  0.367839
  2.369083 -0.113211 -0.156582            4.491854  0.032341 -0.049893
  2.348367 -0.207160 -0.128413            4.492094  0.002405  0.153038
  1.640793 -0.159006 -0.192482            4.247415 -0.054984 -0.141642
  0.926558 -0.274495 -0.315215            3.883215 -0.139969 -0.310629
  0.613980 -0.156289 -0.156289            3.836247 -0.023484 -0.023484
           -0.250062                                -0.037574
```

```
These are wavefront aberrations with simple scaling factors
    SA3     COMA3      AST3     PETZ3     DIST3    LCOLOR    TCOLOR    PUP-SA    PUP-CO
    -S1       -S2       -S3       -S4       -S5        C1        C2    -S1bar    -S2bar
 0.02216 -0.01953  0.01721  0.00998 -0.0240   0.00000   0.00000   0.01044  -0.01185
 0.42360 -0.03806  0.00342  0.00945 -0.0012   0.00000   0.00000  -0.00000   0.00003
-0.27057 -0.01348 -0.00067 -0.00193 -0.0001   0.00000   0.00000   0.00001   0.00019
-0.03561  0.01454 -0.00594 -0.00725  0.0054   0.00000   0.00000  -0.00094   0.00231
-0.09958 -0.00987 -0.00098 -0.00580 -0.0007   0.00000   0.00000   0.00005   0.00047
-0.14733  0.05115 -0.01776 -0.00310  0.0072   0.00000   0.00000  -0.00078   0.00225
 0.00442  0.00084  0.00016 -0.01053 -0.0020   0.00000   0.00000   0.00013   0.00068
-0.29764  0.09369 -0.02949  0.00136  0.0089   0.00000   0.00000  -0.00055   0.00175
 0.00000  0.00000  0.00000  0.00000  0.0000   0.00000   0.00000   0.00000   0.00000
 0.42621 -0.07113  0.01187  0.00889 -0.0035   0.00000   0.00000  -0.00043   0.00258
 0.00714  0.00689  0.00666  0.02334  0.0290   0.00000   0.00000   0.00740   0.00766
 0.00015  0.00129  0.01073 -0.01512 -0.0366   0.00000   0.00000  -0.08720  -0.01045
-0.02028 -0.00646 -0.00206 -0.00602 -0.0026   0.00000   0.00000  -0.00051  -0.00160
 0.00417 -0.00497  0.00593 -0.01102  0.0061   0.00000   0.00000  -0.00387   0.00325
-0.02725 -0.02005 -0.01476 -0.00671 -0.0158   0.00000   0.00000  -0.02304  -0.03131
 0.01628  0.01604  0.01581  0.01445  0.0298   0.00000   0.00000   0.04695   0.04764
-0.00585 -0.00088 -0.00013  0.00000 -0.0000   0.00000   0.00000  -0.00012  -0.00083
```

Total aberration

0.00001	-0.00000	0.00000	0.00000	-0.0000	0.00000	0.00000	-0.05248	0.01279

SA5	COMA5	LCOMA	AST5	PETZ5	SOBSA	TOBSA	DIST5	SA7
0.00068	-0.00277	-0.00312	0.00022	-0.00020	0.00106	0.00433	-0.0000	0.00002
0.07971	-0.06136	-0.00939	0.00022	-0.00007	0.00931	0.03033	-0.0001	0.01548
-0.06115	0.02561	0.00340	0.00007	0.00014	-0.00549	-0.01357	0.0000	-0.01381
-0.00147	0.00035	-0.00004	-0.00002	0.00016	-0.00019	0.00027	-0.0000	-0.00022
-0.01582	0.00558	0.00096	0.00003	0.00019	-0.00163	-0.00253	0.0001	-0.00319
-0.02075	0.01810	0.00469	-0.00024	0.00004	-0.00284	-0.00993	0.0001	-0.00425
0.00027	0.00123	0.00021	-0.00002	0.00016	-0.00045	0.00016	0.0001	-0.00001
-0.06616	0.06292	0.01487	-0.00060	-0.00013	-0.00936	-0.03639	0.0002	-0.01528
0.00000	0.00000	0.00000	0.00000	0.00000	0.00000	0.00000	0.0000	0.00000
0.08651	-0.04792	-0.01114	0.00040	0.00011	0.00826	0.02517	-0.0001	0.01748
0.00055	0.00304	0.00409	0.00003	0.00030	0.00229	0.00601	0.0003	0.00005
0.00007	0.00066	-0.00248	0.00021	-0.00050	-0.00070	-0.00118	-0.0014	0.00001
-0.00090	-0.00197	-0.00297	-0.00026	-0.00012	-0.00075	-0.00211	-0.0004	-0.00003
-0.00002	-0.00024	0.00041	-0.00029	-0.00010	-0.00004	0.00046	0.0002	-0.00002
-0.00270	-0.00941	-0.01134	-0.00129	-0.00066	-0.00283	-0.01263	-0.0022	-0.00026
0.00158	0.00647	0.01224	0.00157	0.00073	0.00324	0.01129	0.0033	0.00014
-0.00038	-0.00029	-0.00038	-0.00002	-0.00003	0.00012	0.00031	-0.0000	-0.00002

Total aberration

0.00000	-0.00000	-0.00000	0.00000	0.00000	0.00000	0.00000	-0.0000	-0.00390

```
Total Aberration Coefficients converted to Transverse Aberrations
   SA3  0.00002  COMA3 -0.00001    AST3  0.00000  PETZ3  0.00000  DIST3 -0.00001
LCOLOR  0.00000 TCOLOR  0.00000
   SA5  0.00001  COMA5 -0.00000   LCOMA -0.00000    AST5  0.00000  PETZ5  0.00000
 SOBSA  0.00000  TOBSA  0.00000   DIST5 -0.00000
   SA7 -0.00780
```

It is interesting to compare this design with a similar system that is simply optimized in the normal way, with a ray-based merit function that includes enough aberrations to control high-order terms. We find, not surprisingly, that the aberra-

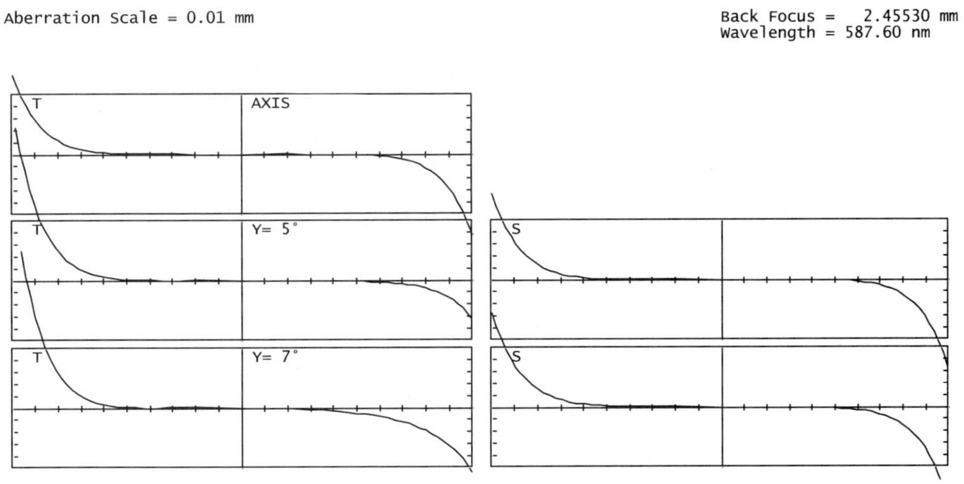

Figure 2.7 Transverse ray aberrations of Shafer lens with zero-, third-, and fifth-order aberrations.

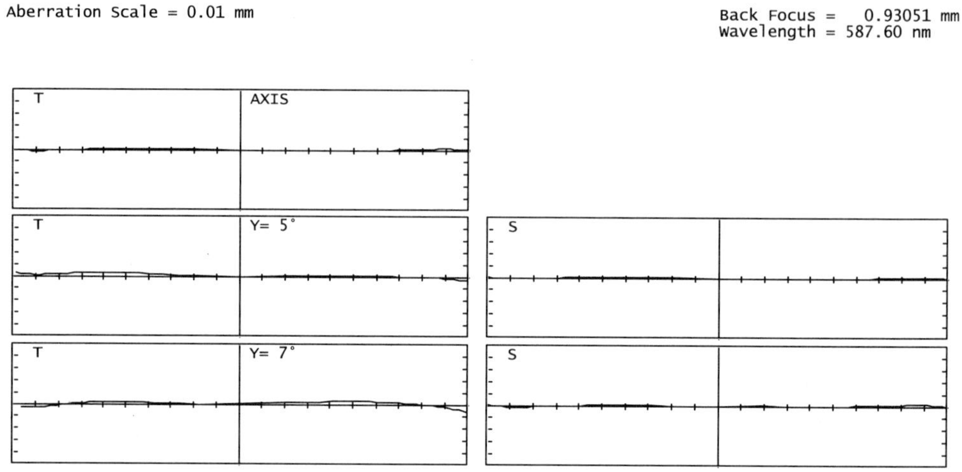

Figure 2.8 Transverse ray aberrations of Shafer lens reoptimized with exact ray aberrations.

tion balance is much better, as shown in Fig. 2.8. It is for this reason that, except perhaps in the early stages of a design, it is almost always preferable to optimize with sets of real rays, rather than with Buchdahl aberrations.

```
EFL =     30.500
WAVELENGTHS  [nm]
      587.60
SURFACES
```

#	SURF SPACE	RADIUS	SEPN	INDEX1	V	CLR RAD	GLASS
0			0.00000	1.000000			
1	S	-37.31650	2.00000	1.600000		10.283	User-FICTIONL
2	S	39.03369	16.50000	1.000000		10.442	
3	S	147.31895	9.40000	1.600000		14.088	User-FICTIONL
4	S	-49.94818	0.10000	1.000000		14.766	
5	S	73.85676	10.70000	1.600000		14.702	User-FICTIONL
6	S	-89.92541	0.10000	1.000000		14.001	
7	S	31.37205	9.00000	1.600000		13.025	User-FICTIONL
8	S	171.95476	6.00000	1.000000		10.968	
9#	S	Plane	7.50000	1.000000		8.456	
10	S	-51.34774	2.40000	1.600000		6.603	User-FICTIONL
11	S	18.37541	19.00000	1.000000		6.180	
12	S	18.46852	6.10000	1.600000		6.924	User-FICTIONL
13	S	-62.23595	0.10000	1.000000		6.368	
14	S	25.06957	4.45000	1.600000		6.049	User-FICTIONL
15	S	45.71355	2.55000	1.000000		4.998	
16	S	-25.09858	2.00000	1.600000		4.307	User-FICTIONL
17	S	Plane	0.93051	1.000000		3.963	
18	S	Plane				3.783	

```
Total Aberration Coefficients converted to Transverse Aberrations
   SA3 -0.00174  COMA3  0.00311   AST3 -0.00044  PETZ3  0.00026  DIST3 -0.00088
LCOLOR  0.00000 TCOLOR  0.00000
   SA5  0.00476  COMA5 -0.00525  LCOMA  0.00019   AST5  0.00035  PETZ5 -0.00006
 SOBSA  0.00025  TOBSA -0.00158  DIST5  0.00115
   SA7 -0.00182
```

References

1. F.D. Cruickshank and G.A. Hills, "Use of optical aberration coefficients in optical design," *JOSA* **50**, 379–387 (1960).
2. H.A. Buchdahl, *Optical Aberration Coefficients*, Oxford Univ. Press (1954).
3. M.P. Rimmer, "Optical Aberration Coefficients," Master's thesis, Univ. of Rochester (1963).
4. W.T. Welford, *Aberrations of Optical Systems*, Adam Hilger (1987).
5. H.H. Hopkins, *The Wave Theory of Aberrations,* Oxford Univ. Press (1950).
6. R. Kingslake, *Lens Design Fundamentals*, Academic Press, (1978).
7. E. Glatzel, "New lenses for microlithography," in *SPIE Proc.* Vol. 237, Bellingham, WA (1980).
8. D. Shafer, "I plead the fifth," in *SPIE Proc.* Vol. 1049, Bellingham, WA (1989).
9. D. Shafer, "Design challenges for the 1990s," in *SPIE Proc.* Vol. 1354, Bellingham, WA (1990).

CHAPTER 3
SYNTHESIS OF NEW LENS DESIGNS

We already know that DLS lens design optimization programs are, in most practical cases, unable to do more than find a local minimum in parameter space.[1] Even in the simple case of the split doublet, a program can converge on the poor Gauss solution, even though the Fraunhofer solution is usually greatly superior. However, we also know that the capability of such programs to find a local minimum is of great value in lens design, as the optimization routines described in Chapter 1 are usually able to find this minimum quickly and effectively.

While there has been considerable progress in the development of so-called global optimization programs, they do not yet guarantee that the true global solution has really been found. Forbes[2] has shown that a global optimization program, running on a computer tracing 1 million ray surfaces per second could take 10^{11} seconds (about 10,000 years) to have any reasonable certainty of finding the true global minimum for the 15-parameter problem that was proposed for the 1990 International Lens Design conference. Since that time, global search programs have found solutions similar to the best one shown in Fig. 1.3, but not a better one.[3,4,5] Lens systems often have many more than 15 variables; for practical purposes, therefore, the result of a lens design optimization remains critically dependent on the starting point given to the program.

3.1 Choice of a starting point

It is possible to identify four distinct methods of finding a starting point, some of which have been discussed by Kingslake.[6]

3.1.1 Modification of an existing design

A design might come from a company's files, or from the patent literature, or from a collection of lens designs. This is probably the usual way to start a new design problem. Where the new design is a relatively minor variation on an existing design, this approach is almost always the most sensible, although it does almost guarantee that a novel solution will not be found. Most companies are not concerned with the whole range of possible lens design types, but only with a small subset. For example, some designers have extensive files describing photographic objectives, others design lenses for laser applications, and so on. Therefore, most designers become experienced in the particular types of system that they are mainly concerned with, and modification of existing designs is eminently sensible.

If a simple reoptimization of an existing design is not successful, it is often possible to split cemented components and to add components, so it is very likely that the new design will differ greatly from the starting design, particularly if the specifications are quite different. Of course, if the starting design was taken from the patent literature, and if the patent is still valid, it is essential that the final design should not infringe the claims of the patent.

3.1.2 Purchase of a competing lens

According to Kingslake, the purchase of a competing lens, followed by an analysis of its structure, has often been done. In most cases, this is not a realistic option, as either the competitor's lenses are generally not available, or, quite often, there is no direct competitor with an existing design.

3.1.3 Analytic solutions

In some cases, e.g., doublets, triplets, and some monocentric systems, it is possible to write down a set of equations that can be analytically solved to find at least a thin-lens solution. However, these cases are generally simple and therefore should be well known, so it is extremely uncommon to find that an analytic solution is an effective way of solving a new problem.

3.1.4 Nonanalytic synthesis of new design forms

Nonanalytic synthesis of new design forms the subject of this chapter. Suppose that we were unaware of the existence of the double-Gauss objective, for example; what thought processes might result in us discovering it for ourselves?

The starting point for the synthesis of any new lens design must be an understanding of paraxial optics and aberration theory.[7,8] In practice, this means first- and third-order (Seidel) theory, since fifth-order theory is very much more complicated. As we have discussed in the previous chapters, high-order aberrations are induced because a surface is operating in an aberrated beam. It is usually considered that induced higher-order aberrations are much larger than intrinsic higher-order aberrations, and this implies that it is generally preferable to design systems in which the aberrations of individual components (or surfaces) are small, so that surfaces that follow them are not operating in a highly aberrated beam. Since this also assists in making the lens less sensitive in manufacture, it is almost always desirable.

3.2 Examples

3.2.1 A unit magnification telecentric doublet pair

A common problem in lens design is the telecentric doublet pair. If a pair of identical doublets of focal length *f* is separated by a distance 2*f*, they form a telecentric system. If they are working at unit magnification, they can have front-back symmetry, and therefore be free from coma, distortion, and lateral color. We therefore consider a telecentric pair of doublets, with a numerical aperture of 0.10, with an image radius of 2 mm, with an (arbitrary) object-image distance of about 67 mm.

Initially we will represent each doublet by a nondispersive aspheric single lens. Since the dispersion is zero, the chromatic aberration is automatically zero. Like a doublet, an aspheric single lens can be corrected for spherical aberration and coma; the doublet can be corrected for axial color, so, apart from higher-order aberrations and secondary spectrum, an aspheric singlet might usefully represent a doublet during the early stages of the design.

We start by setting up a pair of plano-convex spherical singlets, as shown in Fig. 3.1. The system is evaluated only at the central wavelength, to avoid chromatic aberrations. The focal length of each lens is chosen to make the system telecentric, given a central air space of 32 mm, so the lens is telecentric in object space (the chief ray is traveling parallel to the optical axis as it leaves the object from an entrance pupil at infinity). The refractive index is quite arbitrary at this stage, as it will be changed later when an achromatic doublet replaces the singlet.

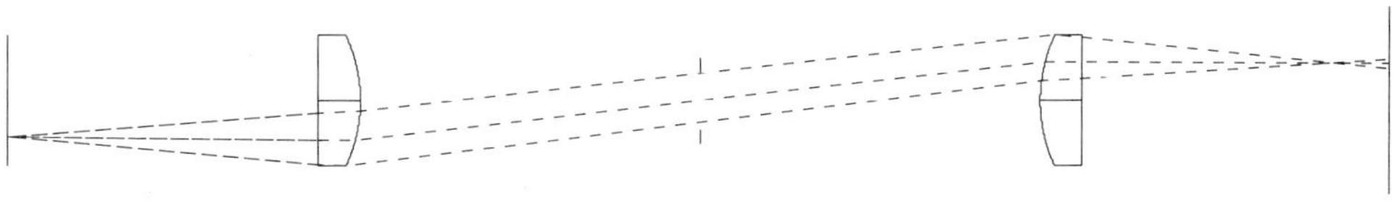

Figure 3.1 Representation by a fictional pair of spherical singlets.

```
 EFL = -5576.885
OBJECT DISTANCE = 14.7351, CURVATURE = 0, TILT ANGLE = 0
WAVELENGTHS  [nm]
      587.60       656.30       486.10
SURFACES
  # SURF SPACE    RADIUS       SEPN      INDEX1      V      CLR RAD  GLASS
  0                          0.00000   1.000000
  1 S             Plane      2.00000   1.610248   56.69     3.600  S-SK1
  2 S           -9.75000    16.00000   1.000000             3.600
  3#S             Plane     16.00000   1.000000             1.584
  4 S            9.75000     2.00000   1.610248   56.69     3.600  S-SK1
  5 S             Plane     14.73507   1.000000             3.600
  6 S             Plane                                     2.285

LAGRANGE INVARIANT =    -0.2000

  #     H        U        HBAR       UBAR     D(U/N)      A        ABAR
  1   1.47351  0.10000  -2.12501  -0.00848  -0.06143  0.10000  -0.00848
```

2	1.59771	0.06210	-2.13554	-0.00527	-0.03857	-0.16387	0.34421
3	1.59771	-0.00000	-0.13268	0.12518	0.00000	-0.00000	0.12518
4	1.59771	-0.00000	1.87019	0.12518	-0.03857	0.16387	0.31699
5	1.47351	-0.06210	1.88028	0.00505	-0.06143	-0.10000	0.00812
		-0.10000		0.00812			

#	S1	S2	S3	S4	S5	C1	C2
1	0.000905	-0.000077	0.000007	0.000000	-0.000001	0.000985	-0.000084
2	0.001655	-0.003476	0.007301	0.001555	-0.018601	0.001750	-0.003676
3	0.000000	0.000000	0.000000	0.000000	0.000000	0.000000	0.000000
4	0.001655	0.003201	0.006192	0.001555	0.014985	0.001750	0.003386
5	0.000905	-0.000074	0.000006	0.000000	-0.000000	0.000985	-0.000080
Sum	0.005120	-0.000425	0.013505	0.003110	-0.003617	0.005471	-0.000454

We can immediately see from the prescription above that the clear diameters of the lenses are almost twice the image diameter. This property makes this configuration unsuitable for endoscopes, but quite useful for many other purposes. Note also that the focal length is quite large. This is correct; since the system is almost telecentric in the image space as well as in the object space, it is close to being an afocal system, with zero power, and therefore infinite focal length.

If we optimize using the two curves of the first lens and the conic constant of the conic to control spherical aberration (the system has front-back symmetry, so the second lens will be a copy of the first lens) while maintaining telecentricity in the image space, we obtain the design shown in Fig. 3.2. The coma of the complete system is automatically zero, because it has front-back symmetry, but the coma of each lens is used in optimization to introduce astigmatism to balance the Petzval sum, which is uncorrectable by two simple positive-powered singlets.

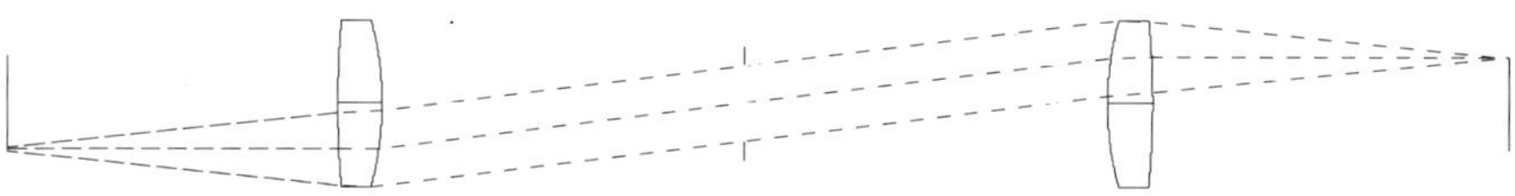

Figure 3.2 Aspheric singlets.

```
EFL =     487.996
OBJECT DISTANCE = 14.7358, CURVATURE = 0, TILT ANGLE = 0
WAVELENGTHS   [nm]
      587.60        656.30        486.10
SURFACES
  # SURF SPACE     RADIUS        SEPN       INDEX1       V      CLR RAD  GLASS
  0                             0.00000    1.000000
  1 E             34.34350      2.00000    1.610248    56.69     3.600  S-SK1
  2 S            -12.94020     15.05610    1.000000              3.600
  3#S              Plane       15.05610    1.000000              1.571
  4 S             12.94020      2.00000    1.610248    56.69     3.600  S-SK1
  5 E            -34.34350     14.73575    1.000000              3.600
  6 S              Plane                                         2.020
CONIC    SURFACE    1 CC = -41.881935
CONIC    SURFACE    5 CC = -41.881935
```

LAGRANGE INVARIANT = -0.2000

#	H	U	HBAR	UBAR	D(U/N)	A	ABAR
1	1.47358	0.10000	-2.00184	-0.00012	-0.07153	0.14291	-0.05841
2	1.56526	0.04584	-1.95781	0.02201	-0.02847	-0.12096	0.27907
3	1.56526	-0.00000	-0.03403	0.12777	0.00000	-0.00000	0.12777
4	1.56526	-0.00000	1.88976	0.12777	-0.02847	0.12096	0.27381
5	1.47358	-0.04584	1.93777	0.02401	-0.07153	-0.14291	-0.05220
		-0.10000		0.00422			

#	S1	S2	S3	S4	S5	C1	C2
1	0.002153	-0.000880	0.000360	0.000441	-0.000327	0.001408	-0.000575
2	0.000652	-0.001504	0.003470	0.001171	-0.010709	0.001266	-0.002920
3	0.000000	0.000000	0.000000	0.000000	0.000000	0.000000	0.000000
4	0.000652	0.001476	0.003341	0.001171	0.010214	0.001266	0.002865
5	0.002153	0.000786	0.000287	0.000441	0.000266	0.001408	0.000514
	ASPHERIC SURFACE 1						
	-0.002975	0.004042	-0.005490	0.000000	0.007459		
	ASPHERIC SURFACE 5						
	-0.002975	-0.003912	-0.005145	0.000000	-0.006765		
Sum	-0.000341	0.000007	-0.003177	0.003226	0.000137	0.005347	-0.000116

Transverse ray aberrations of this design are shown in Fig. 3.3.

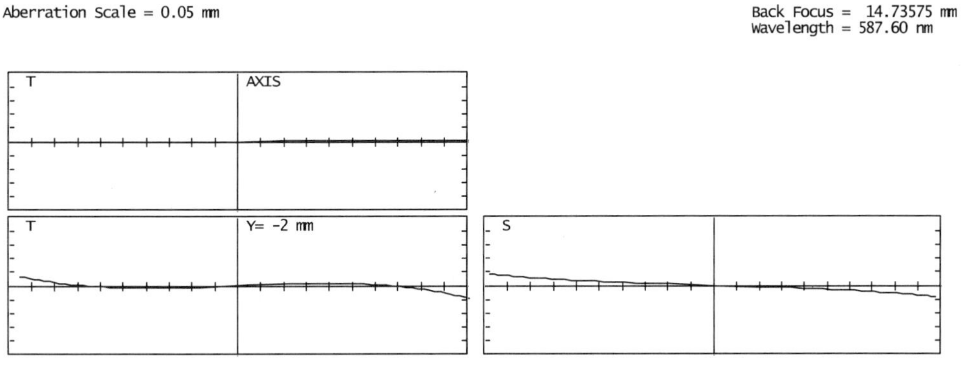

Figure 3.3 Transverse ray aberrations of two aspheric singlets.

3.2.1.1 Replacement of the fictional aspheric singlets by real doublets

It is of course possible to contemplate the use of either crowns or flints as the outer glasses. Since we have no *a priori* way of knowing which is preferable, we will use flints as the outer glasses, but the student can try a crown glass instead. When we replace the aspheric singlet by a real doublet, with BK7 and F8 as the glass types, we cannot expect that correction of astigmatism will necessarily be possible, because the glass choice will be critical. It is therefore necessary to let the optimization routine vary one or both of the glass types. The resulting design is shown in Fig. 3.4.

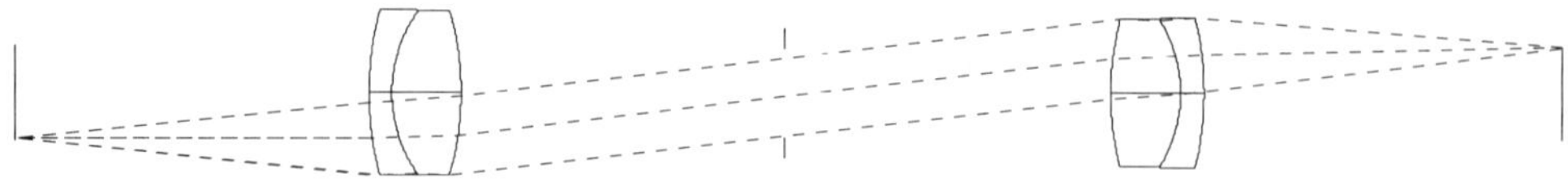

Figure 3.4 Achromatic doublet pair.

EFL = 82.610

OBJECT DISTANCE = 15.5018, CURVATURE = 0, TILT ANGLE = 0

WAVELENGTHS [nm]

587.60 656.30 486.10

SURFACES

#	SURF SPACE	RADIUS	SEPN	INDEX1	V	CLR RAD	GLASS
0			0.00000	1.000000			
1	S	14.32580	1.00000	1.672693	32.20	3.600	S-SF5
2	S	6.15050	3.00000	1.516798	64.14	3.600	S-BK7
3	S	-13.12050	14.01430	1.000000		3.600	
4#	S	Plane	14.01430	1.000000		1.701	
5	S	13.12050	3.00000	1.516798	64.14	3.600	S-BK7
6	S	-6.15050	1.00000	1.672693	32.20	3.600	S-SF5
7	S	-14.32580	15.50177	1.000000		3.600	
8	S	Plane				2.020	

LAGRANGE INVARIANT = -0.2000

#	H	U	HBAR	UBAR	D(U/N)	A	ABAR
1	1.55018	0.10000	-1.83968	0.01034	-0.09028	0.20821	-0.11808
2	1.56644	0.01627	-1.78185	0.05783	0.01936	0.45322	-0.38787
3	1.69879	0.04411	-1.67987	0.03399	-0.02908	-0.12948	0.24577
4	1.69879	0.00000	-0.02995	0.11773	0.00000	0.00000	0.11773
5	1.69879	0.00000	1.61997	0.11773	-0.02908	0.12948	0.24120
6	1.56644	-0.04411	1.72662	0.03555	0.01936	-0.45322	-0.37189
7	1.55018	-0.01627	1.78502	0.05840	-0.09028	-0.20821	-0.11073
		-0.10000		0.01387			

#	S1	S2	S3	S4	S5	C1	C2
1	0.006067	-0.003440	0.001951	0.001123	-0.001743	0.004032	-0.002286
2	-0.006229	0.005331	-0.004562	-0.000400	0.004246	-0.005097	0.004362
3	0.000828	-0.001572	0.002984	0.001039	-0.007636	0.001168	-0.002218
4	0.000000	0.000000	0.000000	0.000000	0.000000	0.000000	0.000000
5	0.000828	0.001543	0.002874	0.001039	0.007290	0.001168	0.002177
6	-0.006229	-0.005111	-0.004194	-0.000400	-0.003769	-0.005097	-0.004182
7	0.006067	0.003226	0.001716	0.001123	0.001510	0.004032	0.002144
Sum	0.001332	-0.000023	0.000770	0.003524	-0.000103	0.000207	-0.000004

We can see in Fig. 3.5 that spherical aberration and astigmatism are corrected, but there is a considerable amount of tangential oblique spherical aberration, which we can assume is coming from the cemented surfaces.

These are wavefront aberrations with simple scaling factors

SA3	COMA3	AST3	PETZ3	DIST3	LCOLOR	TCOLOR	PUP-SA	PUP-CO
-S1	-S2	-S3	-S4	-S5	C1	C2	-S1bar	-S2bar
-0.00607	0.00333	-0.00183	-0.00112	0.0016	0.00403	-0.00221	-0.00054	0.00098
0.00623	-0.00522	0.00438	0.00040	-0.0040	-0.00510	0.00427	0.00299	-0.00357
-0.00083	0.00156	-0.00293	-0.00104	0.0075	0.00117	-0.00220	-0.00927	0.00493

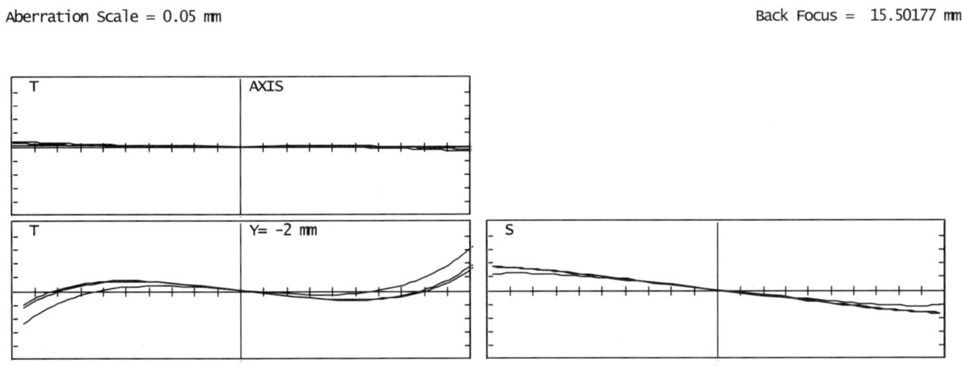

Figure 3.5 Transverse ray aberrations of achromatic doublet pair.

0.00000	0.00000	0.00000	0.00000	0.0000	0.00000	0.00000	0.00000	0.00000
-0.00083	-0.00156	-0.00293	-0.00104	-0.0075	0.00117	0.00220	-0.00927	-0.00493
0.00623	0.00522	0.00438	0.00040	0.0040	-0.00510	-0.00427	0.00299	0.00357
-0.00607	-0.00333	-0.00183	-0.00112	-0.0016	0.00403	0.00222	-0.00054	-0.00098
Total aberration								
-0.00133	-0.00000	-0.00077	-0.00352	-0.0000	0.00021	-0.00000	-0.01364	-0.00000

SA5	COMA5	LCOMA	AST5	PETZ5	SOBSA	TOBSA	DIST5	SA7
-0.00007	0.00030	0.00030	-0.00002	-0.00000	-0.00012	-0.00044	0.0000	-0.00000
0.00037	-0.00169	-0.00256	0.00022	0.00001	0.00063	0.00298	-0.0002	0.00003
-0.00001	0.00004	0.00030	-0.00007	-0.00003	-0.00004	-0.00014	0.0002	0.00000
0.00000	0.00000	0.00000	0.00000	0.00000	0.00000	0.00000	0.0000	0.00000
0.00001	-0.00000	-0.00021	-0.00006	-0.00007	0.00000	-0.00010	-0.0005	-0.00001
0.00032	0.00181	0.00295	0.00036	0.00017	0.00055	0.00305	0.0006	0.00004
-0.00003	-0.00045	-0.00080	-0.00012	-0.00011	-0.00007	-0.00050	-0.0002	-0.00002
Total aberration								
0.00060	-0.00000	-0.00001	0.00030	-0.00004	0.00095	0.00486	-0.0001	0.00005

```
Total Aberration Coefficients converted to Transverse Aberrations
   SA3 -0.00666  COMA3 -0.00000   AST3 -0.00385  PETZ3 -0.01762  DIST3 -0.00000
LCOLOR  0.00205 TCOLOR -0.00000

   SA5  0.00300  COMA5 -0.00001  LCOMA -0.00007   AST5  0.00150  PETZ5 -0.00021
 SOBSA  0.00475  TOBSA  0.02429  DIST5 -0.00050

   SA7  0.00025
```

Note that the value of TOBSA (underlined above), when converted to a transverse ray aberration, is 0.02429. This is reasonably close to the values seen in the aberration curves shown above (allowing for the fact that this aberration can be balanced against Seidel astigmatism and field curvature), but there are clearly no more free variables that we can use to control this fifth-order term.

3.2.1.2 Addition of an extra crown component

It is difficult to see whether there is any way of controlling the tangential oblique spherical aberration by any simple means. The obvious approach is to reduce the

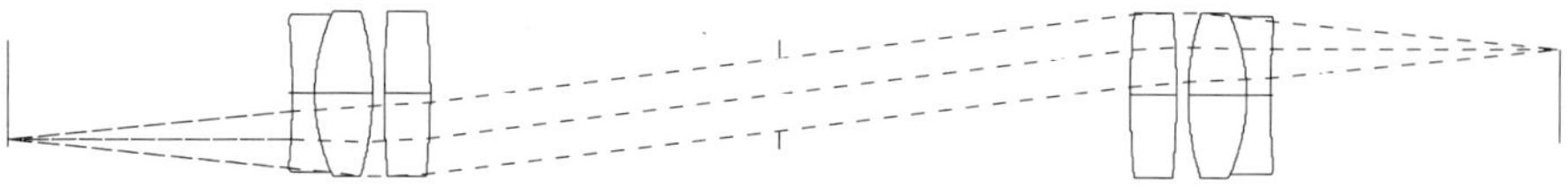

Figure 3.6 Extra crown singlets.

angles of incidence at the cemented surfaces, replacing the doublet by a triplet with an extra crown element, and using a higher refractive index crown glass. This is shown after reoptimization in Fig.3.6. We can see in Fig. 3.7 that the tangential oblique spherical aberration is very much smaller when the extra component is added.

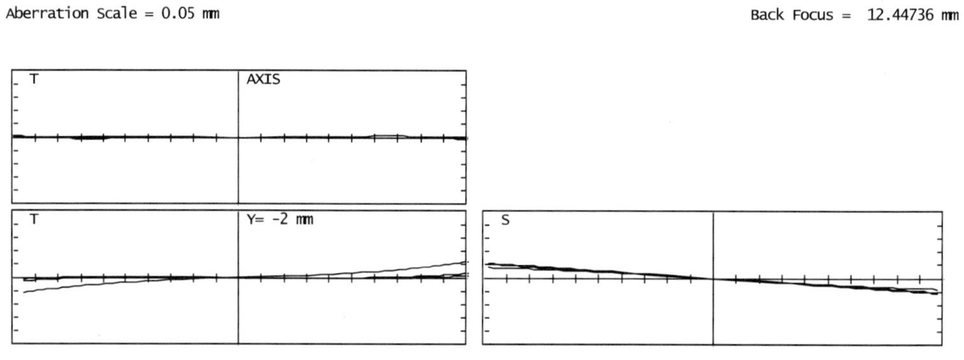

Figure 3.7 Transverse ray aberrations with extra singlets.

EFL = 320.063
OBJECT DISTANCE = 12.4474, CURVATURE = 0, TILT ANGLE = 0
WAVELENGTHS [nm]
587.60 656.30 486.10
SURFACES

# SURF SPACE	RADIUS	SEPN	INDEX1	V	CLR RAD	GLASS
0		0.00000	1.000000			
1 S	-32.20930	1.00000	1.672693	32.20	3.600	S-SF5
2 S	8.53770	2.50000	1.642496	57.94	3.600	S-LAKN6
3 S	-12.45820	0.50000	1.000000		3.600	
4 S	39.84010	2.00000	1.642496	57.94	3.600	S-LAKN6
5 S	-36.99690	15.05530	1.000000		3.600	
6#S	Plane	15.05530	1.000000		1.583	
7 S	36.99690	2.00000	1.642496	57.94	3.600	S-LAKN6
8 S	-39.84010	0.50000	1.000000		3.600	
9 S	12.45820	2.50000	1.642496	57.94	3.600	S-LAKN6
10 S	-8.53770	1.00000	1.672693	32.20	3.600	S-SF5
11 S	32.20930	12.44736	1.000000		3.600	
12 S	Plane				2.020	

LAGRANGE INVARIANT = -0.2000

#	H	U	HBAR	UBAR	D(U/N)	A	ABAR
1	1.24474	0.10000	-2.00154	-0.00012	-0.05497	0.06135	0.06202
2	1.32006	0.07533	-2.02660	-0.02506	0.00340	0.38462	-0.43897
3	1.51894	0.07955	-2.10132	-0.02989	0.00390	-0.06959	0.22795

4	1.54511	0.05233	-2.07169	0.05928	-0.04217	0.09111	0.00728
5	1.57849	0.01669	-1.95883	0.05643	-0.01016	-0.04267	0.17965
6	1.57849	-0.00000	-0.05126	0.12670	0.00000	-0.00000	0.12670
7	1.57849	-0.00000	1.85630	0.12670	-0.01016	0.04267	0.17688
8	1.54511	-0.01669	1.97132	0.05751	-0.04217	-0.09111	0.01319
9	1.51894	-0.05233	2.00266	0.06268	0.00390	0.06959	0.22343
10	1.32006	-0.07955	1.94086	-0.02472	0.00340	-0.38462	-0.41399
11	1.24474	-0.07533	1.92068	-0.02017	-0.05497	-0.06135	0.06600
		-0.10000		0.00637			

#	S1	S2	S3	S4	S5	C1	C2
1	0.000258	0.000260	0.000263	-0.000499	-0.000239	0.000954	0.000964
2	-0.000664	0.000758	-0.000865	-0.000051	0.001046	-0.002914	0.003326
3	-0.000029	0.000094	-0.000307	0.001256	-0.003107	0.000714	-0.002338
4	0.000541	0.000043	0.000003	0.000393	0.000032	0.000950	0.000076
5	0.000029	-0.000123	0.000518	0.000423	-0.003960	0.000455	-0.001915
6	0.000000	0.000000	0.000000	0.000000	0.000000	0.000000	0.000000
7	0.000029	0.000121	0.000502	0.000423	0.003834	0.000455	0.001885
8	0.000541	-0.000078	0.000011	0.000393	-0.000059	0.000950	-0.000138
9	-0.000029	-0.000092	-0.000295	0.001256	0.003084	0.000714	0.002291
10	-0.000664	-0.000715	-0.000770	-0.000051	-0.000884	-0.002914	-0.003137
11	0.000258	-0.000277	0.000298	-0.000499	0.000217	0.000954	-0.001026
Sum	0.000269	-0.000009	-0.000642	0.003041	-0.000036	0.000317	-0.000010

3.2.2 A simple zoom lens

A more complex example discussed below is a simple zoom lens, in principle not very different from the lens by H.H. Hopkins that is discussed later in Chapter 9. The specification is focal lengths 225 mm, 100 mm, and 500 mm; aperture *f*/5, wavelengths 546 nm, 625 nm, and 500 nm, to cover a 35-mm negative with a semi-diameter of 20 mm.

3.2.2.1 Paraxial layout at 225-mm focal length

As we shall see in Chapter 9, a system with a unit-magnification telescope in front of a lens with a focal length of 225 mm will be a reasonable starting point, although in this example we will keep the aperture stop in a fixed position, in the rear fixed group of lenses. The layout for this lens is shown in Fig. 3.8 (ZOOM1):

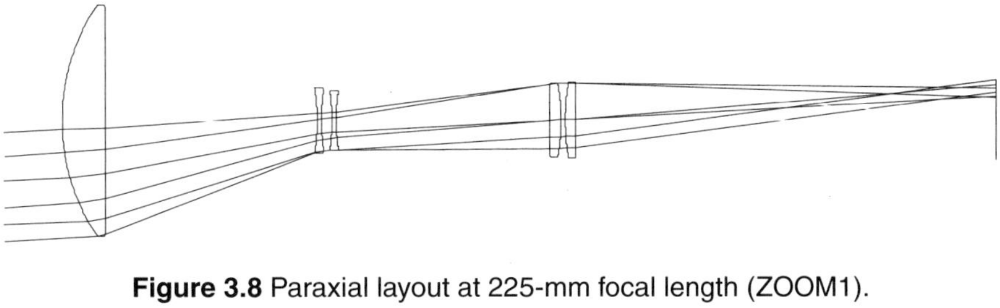

Figure 3.8 Paraxial layout at 225-mm focal length (ZOOM1).

SURFACES

#	SURF	SPACE	RADIUS	SEPN	INDEX1	V	CLR RAD	GLASS
0#				0.00000	1.000000			
1	S		100.00000	20.00000	1.500000	Inf	54.997	FICTITIOUS
2	S	Z	Plane	100.00000	1.000000		54.122	
3	S		-100.00000	2.00000	1.500000	Inf	15.737	FICTITIOUS
4	S	Z	100.00000	5.00000	1.000000		14.920	
5	S		-100.00000	2.00000	1.500000	Inf	14.335	FICTITIOUS
6	S	Z	100.00000	100.00000	1.000000		13.995	
7	S		Plane	5.00000	1.500000	Inf	17.422	FICTITIOUS
8	S		-100.00000	2.00000	1.000000		17.598	
9#	S		112.50000	5.00000	1.500000	Inf	18.119	FICTITIOUS
10	S		Plane	198.47960	1.000000		18.166	
11	S		Plane				14.348	

The four lenses that form the telescope have focal lengths of 200, –100, –100, and 200 mm; and the last lens has a focal length of 225 mm. Note that lens 1 has a diameter of 110 mm and a focal length of 200 mm, which means that it is operating at *f*/1.8. This is much too high for a doublet and this component will have to be at least a triplet, as we shall see. However, we will use this as a starting point. The overall length is 440 mm.

The next stage is to get a system that zooms correctly to give focal lengths of 100 and 500 mm, with a constant back focus of 200 mm. At this stage we are just looking at the paraxial properties and all calculations are monochromatic. After optimization, design ZOOM2A is shown in Fig. 3.9.

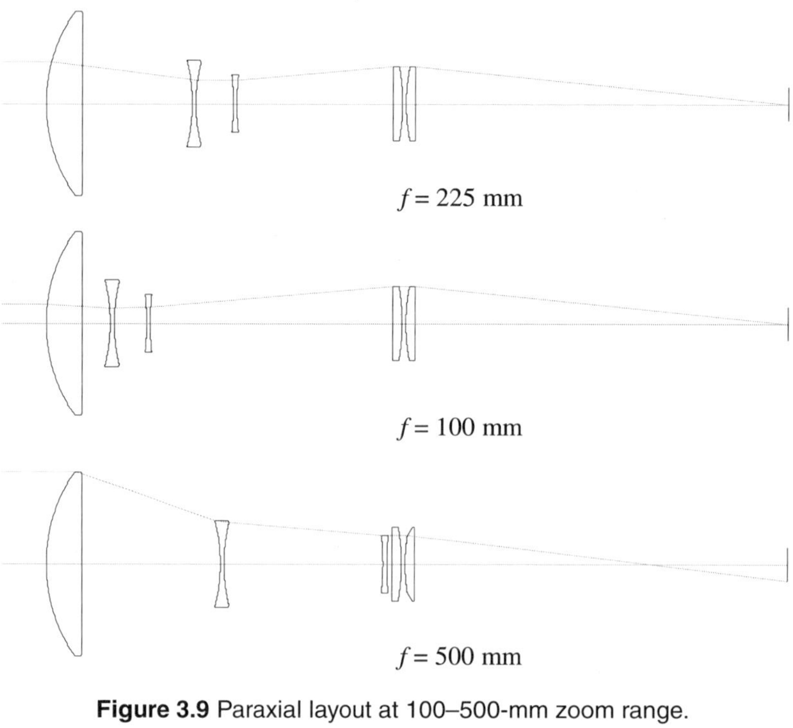

Figure 3.9 Paraxial layout at 100–500-mm zoom range.

SURFACES

#	SURF	SPACE	RADIUS	SEPN	INDEX1	V	CLR RAD	GLASS
0#				0.00000	1.000000			
1	S		85.60149	20.00000	1.500000	Inf	50.000	FICTITIOUS
2	S	Z	Plane	59.59851	1.000000		49.117	
3	S		-99.13358	2.00000	1.500000	Inf	23.678	FICTITIOUS
4	S	Z	99.13358	20.00000	1.000000		22.674	
5	S		-147.57856	2.00000	1.500000	Inf	15.700	FICTITIOUS
6	S	Z	147.57856	84.40148	1.000000		15.626	
7	S		Plane	5.00000	1.500000	Inf	20.232	FICTITIOUS
8	S		-118.55124	2.00000	1.000000		20.418	
9#	S		102.13945	5.00000	1.500000	Inf	20.200	FICTITIOUS
10	S		Plane	200.00003	1.000000		20.213	
11	S		Plane				11.857	

ZOOM POSITION	EFL	D(2)	D(4)	D(6)
1	225	59.59851	20.00000	84.40148
2	100	15.06008	17.56148	131.37844
3	500	74.92767	86.24325	2.82907

This lens has the correct paraxial properties, but the NA of lens 1 will again be too large. For a curvature of 0.01168 and a diameter of 100 mm, the aperture is effectively *f*/1.71. The overall length is 400 mm. To reduce this aperture for lens 1, the curvature of surface 1 has been reduced and the system reoptimized to that shown in Fig. 3.10 (ZOOM2C). At this stage it is helpful to remember that we are controlling eight conditions: focal length at three zoom positions, back focus at three zoom positions, and lens length at two zoom positions. It is therefore neces-

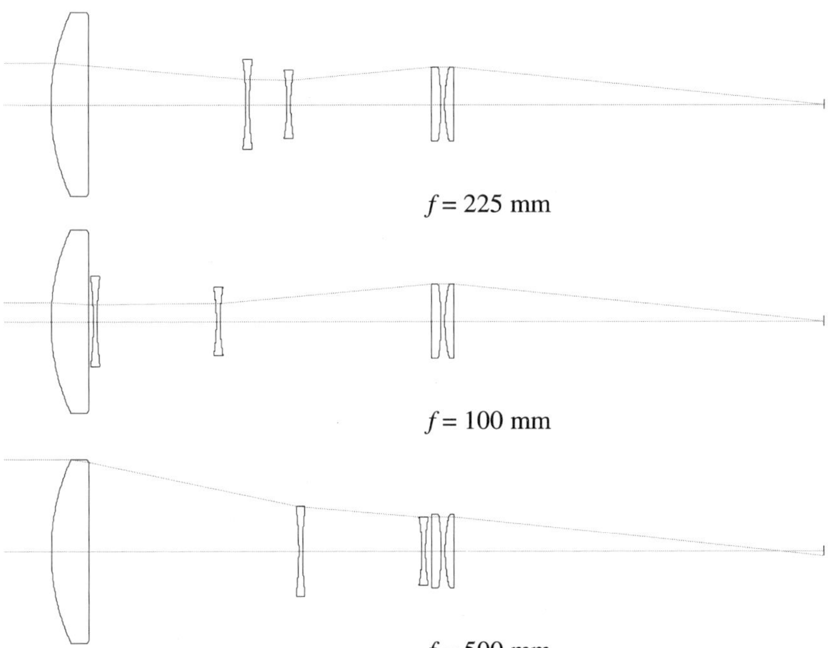

Figure 3.10 Reoptimized paraxial layout at 100–500-mm zoom range.

sary for there to be at least eight suitable variables. Here we have used four lens powers (but lenses 4 and 5 are effectively in contact) and six separations, so there are nine useful variables.

SURFACES

#	SURF	SPACE	RADIUS	SEPN	INDEX1	V	CLR RAD	GLASS
0				0.00000	1.000000			
1	S		125.00000	20.00000	1.500000	Inf	50.000	FICTITIOUS
2	S	Z	Plane	84.65626	1.000000		48.637	
3	S		-199.57208	2.00000	1.500000	Inf	24.540	FICTITIOUS
4	S	Z	199.57208	20.00000	1.000000		24.048	
5	S		-118.80425	2.00000	1.500000	Inf	18.655	FICTITIOUS
6	S	Z	118.80425	75.88170	1.000000		18.630	
7	S		Plane	5.00000	1.500000	Inf	20.142	FICTITIOUS
8	S		-135.55934	2.00000	1.000000		20.350	
9#	S		88.75141	5.00000	1.500000	Inf	20.371	FICTITIOUS
10	S		Plane	200.00001	1.000000		20.258	
11	S		Plane				11.083	

ZOOM POSITION	EFL	D(2)	D(4)	D(6)
1	225	84.65626	20.00000	75.88170
2	100	2.62878	64.29425	113.61493
3	500	112.77913	64.29425	3.46458

This design also satisfies all of the paraxial conditions, but lens 1 is now effectively *f*/2.5, which is more reasonable for a triplet. In order to maintain some of the symmetry conditions in the H.H. Hopkins design, we have also set a condition that D(4) in zoom 2 equals D(4) in zoom 3. This helps to ensure that the zoom movements are more or less symmetrical. The overall length is now 417 mm.

3.2.2.2 Aberration correction with fictitious aspheric lenses

Now that we have a design that has reasonable curves and the correct paraxial properties, we can make one surface of each lens aspheric and optimize at the correct field angles for an image radius of 20 mm. The results (ZOOM2D) are shown in Fig. 3.11.

SURFACES

#	SURF	SPACE	RADIUS	SEPN	INDEX1	V	CLR RAD	GLASS
0				0.00000	1.000000			
1	E		151.65382	12.00000	1.500000	Inf	58.395	FICTITIOUS
2	S	Z	1119.18792	124.81078	1.000000		58.021	
3	S		345.41885	2.00000	1.500000	Inf	37.090	FICTITIOUS
4	E	Z	109.67074	20.00000	1.000000		35.578	
5	S		-134.30722	2.00000	1.500000	Inf	21.749	FICTITIOUS
6	E	Z	184.32534	105.66750	1.000000		21.437	
7	S		-4884.57744	5.00000	1.500000	Inf	20.589	FICTITIOUS
8	E		-137.97639	2.00000	1.000000		20.651	
9#	S		111.82126	5.00000	1.500000	Inf	20.305	FICTITIOUS
10	E		-6786.42444	200.05090	1.000000		20.852	
11	S		Plane				19.519	

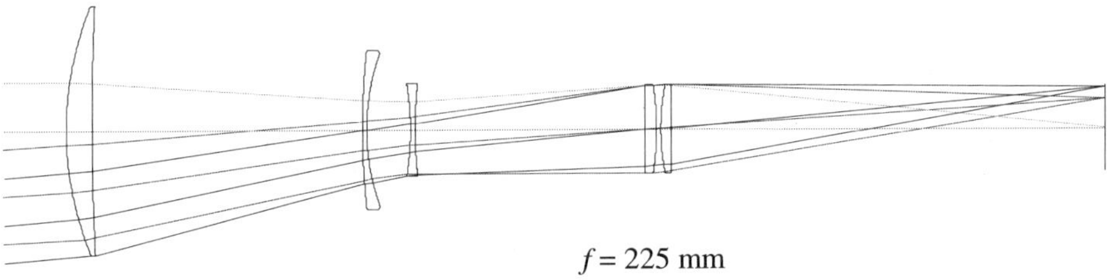

$f = 225$ mm

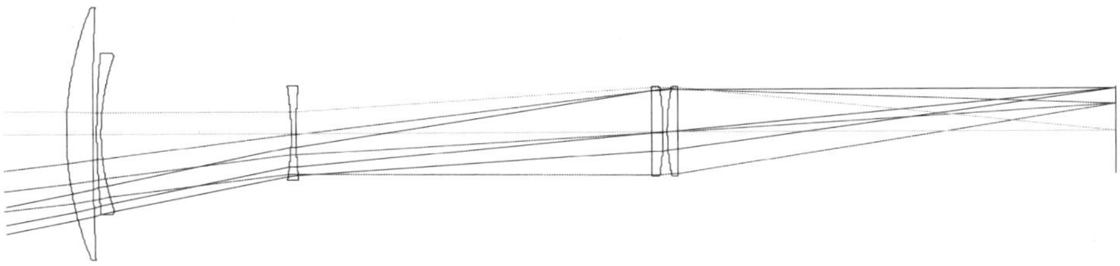

$f = 100$ mm

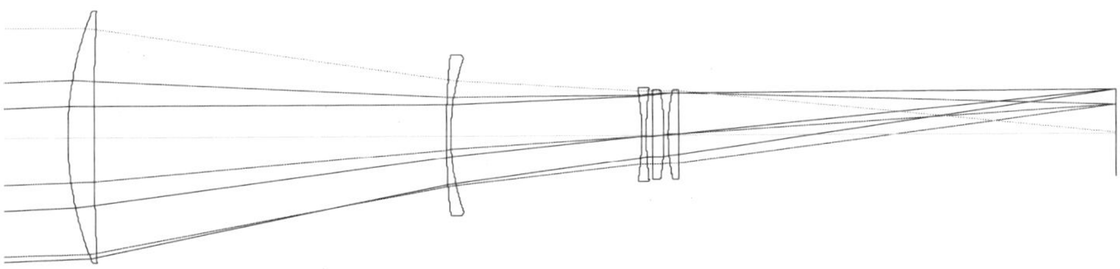

$f = 500$ mm

Figure 3.11 Fictitious aspheric elements at 100–500-mm zoom range.

CONIC	SURFACE	1	CC =	-0.319690	(ELLIPSE)
CONIC	SURFACE	4	CC =	0.738563	
CONIC	SURFACE	6	CC =	-6.624520	
CONIC	SURFACE	8	CC =	-1.529800	
CONIC	SURFACE	10	CC =	0.032467	

ZOOM POSITION	EFL	D(2)	D(4)	D(6)
1	225	124.81078	20.00000	105.66750
2	100	1.95369	87.00367	161.52091
3	500	160.36277	87.00367	3.11184

This system is getting physically longer (overall length 478 mm), but the aberration correction is fine. Lens 1 has a focal length of 350 mm, so it is *f*/3, which is reasonable for a triplet (or even possibly a doublet). The aberration curves of Fig. 3.12 show the best that we can hope for, when we replace the aspheric monochromatic lenses with real doublets.

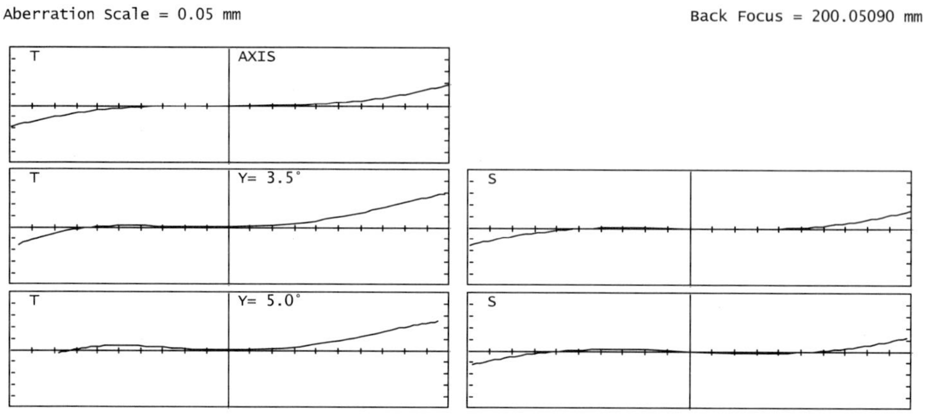

f = 225 mm

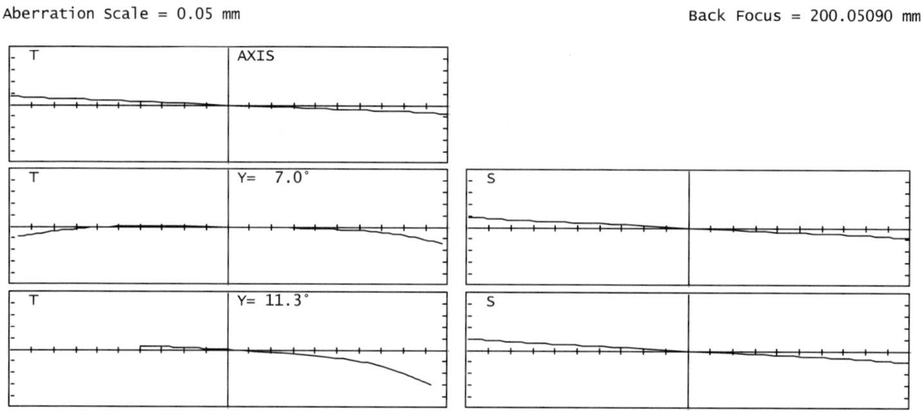

f = 100 mm

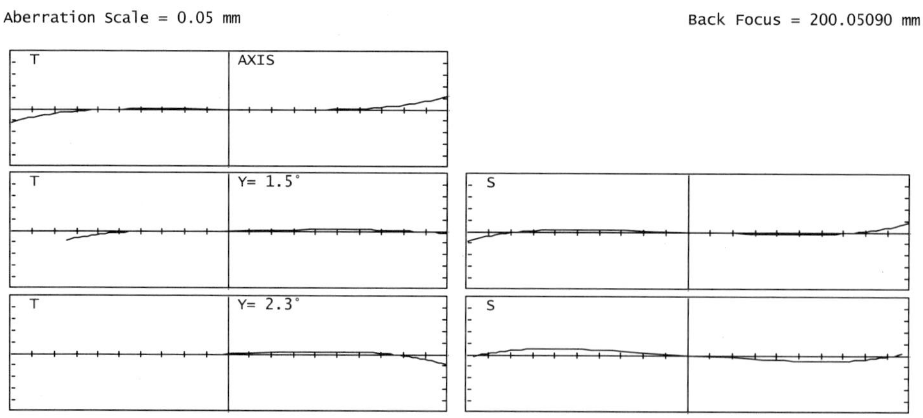

f = 500 mm

Figure 3.12 Transverse aberrations with fictitious aspheric elements at 100–500-mm zoom range.

3.2.2.3 Aberration correction with triplet and cemented doublets

We now replace lens 1 with a triplet, and the other lenses with cemented doublets. After optimizing again we get ZOOM4, shown in Fig. 3.13, with an overall length of 450 mm. Transverse ray aberrations of the three zoom positions are shown in Fig. 3.14. The overall length can be reduced, but with a loss of performance.

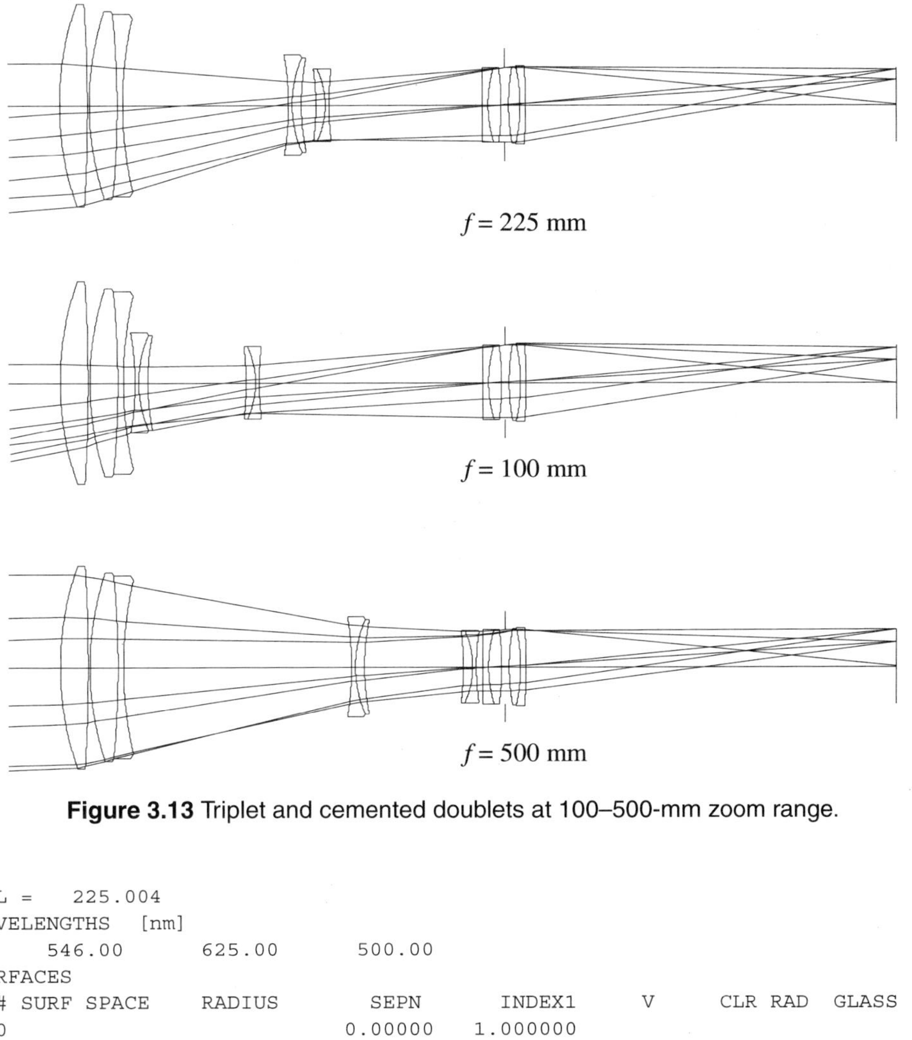

Figure 3.13 Triplet and cemented doublets at 100–500-mm zoom range.

```
EFL =    225.004
WAVELENGTHS   [nm]
      546.00        625.00         500.00
SURFACES
  # SURF SPACE     RADIUS          SEPN        INDEX1        V        CLR RAD  GLASS
  0                               0.00000    1.000000
  1 S             168.81173     15.00000    1.518726    85.69     55.126  S-BK7
  2 S            -940.27823      1.00000    1.000000              54.402
  3 S             165.26123     15.00000    1.518726    85.69     51.169  S-BK7
  4 S            -526.43936      3.00000    1.677649    43.09     49.927  S-SF5
  5 S     Z       219.08728     89.12227    1.000000              47.418
  6 S            -161.67210      2.00000    1.518726    85.69     27.454  S-BK7
  7 S              64.53811      5.00000    1.677649    43.09     25.899  S-SF5
  8 S     Z       132.08792      7.76387    1.000000              25.518
```

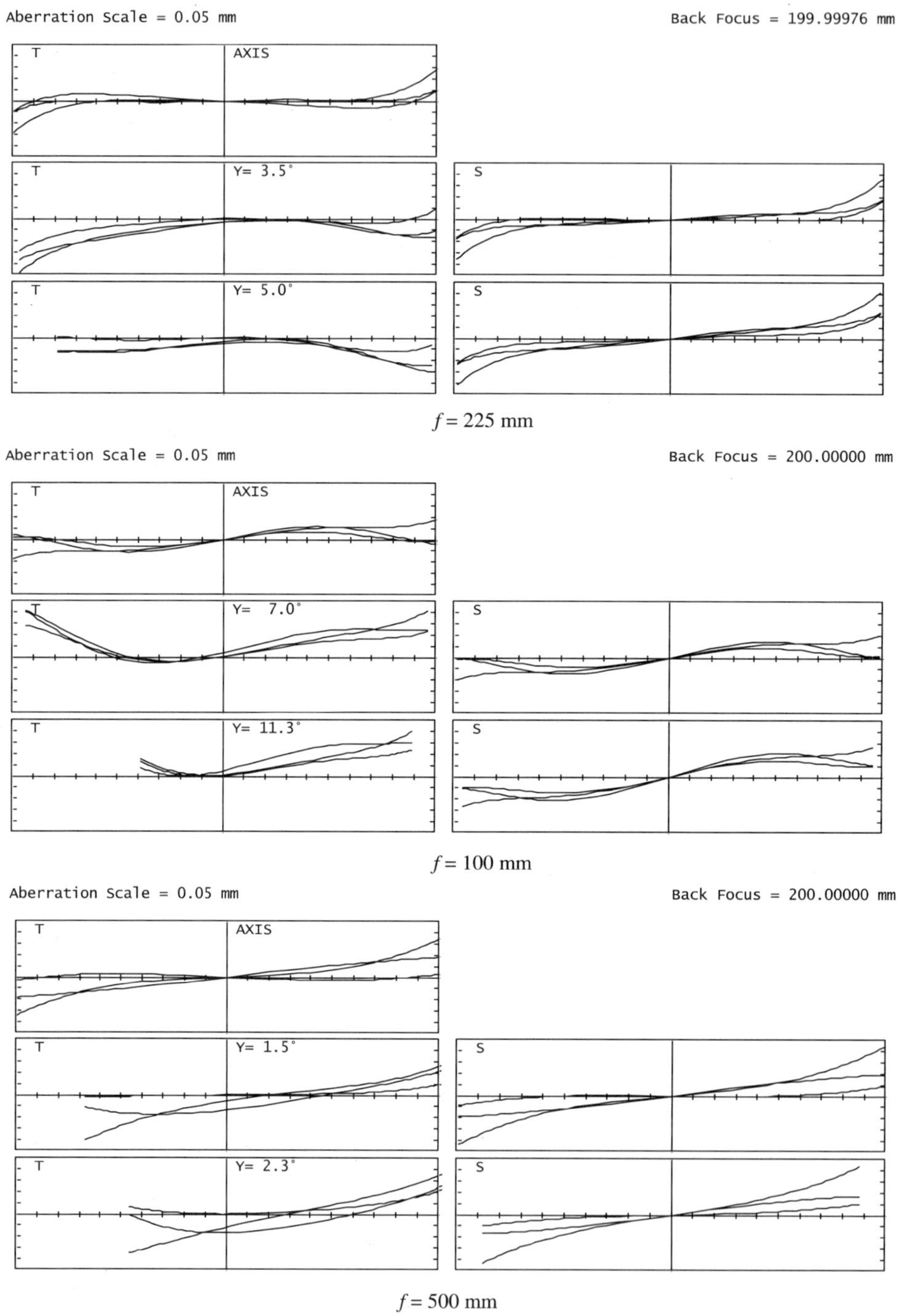

Figure 3.14 Transverse aberrations with triplet and doublets at 100–500-mm zoom range.

9 S		-154.21725	5.00000	1.668862	47.90	19.578	S-BASF2
10 S		-47.04288	2.00000	1.609731	79.36	19.485	S-SK7
11 S	Z	152.76257	82.11404	1.000000		19.118	
12 S		-626.24623	3.00000	1.630044	47.73	19.966	S-F1
13 S		70.20254	7.00000	1.521111	76.63	20.101	S-K4
14 S		-148.52615	2.00000	1.000000		20.221	
15#S		Plane	2.00000	1.000000		20.130	
16 S		123.44408	6.00000	1.654260	78.11	20.803	S-LAKN7
17 S		-101.15918	3.00000	1.671591	44.15	20.859	S-SF19
18 S	Z	-239.48058	200.00000	1.000000		20.982	
19 S		Plane				19.624	

ZOOM POSITION	EFL	D(5)	D(8)	D(11)
1	225	89.12227	7.76387	82.11404
2	100	6.13460	52.93415	119.83143
3	500	122.57596	52.93415	3.49007

Of course, this is not a finished design, but it does illustrate how we can get a configuration for a fairly complex system by starting from a paraxial layout, then simulating real lenses with monochromatic aspheric single lenses. Most important, in the context of this chapter, it shows that it is possible to achieve essentially the same monochromatic performance after replacing the aspheric lenses with doublets.

To complete this design we would have to do at least all of the following:

1. Check that lenses do not get too close at other zoom positions,
2. Look at trade-offs between overall length and performance,
3. Look at trade-offs between back focus and performance,
4. Look at vignetting,
5. Look at the effect of reversing the order of crown and flint glasses,
6. Tolerance analysis.

3.3 The use of catalog components

Several manufacturers supply ready-made lenses that can be ordered from catalogs; many of them also supply data for most of the lenses in their catalogs, although not for all. These data can be used to set up systems consisting of catalog components, and this section describes some possibilities of such systems.

In general, the number of variables in assembling a high-performance system is somewhat limited, and the major use of catalog components is in situations where high resolution over an extended field is not required. There are, of course, many cases where low cost and short development times are the higher priorities and the use of catalog components is an appropriate course of action.

3.3.1 Singlets

Singlets are available in a very large range of focal lengths, from a few millimeters up to, typically, 1000 mm. Most lenses are either plano-convex, plano-concave, equi-convex, or equi-concave, and most are made of BK7. Of course, single BK7 lenses may have large aberrations, and they are mostly used when apertures are quite small or when aberrations are not important. When UV transmission is required, fused silica lenses are available, and for a few special applications, sapphire lenses are available.

3.3.2 Doublets and triplets

For applications where improved resolution is desirable, achromats are clearly useful. Spherical aberration, coma, and chromatic aberration are corrected; even for monochromatic purposes, the correction of spherical aberration can be very important. As an example, look at Melles Griot LAO028, in Fig. 3.15, with aberrations shown in Fig. 3.16.

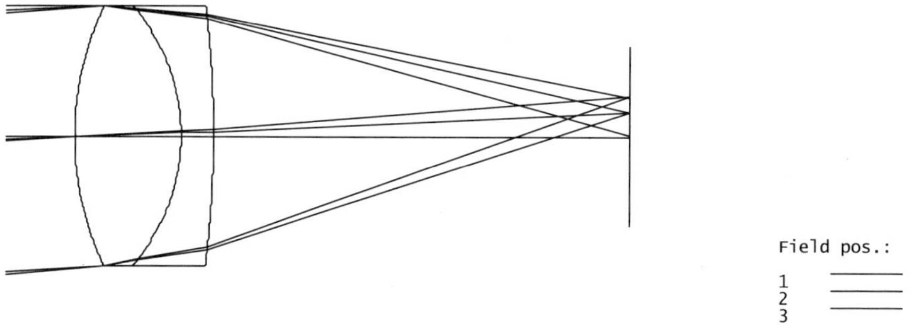

Figure 3.15 Achromatic doublet, Melles Griot LAO 028.

EFL = 30.995
WAVELENGTHS [nm]
546.10 656.30 486.10
SURFACES

# SURF	SPACE	RADIUS	SEPN	INDEX1	V	CLR RAD	GLASS
0#			0.00000	1.000000			
1 S		21.24000	6.79000	1.620317	55.35	8.750	A-SSK4
2 S		-13.70000	2.06000	1.694154	31.40	8.750	S-SF8
3 S		-75.57000	26.27930	1.000000		8.750	
4 S		Plane				6.000	

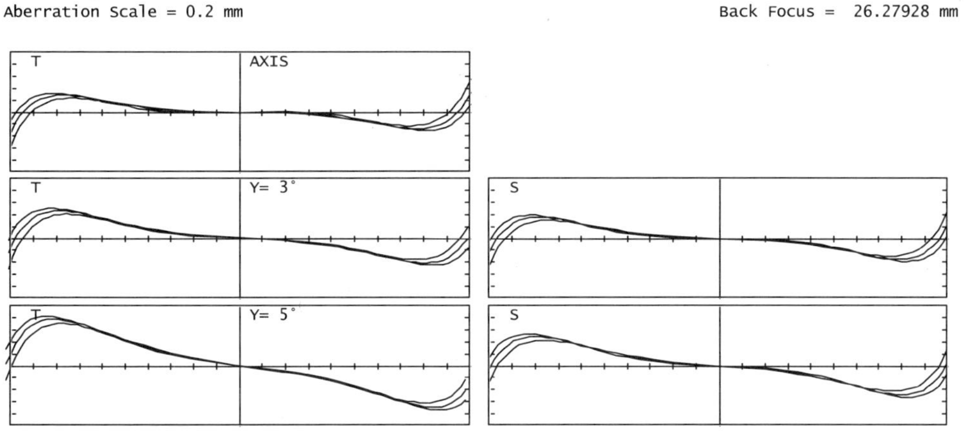

Figure 3.16 Transverse ray aberrations of Melles Griot LAO 028.

3.3.3 Meniscus singlets

Meniscus single lenses are also available, possibly approximating an aplanatic condition on the front surface and a rear surface that is almost concentric with the marginal ray when used in the converging beam after a doublet. The combination shown in Fig. 3.17 has significantly better resolution than the equivalent doublet in Fig. 3.15. However, the meniscus lens is not color corrected. Aberrations are shown in Fig. 3.18.

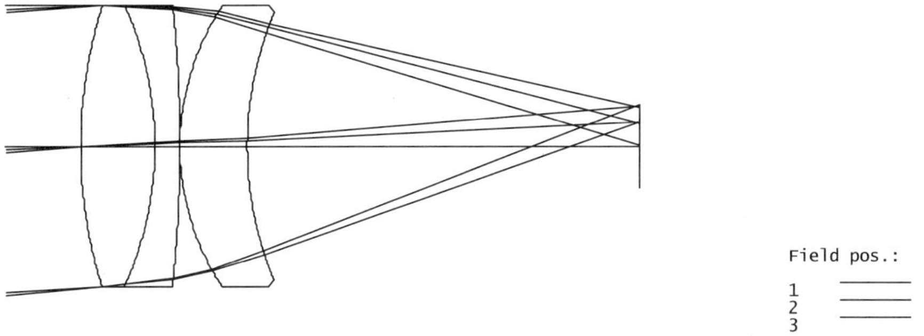

Figure 3.17 Melles Griot LAO 059 + LAM 059.

EFL = 29.265
WAVELENGTHS [nm]
546.10 656.30 486.10
SURFACES

#	SURF SPACE	RADIUS	SEPN	INDEX1	V	CLR RAD	GLASS
0#			0.00000	1.000000			
1	S	32.16000	4.46000	1.566050	61.01	9.000	S-SK11
2	S	-22.47000	1.50000	1.677636	32.43	9.000	S-SF5
3	S	-89.36000	0.00000	1.000000		9.000	
4	S	17.10000	4.00000	1.694154	31.40	9.000	S-SF8
5	S	23.25000	23.44704	1.000000		8.500	
6	S	Plane				2.687	

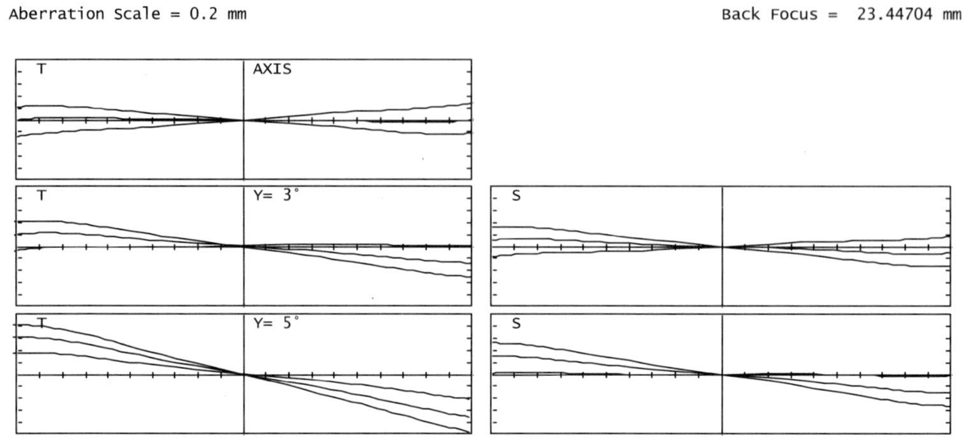

Figure 3.18 Transverse ray aberrations of Melles Griot LAO 059 + LAM 059.

3.3.4 Field flatteners

Plano-concave singlets for field flatteners are also available, and although they do not correct astigmatism, the improvement in field curvature is significant. Compare Melles Griot LAO123 in Figs. 3.19 and 3.20 with the combination of LAO123 and the LFF123 field flattener shown in Figs. 3.21 and 3.22.

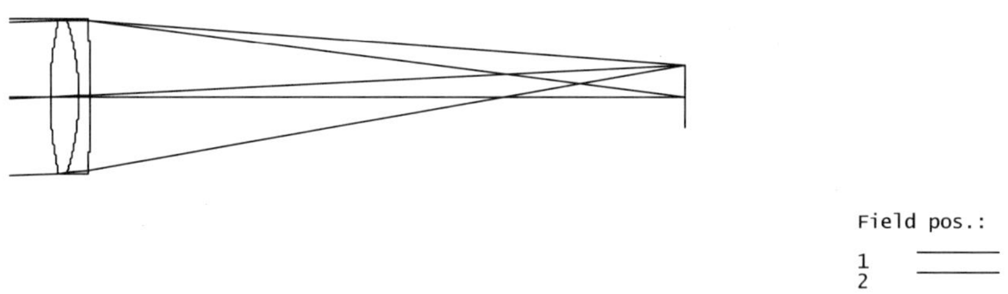

Figure 3.19 Melles Griot LAO 123.

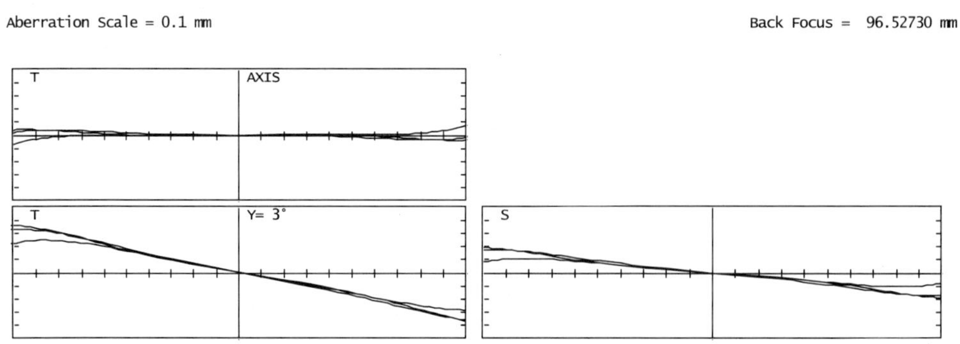

Figure 3.20 Transverse ray aberrations of Melles Griot LAO 123.

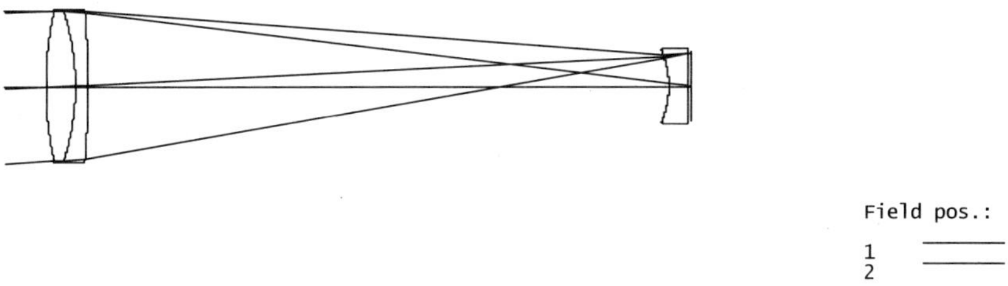

Figure 3.21 Melles Griot LAO 123 + LFF 123.

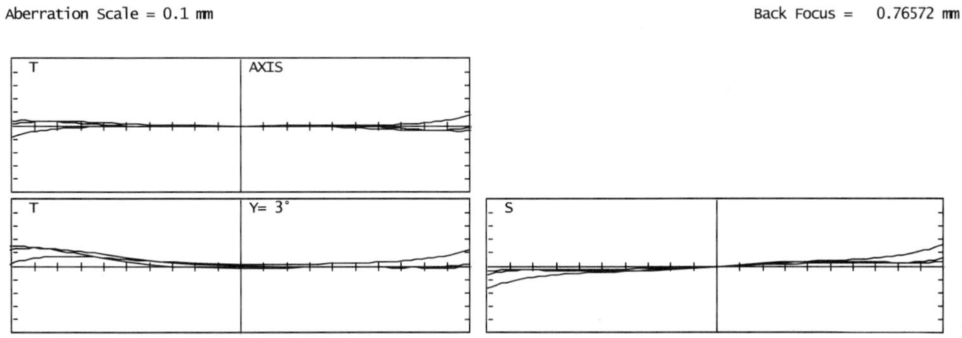

Figure 3.22 Transverse ray aberrations of Melles Griot LAO 123 + LFF 123.

EFL = 108.460
WAVELENGTHS [nm]
546.10 656.30 486.10
SURFACES

# SURF SPACE	RADIUS	SEPN	INDEX1	V	CLR RAD	GLASS
0#		0.00000	1.000000			
1 S	64.04000	4.80000	1.566050	61.01	13.250	S-SK11
2 S	-44.88000	2.00000	1.677636	32.43	13.250	S-SF5
3 S	-182.72000	94.00000	1.000000		13.250	
4 S	-16.81000	3.00000	1.518721	64.38	6.200	S-BK7
5 S	Plane	0.76572	1.000000		6.500	
6 S	Plane				5.824	

3.3.5 Cemented triplets

Symmetrical triplets are also available and clearly have zero coma, distortion, and lateral color at unit magnification. Figure 3.23 shows one such triplet used at infinite conjugates, where the aberrations may be small enough for some applications, as shown in Fig. 3.24.

EFL = 22.518
WAVELENGTHS [nm]
546.10 656.30 486.10
SURFACES

# SURF SPACE	RADIUS	SEPN	INDEX1	V	CLR RAD	GLASS
0		0.00000	1.000000			
1#S	16.20000	1.20000	1.652216	34.07	6.250	S-SF2

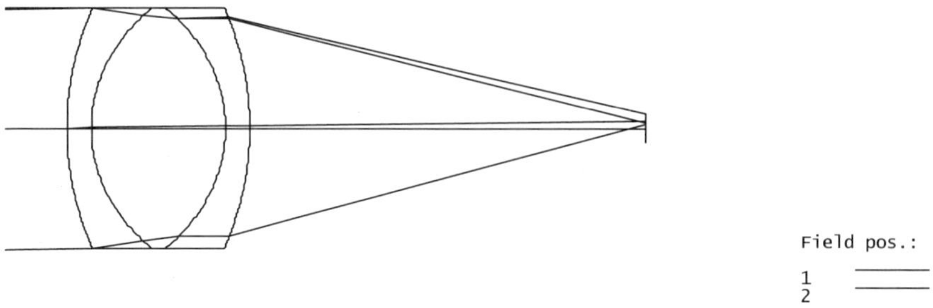

Figure 3.23 Melles Griot LAT 013.

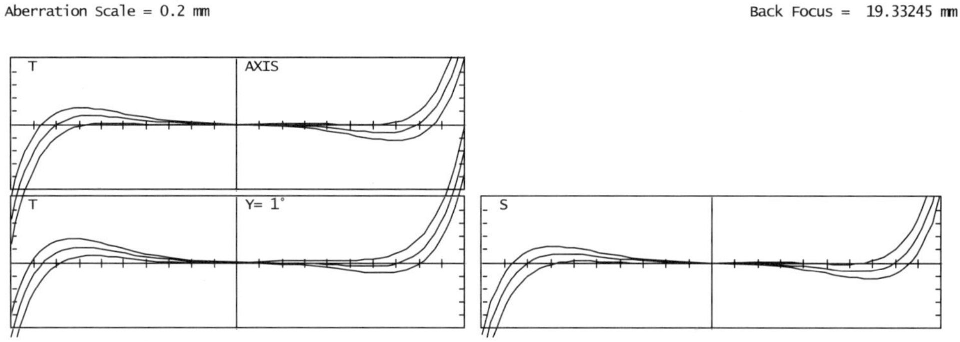

Figure 3.24 Transverse ray aberrations of Melles Griot LAT 013.

2	S	8.06000	6.60000	1.518721	64.38	6.250	S-BK7
3	S	-8.06000	1.20000	1.652216	34.07	6.250	S-SF2
4	S	-16.20000	19.33245	1.000000		6.250	
5	S	Plane				0.740	

References

1. M.J. Kidger and P.T. Leary, "Existence of local minima in lens design," in *SPIE Proc.* Vol. 1354, Bellingham, WA (1990).
2. G.W. Forbes, "Towards global optimization with adaptive simulated annealing," in *SPIE Proc.* Vol. 1354, Bellingham, WA (1990).
3. T.G. Kuper, T.I. Harris, "Global optimization for lens design: an emerging technology," in *SPIE Proc.* Vol. 1780, Bellingham, WA (1993).
4. A.E.W. Jones and G.W. Forbes, "Application of adaptive simulated annealing to lens design," in *OSA Proc.* Vol. 22, Washington, D.C. (1994).
5. K.E. Moore, "Algorithm for global optimization of optical systems based on genetic competition," in *SPIE Proc.* 3780, Bellingham, WA (1999).
6. R. Kingslake, *Lens Design Fundamentals*, 7, Academic Press, (1978).
7. R.G. Bingham and M.J. Kidger, "The principles of aberration-corrected optical systems," in *SPIE Proc.* Vol. 554, Bellingham, WA (1985).
8. M.J. Kidger, "The generation of lens designs for optimization," in *SPIE Proc.* Vol. 1780, Bellingham, WA (1993).

CHAPTER 4
LENSES FOR 35-MM CAMERAS

The Leica, from Leitz, and the Contax, from Zeiss Ikon, were the most important 35-mm cameras in the period before World War II, and lenses for these cameras illustrate the performance available at that time. In most cases, modern computer-optimized designs are far superior.[1–9]

In this chapter, most of the performance data and lens drawings are shown at the same scale. MTF curves show the performance at 30, 15, and 7.5 cycles/mm, plotted as a function of field angle. Bearing in mind that the 35-mm film format is 24 × 36 mm, 30 cycles/mm represents a resolution of 1200 pixels across the image. Transverse ray aberration curves are plotted with a scale of 0.10 mm. Longitudinal astigmatism plots are plotted on a scale from –0.4 mm to +0.4 mm. Distortion plots are scaled automatically, so the scale is different for each lens.

The examples discussed here are

Triplet	*f*/4	50 mm	22 deg	3 glasses
Tessar	*f*/4	50 mm	22 deg	4 glasses
Double Gauss (Modern)	*f*/2	50 mm	22 deg	6 glasses
Zeiss Sonnar (1931)	*f*/2	50 mm	22 deg	6 glasses
Wide-angle (Zeiss Biogon,1934)	*f*/2.8	35 mm	28 deg	6 glasses
Wide-angle (Super-Angulon)	*f*/4	20 mm	45 deg	9 glasses
Wide-angle inverse telephoto for SLR	*f*/2.8	28.5 mm	36 deg	7 glasses
Telephoto	*f*/2.8	100 mm	12 deg	4 glasses
Longer focal length telephoto	*f*/3.5	170 mm	7 deg	9 glasses
Lens for compact point-and-shoot camera	*f*/2.8	35 mm	31 deg	6 glasses
Single lens for disposable camera	*f*/2.8	30 mm	35 deg	1 glass

In almost all cases, these examples are derived from patent data and have been reoptimized to use available glasses.

4.1 The triplet

The simplest design that is reasonable for 35-mm cameras is the triplet configuration. Once the glasses are chosen, there are just enough variables (three powers, three shapes and two separations) to correct all seven Seidel aberrations, as well as the focal length (making eight conditions).

Typically, the focal length of triplets would be 45 or 50 mm, as shown in this example in Fig. 4.1.

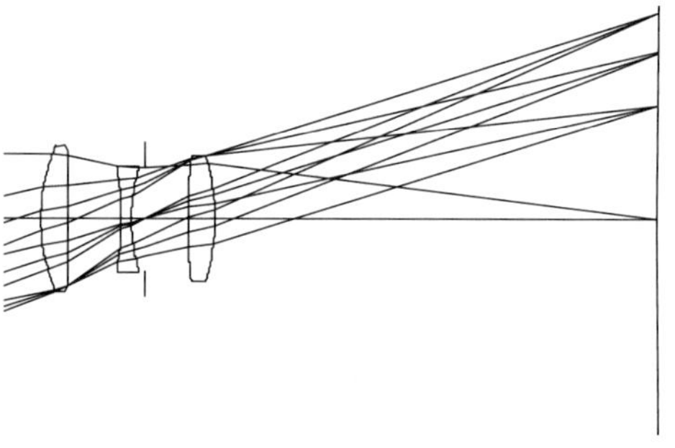

Figure 4.1 The triplet.

EFL = 49.999
WAVELENGTHS [nm]
587.60 656.30 486.10
SURFACES

# SURF SPACE	RADIUS	SEPN	INDEX1	V	CLR RAD	GLASS
0		0.00000	1.000000			
1 S	18.60900	2.50000	1.620408	60.30	7.001	S-SK16
2 S	460.96700	5.15700	1.000000		6.524	
3 S	-29.93400	1.00000	1.636355	35.33	5.058	S-F6
4 S	18.60500	1.20700	1.000000		4.936	
5#S	Plane	4.00700	1.000000		4.979	
6 S	67.59400	2.50000	1.620408	60.30	5.797	S-SK16
7 S	-22.16200	41.60524	1.000000		6.070	
8 S	Plane				20.178	

LAGRANGE INVARIANT = -2.5251

#	H	U	HBAR	UBAR	D(U/N)	A	ABAR
1	6.24991	0.00000	-5.12787	0.40403	-0.07936	0.33585	0.12847
2	5.92844	-0.12859	-4.24077	0.35484	-0.12103	-0.18753	0.56008
3	4.89504	-0.20039	-1.30500	0.56928	0.16441	-0.36392	0.61287
4	4.83617	-0.05887	-0.97406	0.33094	0.10506	0.32903	0.45586
5	4.91956	0.06909	-0.36065	0.50822	0.00000	0.06909	0.50822
6	5.19640	0.06909	1.67579	0.50822	-0.06094	0.14596	0.53301
7	5.22941	0.01320	2.43615	0.30414	-0.13315	-0.36096	0.31471
		-0.12500		0.42464			

#	S1	S2	S3	S4	S5	C1	C2
1	0.055944	0.021399	0.008185	0.131189	0.053312	0.013328	0.005098
2	0.025233	-0.075362	0.225080	-0.005296	-0.656417	0.007059	-0.021083
3	-0.106585	0.179501	-0.302298	-0.082837	0.648610	-0.019610	0.033025
4	-0.055006	-0.076210	-0.105589	-0.133278	-0.330947	-0.017517	-0.024269
5	0.000000	0.000000	0.000000	0.000000	0.000000	0.000000	0.000000
6	0.006747	0.024637	0.089966	0.036117	0.460411	0.004816	0.017586
7	0.090721	-0.079098	0.068964	0.110157	-0.156171	0.011985	-0.010450
Sum	0.017054	-0.005133	-0.015692	0.056051	0.018798	0.000062	-0.000092

At an aperture of *f*/4, as shown in Fig. 4.2, the major transverse ray aberrations are astigmatism and field curvature. Spherical aberration is well corrected, and coma is small because of the near front-to-back symmetry. Considering the sagittal curves (labeled S), the performance clearly drops off above about 10 deg off-axis, while the tangential curves (labeled T) are better. The sagittal astigmatism is mainly a result of uncorrected third-order field curvature (S_4).

Overall, the performance shown in Fig. 4.3 is only just acceptable at full aperture, though it will obviously improve when the lens is stopped down to *f*/5.6 or *f*/8.

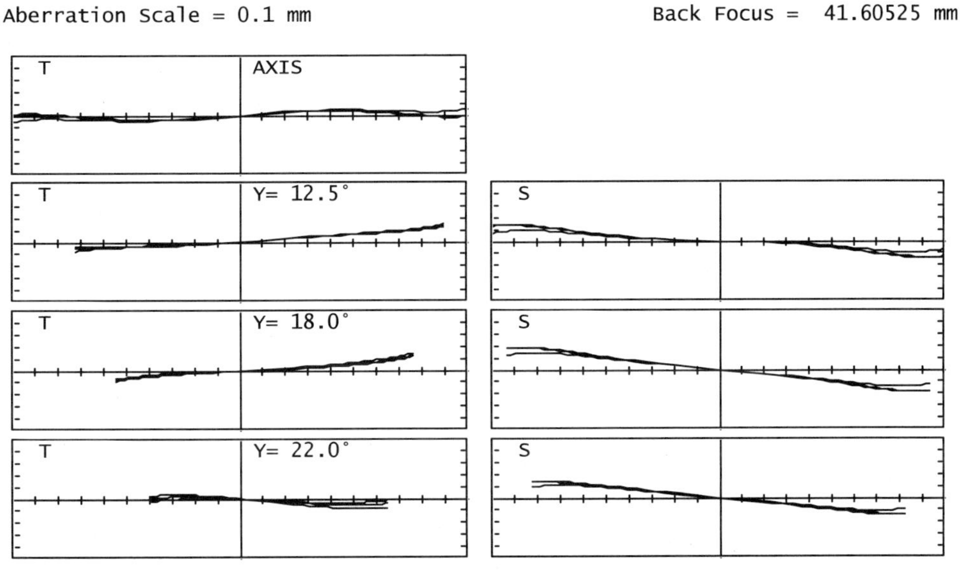

Figure 4.2 Transverse ray aberrations of the triplet.

4.2 The tessar

Paul Rudolph invented the tessar in 1902. Figure 4.4 shows that it is similar to the triplet, with a cemented doublet replacing the rear singlet. The extra variables give better performance, as can be seen from the results shown in Figs. 4.5 and 4.6.

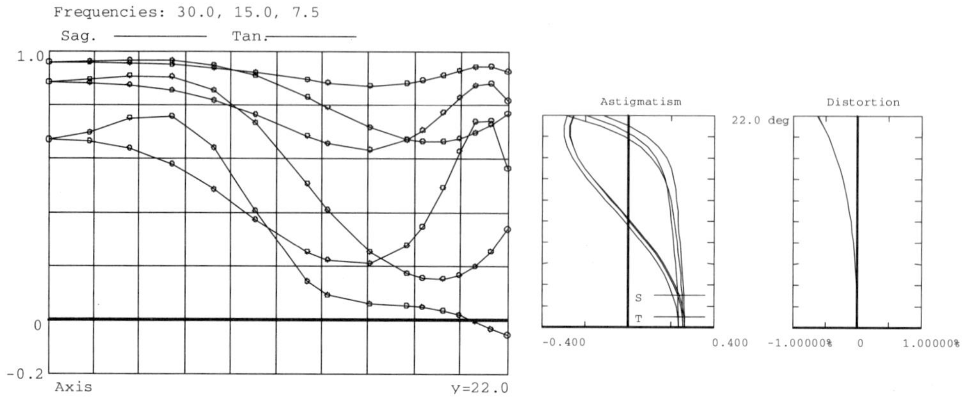

Figure 4.3 MTF as function of field, astigmatism, and distortion of the triplet.

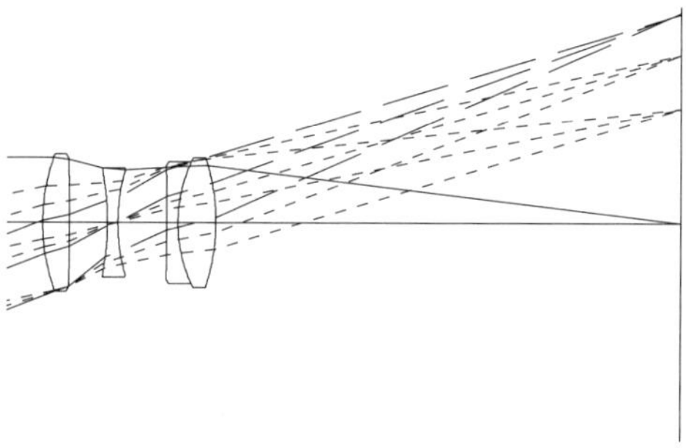

Figure 4.4 The tessar.

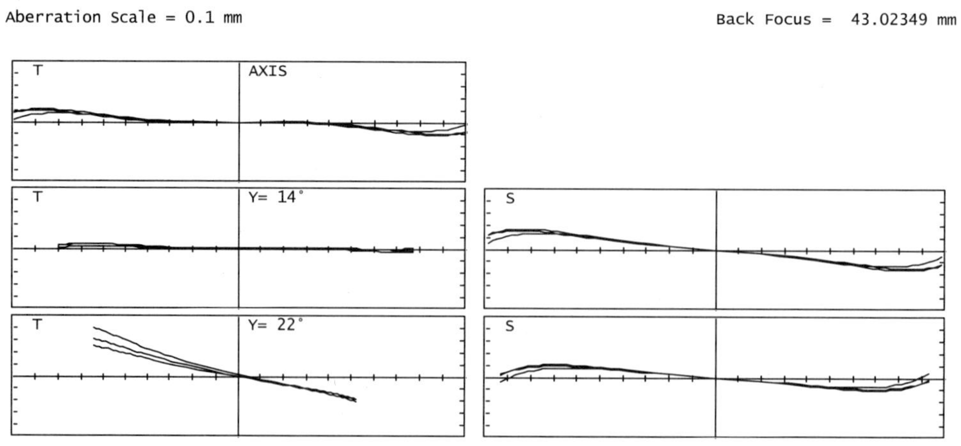

Figure 4.5 Transverse ray aberrations of the tessar.

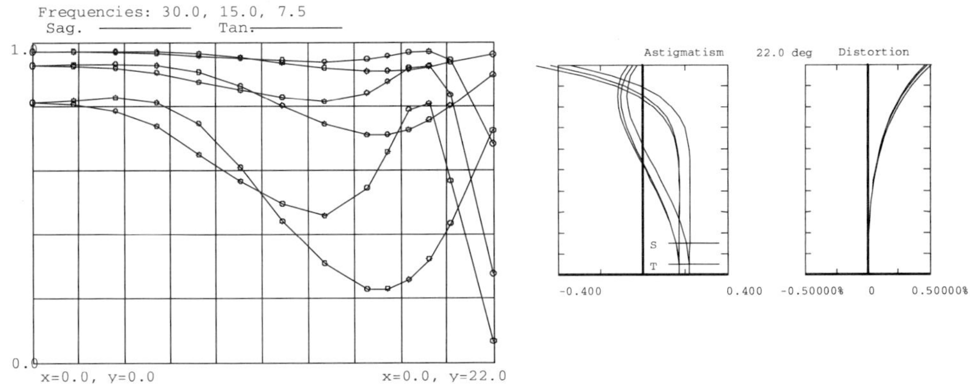

Figure 4.6 MTF as function of field, astigmatism, and distortion of the tessar.

EFL = 49.999

WAVELENGTHS [nm]

587.60 656.30 486.10

SURFACES

#	SURF SPACE	RADIUS	SEPN	INDEX1	V	CLR RAD	GLASS
0			0.00000	1.000000			
1	S	19.60700	2.50000	1.691000	54.69	6.630	S-LAK9
2	S	-192.80200	3.56200	1.000000		6.181	
3	S	-27.30400	1.00000	1.636355	35.33	5.252	S-F6
4#	S	18.22800	4.60800	1.000000		5.088	
5	S	-206.16800	1.00000	1.658438	50.86	5.422	S-SSKN5
6	S	14.61800	3.50000	1.691000	54.69	5.872	S-LAK9
7	S	-20.57200	43.02349	1.000000		6.251	
8	S	Plane				20.500	

LAGRANGE INVARIANT = -2.5251

#	H	U	HBAR	UBAR	D(U/N)	A	ABAR
1	6.24988	0.00000	-3.08633	0.40403	-0.07703	0.31876	0.24662
2	5.92425	-0.13026	-2.32821	0.30325	-0.16447	-0.27222	0.53322
3	5.06404	-0.24149	-0.47191	0.52114	0.19538	-0.42696	0.53842
4	4.98859	-0.07545	-0.16015	0.31176	0.09680	0.32436	0.49577
5	5.22215	0.05069	2.16482	0.50455	-0.02619	0.02536	0.49405
6	5.26277	0.04062	2.47322	0.30840	-0.00503	0.66443	0.79206
7	5.37794	0.03290	3.52044	0.29921	-0.14446	-0.38642	0.21658
		-0.12500		0.38771			

#	S1	S2	S3	S4	S5	C1	C2
1	0.048915	0.037845	0.029280	0.132888	0.125466	0.014886	0.011517
2	0.072202	-0.141427	0.277022	0.013514	-0.569091	0.012051	-0.023604
3	-0.180370	0.227456	-0.286835	-0.090815	0.476238	-0.023802	0.030015
4	-0.050805	-0.077651	-0.118684	-0.136034	-0.389317	-0.017813	-0.027225
5	0.000088	0.001714	0.033388	-0.012279	0.411296	0.001034	0.020139
6	0.011696	0.013942	0.016620	0.005065	0.025850	-0.001166	-0.001390
7	0.116005	-0.065018	0.036441	0.126655	-0.091411	0.015529	-0.008703
Sum	0.017732	-0.003140	-0.012769	0.038994	-0.010969	0.000718	0.000748

4.3 The double-Gauss (planar-type)

The Gauss doublet objective (1817) has been well known for many years, and it was an obvious move for two Gauss doublets to be placed back to back. This was first tried in 1888, but these attempts were unsuccessful. These first double-Gauss lenses were quite thin, but in 1896 P. Rudolph described the Zeiss planar, which used thick meniscus components, with an aperture of *f*/4.5.

The planar is shown in Fig. 4.7, with transverse ray aberrations in Fig. 4.8. As described in U.S. Patent 583,336, it has exact front-back symmetry, which results in some coma, since the lens is used at infinite conjugates. The coma can be removed by a slight departure from front-back symmetry, but this is not stated in the patent. Rudolph found that one component needed to be thick, and the thick component should also be made achromatic. The patent shows two examples, but since the negative component is always the doublet in modern double-Gauss lenses, we have shown this version. There is large oblique spherical aberration, even at the aperture of *f*/4. The planar also uses buried surfaces, in which there is a difference of dispersion (to correct chromatic aberration) but no significant difference in refractive index. Although this is not desirable in itself, it does provide a simple way of correcting chromatic aberration without affecting the monochromatic aberrations.

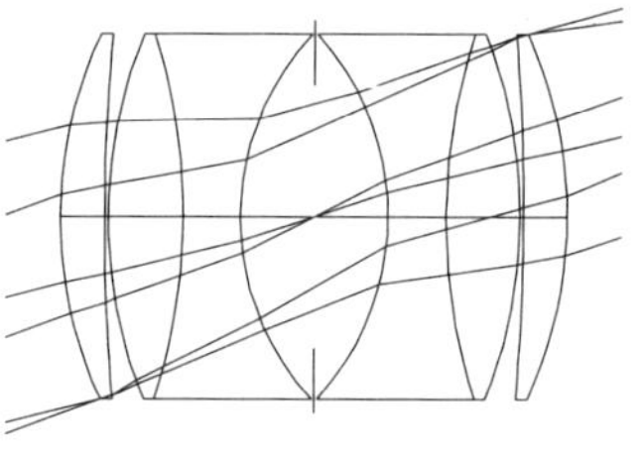

Figure 4.7 The planar.

EFL = 50.017

WAVELENGTHS [nm]

587.60 656.30 486.10

SURFACES

#	SURF SPACE	RADIUS	SEPN	INDEX1	V	CLR RAD	GLASS
0			0.00000	1.000000			
1	S	15.73500	1.54000	1.572440	45.15	6.500	User-USER
2	S	77.12000	0.13000	1.000000		6.500	
3	S	17.99500	2.57000	1.572440	45.15	6.500	User-USER
4	S	-21.85000	2.05500	1.576310	36.11	6.500	User-USER-3
5	S	9.77000	2.57000	1.000000		6.500	
6	#S	Plane	2.57000	1.000000		4.662	
7	S	-9.77000	2.05500	1.576310	36.11	6.500	User-USER-3
8	S	21.85000	2.57000	1.572440	45.15	6.500	User-USER
9	S	-17.99500	0.13000	1.000000		6.500	
10	S	-77.12000	1.54000	1.572440	45.15	6.500	User-USER
11	S	-15.73500	40.60602	1.000000		6.500	
12	S	Plane				18.554	

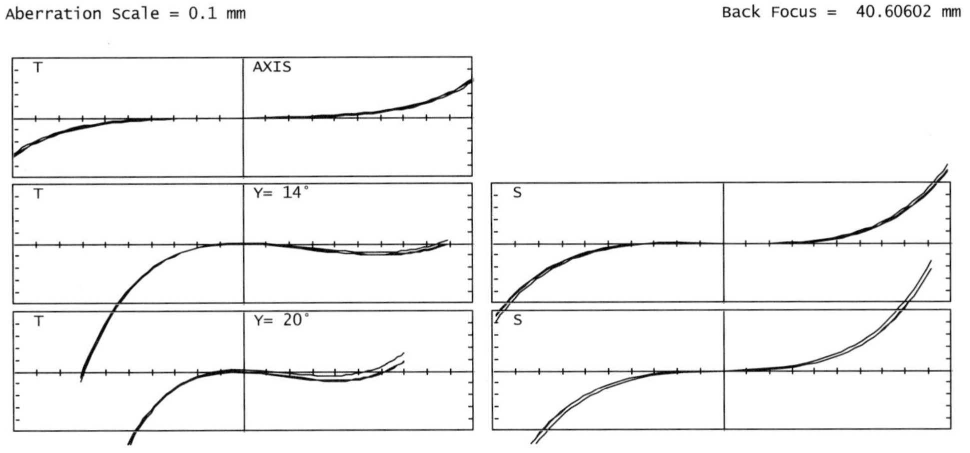

Figure 4.8 Transverse ray aberrations of the planar.

LAGRANGE INVARIANT = -2.2756

#	H	U	HBAR	UBAR	D(U/N)	A	ABAR
1	6.25215	0.00000	-3.59078	0.36397	-0.09199	0.39734	0.13577
2	6.02939	-0.14465	-3.10638	0.31454	-0.09071	-0.10452	0.43127
3	6.00564	-0.18270	-3.04508	0.47155	0.03154	0.15104	0.30233
4	5.39479	-0.23768	-2.11607	0.36148	0.00113	-0.76198	0.72070
5	4.90879	-0.23649	-1.37553	0.36036	0.06680	0.41920	0.34611
6	4.69489	-0.08323	-0.12420	0.48690	0.00000	-0.08323	0.48690
7	4.48098	-0.08323	1.12713	0.48690	0.15611	-0.54188	0.37153
8	4.71707	0.11488	1.84857	0.35106	0.00070	0.52139	0.68674
9	5.01441	0.11570	2.75355	0.35214	-0.05116	-0.25624	0.31310
10	5.01733	0.02241	2.81415	0.46612	0.00171	-0.04264	0.42963
11	5.07575	0.03794	3.29111	0.30971	-0.14913	-0.44758	0.15812
		-0.12500		0.36728			

#	S1	S2	S3	S4	S5	C1	C2
1	0.090803	0.031026	0.010601	0.119806	0.044559	0.020030	0.006844
2	0.005974	-0.024652	0.101721	-0.024444	-0.318862	0.005081	-0.020966
3	-0.004322	-0.008650	-0.017314	0.104760	0.175033	0.007314	0.014640
4	-0.003526	0.003335	-0.003155	-0.000370	0.003334	-0.008475	0.008016
5	-0.057624	-0.047576	-0.039280	-0.193781	-0.192422	-0.020835	-0.017202
6	0.000000	0.000000	0.000000	0.000000	0.000000	0.000000	0.000000
7	-0.205406	0.140834	-0.096561	-0.193781	0.199069	-0.024584	0.016856
8	-0.000894	-0.001177	-0.001551	-0.000370	-0.002530	-0.005071	-0.006679
9	0.016846	-0.020584	0.025151	0.104760	-0.158738	0.010360	-0.012659
10	-0.000016	0.000157	-0.001587	-0.024444	0.262252	-0.001725	0.017381
11	0.151633	-0.053568	0.018924	0.119806	-0.049010	0.018317	-0.006471
Sum	-0.006532	0.019146	-0.003050	0.011942	-0.037315	0.000413	-0.000240

Professor Kingslake has remarked that when a name is attached to a lens design it usually means that the person concerned was not the designer. The planar should be regarded as the prototype of all modern double-Gauss lenses; perhaps this configuration should not be known as the double Gauss but as the planar, or

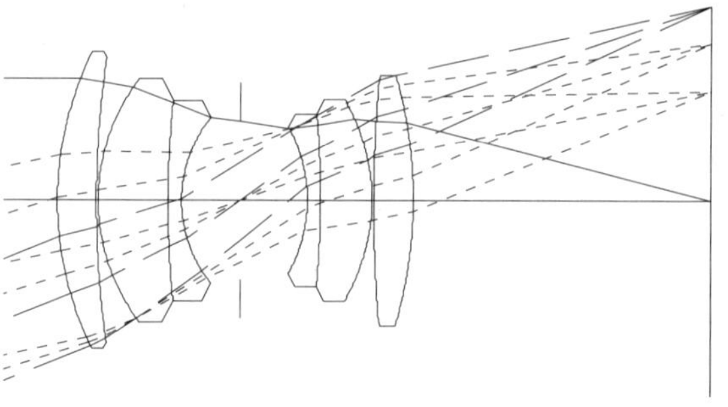

Figure 4.9 Six-glass Mandler double Gauss.

even the Rudolph, in recognition of Paul Rudolph's insight in inventing it more than 100 years ago.

The planar was not immediately developed into a high-aperture lens. H.W. Lee of Taylor Hobson designed the first high-aperture double Gauss in 1920. This lens uses several principles that we know are desirable. It has near front-back symmetry (which makes correction of odd-power aberrations quite simple), but the front-back symmetry is not perfect because the lens is intended for use as a cine-objective, with an object at infinity. It uses much thicker meniscus components, which are helpful in reducing the Petzval sum and the oblique spherical aberration. At an aperture of *f*/2 it has aberrations that are not much worse than the planar at *f*/4. The major limitations of the original Lee design are oblique spherical aberration (both tangential and sagittal) and the high-order sagittal astigmatism.

Since 1920, designers have improved on the original Lee design, and it has been modified in many ways.[8,9] The design shown in Fig. 4.9, by Mandler for the Leica, is typical of modern DLS-optimized double-Gauss designs. The size of the central air space is significantly increased compared with the Lee design, and the Mandler design uses modern high-index glass types. Note from Fig. 4.10 that the

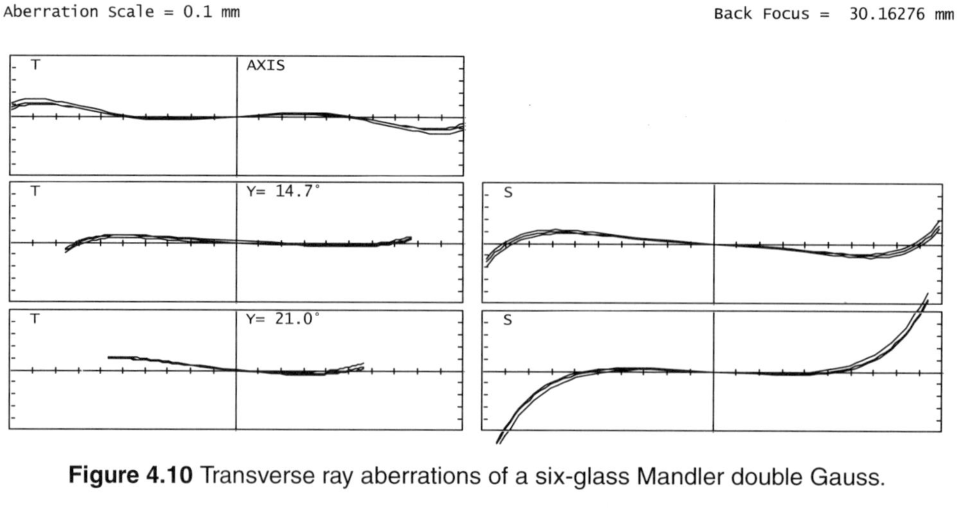

Figure 4.10 Transverse ray aberrations of a six-glass Mandler double Gauss.

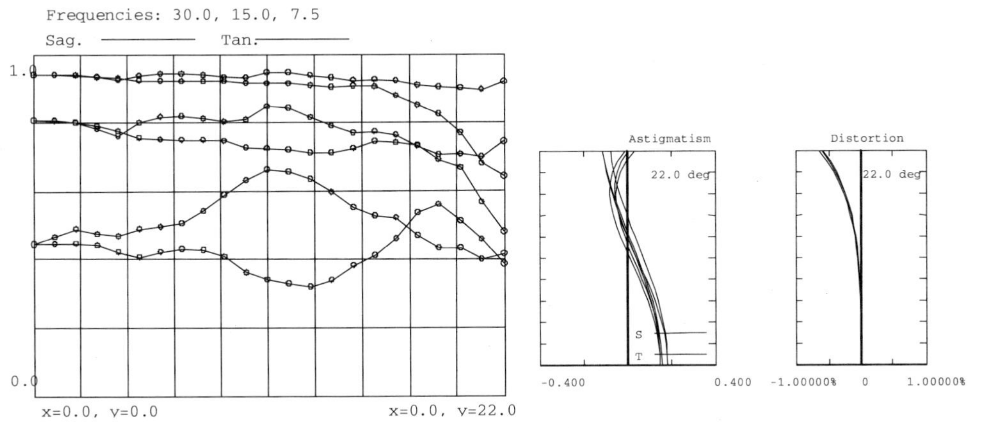

Figure 4.11 MTF as function of field, astigmatism, and distortion of the six-glass Mandler double Gauss.

field curvature and astigmatism are very much smaller than in the tessar, and the tangential oblique spherical aberration (TOBSA) is eliminated within the vignetted aperture, but there is still some sagittal oblique spherical aberration (SOBSA) and some sagittal high-order astigmatism. The variation in MTF across the field, as shown in Fig. 4.11, is considerably reduced from earlier designs. This design almost certainly represents the best performance that can be achieved with a basic six-glass double-Gauss objective.

EFL = 50.001

WAVELENGTHS [nm]

587.60 656.30 486.10

SURFACES

#	SURF SPACE	RADIUS	SEPN	INDEX1	V	CLR RAD	GLASS
0			0.00000	1.000000			
1	S	34.38800	4.00000	1.689001	49.69	15.195	S-LAFN23
2	S	101.75100	0.20000	1.000000		14.583	
3	S	20.18000	7.23000	1.743999	44.70	12.439	S-LAF2
4	S	107.82100	1.30000	1.717355	29.50	10.297	S-SF1
5	S	13.66800	5.91000	1.000000		8.508	
6#	S	Plane	6.89000	1.000000		8.086	
7	S	-15.52400	1.30000	1.717355	29.50	7.536	S-SF1
8	S	-95.46700	5.23000	1.716995	47.94	8.897	S-LAF3
9	S	-20.64500	0.20000	1.000000		10.409	
10	S	159.89500	4.00000	1.788306	47.45	12.402	S-LAFN21
11	S	-46.02500	29.76276	1.000000		12.832	
12	S	Plane				19.115	

LAGRANGE INVARIANT = -4.7961

#	H	U	HBAR	UBAR	D(U/N)	A	ABAR
1	12.49421	0.00000	-9.51783	0.38386	-0.08775	0.36333	0.10709
2	11.90135	-0.14821	-8.15711	0.34018	-0.08199	-0.05278	0.43916
3	11.86740	-0.16975	-8.05324	0.51933	-0.02991	0.41833	0.12026
4	9.34986	-0.34821	-4.66941	0.46803	-0.00546	-0.45604	0.74071
5	8.89191	-0.35227	-4.05241	0.47462	0.06684	0.51229	0.30591

6	8.07468	-0.13828	-0.49226	0.60239	0.00000	-0.13828	0.60239
7	7.12194	-0.13828	3.65824	0.60239	0.20298	-0.59705	0.36674
8	7.26639	0.11111	4.24220	0.44920	0.00002	0.06011	0.69513
9	7.84756	0.11112	6.59197	0.44929	-0.14647	-0.46187	0.22319
10	7.83121	-0.08175	6.70047	0.54249	0.04411	-0.03277	0.58439
11	7.56199	-0.06730	7.83999	0.28488	-0.21224	-0.41418	0.20483
		-0.24988		0.37517			

#	S1	S2	S3	S4	S5	C1	C2
1	0.144735	0.042659	0.012573	0.272870	0.084130	0.037267	0.010984
2	0.002718	-0.022619	0.188200	-0.092220	-0.798610	0.005157	-0.042907
3	0.062128	0.017860	0.005134	0.486270	0.141264	0.047378	0.013620
4	0.010617	-0.017244	0.028008	-0.001898	-0.042409	-0.019682	0.031968
5	-0.155980	-0.093142	-0.055618	-0.702978	-0.452987	-0.064499	-0.038515
6	0.000000	0.000000	0.000000	0.000000	0.000000	0.000000	0.000000
7	-0.515312	0.316536	-0.194436	-0.618932	0.499621	-0.060208	0.036983
8	-0.000000	-0.000005	-0.000063	0.000029	-0.000384	-0.002380	-0.027524
9	0.245194	-0.118484	0.057254	0.465270	-0.252496	0.031571	-0.015256
10	-0.000371	0.006616	-0.117981	0.063415	0.973041	-0.002384	0.042515
11	0.275326	-0.136160	0.067337	0.220309	-0.142252	0.029096	-0.014389
Sum	0.069056	-0.003982	-0.009592	0.092135	0.008917	0.001315	-0.002520

4.4 The Sonnar

Historically, the third type of high-aperture photographic objective was the Sonnar type; this can be thought of as a triplet derivative with a meniscus component in the front air space, although Sonnars are typically more complex than this.

In Fig. 4.12 we show an example of a six-glass Sonnar at an aperture of *f*/2. It should be noted that the Sonnar was introduced by Bertele (Zeiss) in 1931, at a time when antireflection coatings were not yet available. Given that the transmission of a single air-glass surface is typically about 95%, it is clear that a lens with six air-glass surfaces can have a transmission of $0.95^6 = 0.73$. However, a lens with ten air-glass surfaces had a transmission of $0.95^{10} = 0.59$. (The reflection losses at any cemented interface will be much smaller than at air-glass surfaces.)

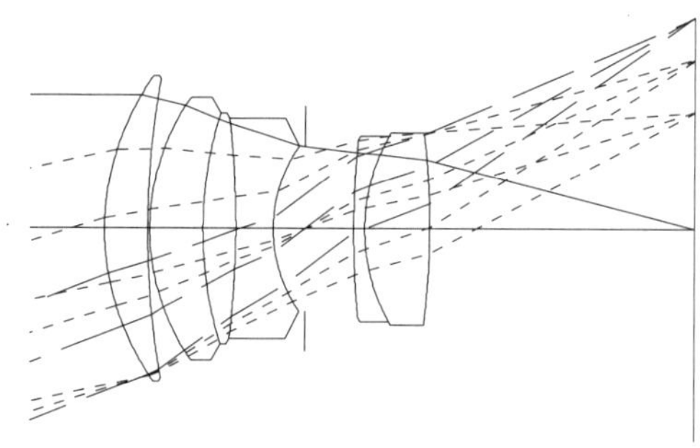

Figure 4.12 The six-glass Sonnar.

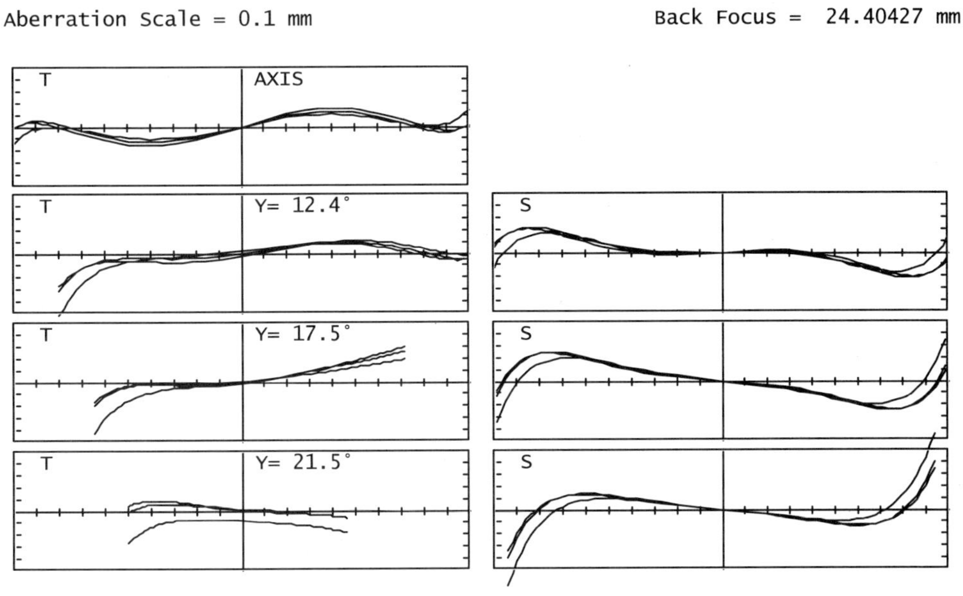

Figure 4.13 Transverse ray aberrations of the six-glass Sonnar.

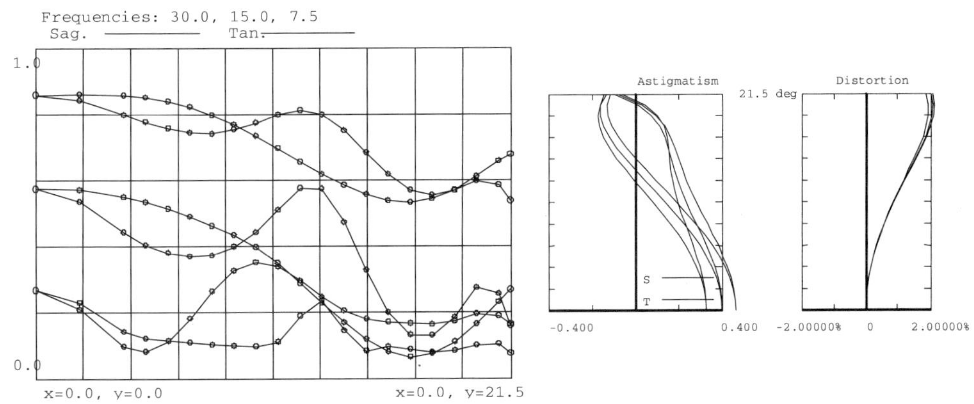

Figure 4.14 MTF as function of field, astigmatism, and distortion of the six-glass Sonnar.

EFL = 50.003
WAVELENGTHS [nm]
587.60 656.30 486.10
SURFACES

# SURF SPACE	RADIUS	SEPN	INDEX1	V	CLR RAD	GLASS
0		0.00000	1.000000			
1 S	24.89000	4.00000	1.620408	60.30	14.315	S-SK16
2 S	80.94700	0.20000	1.000000		14.028	
3 S	21.59600	5.00000	1.670028	47.09	12.351	S-BAFN10
4 S	37.79000	3.00000	1.464499	65.74	10.813	S-FK3
5 S	-90.80700	3.40000	1.688927	31.16	10.351	S-SF8
6 S	13.58300	3.00000	1.000000		7.743	
7#S	Plane	4.50000	1.000000		7.681	

8 S	91.63600	1.00000	1.568828	56.11	8.277	S-BAK4
9 S	18.15000	6.00000	1.670028	47.09	8.681	S-BAFN10
10 S	-73.09800	24.80427	1.000000		8.995	
11 S	Plane				20.285	

LAGRANGE INVARIANT = -4.9194

#	H	U	HBAR	UBAR	D(U/N)	A	ABAR
1	12.48862	0.00000	-9.81660	0.39391	-0.11855	0.50175	-0.00049
2	11.72019	-0.19211	-8.24021	0.39410	-0.10291	-0.07667	0.47365
3	11.67590	-0.22146	-8.12513	0.57544	0.01217	0.31919	0.19921
4	9.92828	-0.34952	-5.64753	0.49552	-0.03769	-0.14496	0.57795
5	8.84316	-0.36171	-4.01527	0.54409	0.06894	-0.67234	0.86157
6	7.82078	-0.30070	-2.43117	0.46591	0.06685	0.46458	0.48460
7	7.48720	-0.11119	-0.44043	0.66358	0.00000	-0.11119	0.66358
8	6.98683	-0.11119	2.54569	0.66358	0.04839	-0.03495	0.69136
9	6.88831	-0.09852	2.95860	0.41291	-0.00639	0.44084	0.90351
10	6.19501	-0.11555	5.22664	0.37801	-0.18057	-0.33451	0.51187
		-0.24976		0.58338			

#	S1	S2	S3	S4	S5	C1	C2
1	0.372745	-0.000363	0.000000	0.372265	-0.000363	0.039787	-0.000039
2	0.007091	-0.043802	0.270579	-0.114466	-0.964360	0.005706	-0.035248
3	-0.014478	-0.009036	-0.005640	0.449593	0.277079	0.031753	0.019817
4	0.007863	-0.031351	0.124994	-0.053816	-0.283784	0.005319	-0.021206
5	-0.275584	0.353149	-0.452545	-0.024181	0.610906	-0.049140	0.062971
6	-0.112843	-0.117704	-0.122774	-0.726759	-0.886125	-0.047560	-0.049608
7	0.000000	0.000000	0.000000	0.000000	0.000000	0.000000	0.000000
8	-0.000413	0.008169	-0.161614	0.095755	1.302867	-0.001578	0.031215
9	0.008555	0.017534	0.035936	0.051503	0.179207	0.006249	0.012808
10	0.125165	-0.191532	0.293090	0.132827	-0.651756	0.017656	-0.027018
Sum	0.118101	-0.014936	-0.017974	0.182721	-0.416330	0.008192	-0.006306

4.5 Wide-angle lenses for rangefinder cameras (Zeiss Biogon)

Early wide-angle lenses were often derived from standard 50-mm lenses. Figure 4.15 shows a famous example that was designed at Zeiss for the Contax camera, and derived from the Sonnar. Note that the back focus is very small, which is accept-

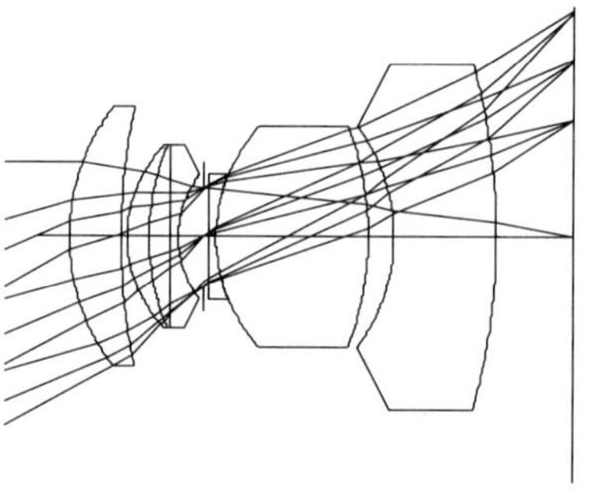

Figure 4.15 The Zeiss Biogon.

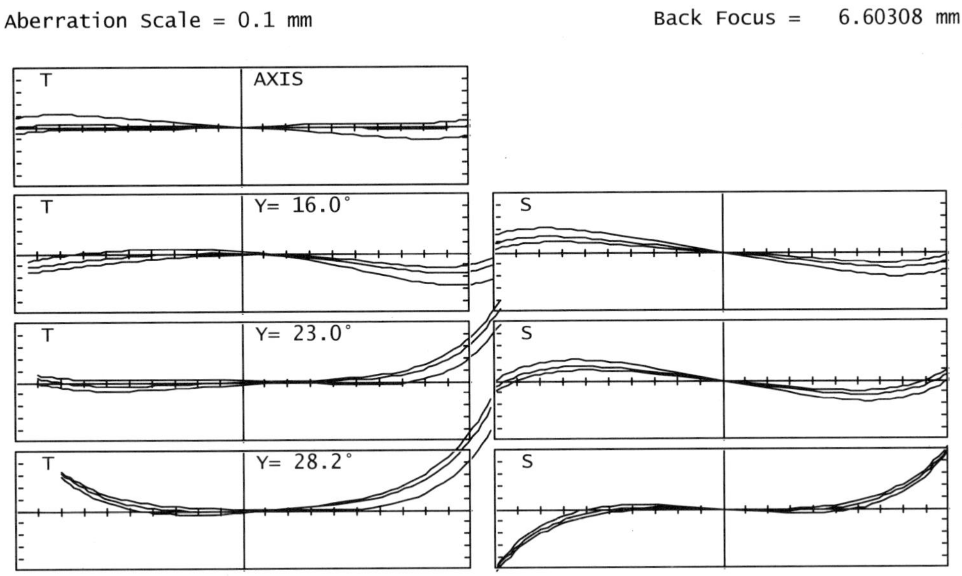

Figure 4.16 Transverse ray aberrations of the Zeiss Biogon.

able, because the Contax was a rangefinder camera, and did not require the long back focus that is necessary in modern SLR cameras. This design is quite asymmetric, and Fig. 4.16 shows significant coma at the edge of the field. Note the effect of the sagittal astigmatism on the MTF curves of Fig. 4.17. Distortion is quite large at 4%, but it is approximately linear and is acceptable.

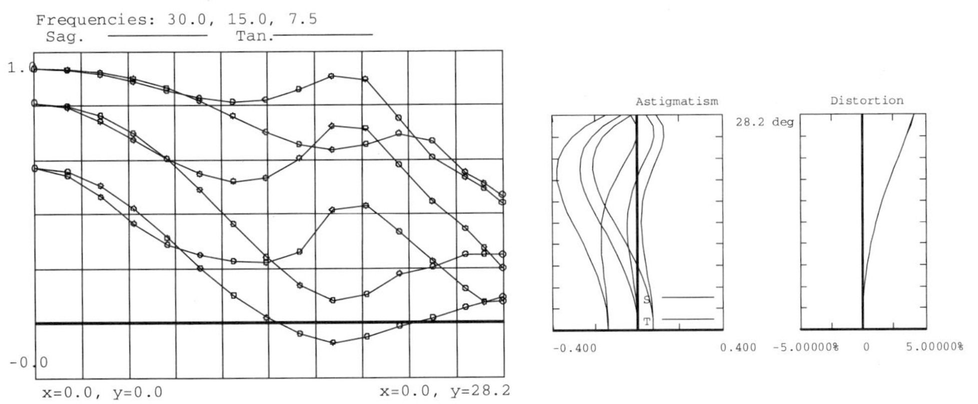

Figure 4.17 MTF as function of field, astigmatism, and distortion of the Zeiss Biogon.

```
EFL =     35.005
WAVELENGTHS   [nm]
      587.60        656.30        486.10
SURFACES
  # SURF SPACE     RADIUS        SEPN       INDEX1       V      CLR RAD  GLASS
  0                            0.00000    1.000000
  1 S             18.54100     4.47000    1.670028    47.09    11.296  S-BAFN10
```

2 S	54.52900	0.54400	1.000000		11.296	
3 S	11.55900	1.89700	1.670028	47.09	7.927	S-BAFN10
4 S	21.81100	1.81000	1.464499	65.74	7.927	S-FK3
5 S	Plane	0.65500	1.688927	31.16	7.927	S-SF8
6 S	8.99600	2.27200	1.000000		5.450	
7#S	Plane	0.34700	1.000000		4.296	
8 S	144.82800	0.65500	1.464499	65.74	5.450	S-FK3
9 S	14.17800	13.08900	1.670028	47.09	9.612	S-BAFN10
10 S	-27.26600	2.18100	1.000000		9.612	
11 S	-16.35900	8.72600	1.531717	48.74	9.612	S-LLF6
12 S	-58.27100	6.60308	1.000000		15.062	
13 S	Plane				19.746	

LAGRANGE INVARIANT = -3.4724

#	H	U	HBAR	UBAR	D(U/N)	A	ABAR
1	6.47594	0.00000	-8.14851	0.53620	-0.08391	0.34928	0.09671
2	5.84954	-0.14013	-5.92516	0.49739	-0.07824	-0.05487	0.64920
3	5.76134	-0.16215	-5.51288	0.75786	-0.01573	0.33628	0.28092
4	5.19780	-0.29707	-4.28903	0.64515	-0.03059	-0.09812	0.74901
5	4.64518	-0.30531	-3.00739	0.70809	0.05172	-0.44713	1.03700
6	4.47178	-0.26474	-2.60522	0.61400	0.05208	0.39241	0.54789
7	4.23396	-0.10467	-0.70244	0.83749	0.00000	-0.10467	0.83749
8	4.19764	-0.10467	-0.41183	0.83749	0.04959	-0.07569	0.83465
9	4.14480	-0.08067	-0.03667	0.57276	-0.00882	0.30999	0.83502
10	2.74797	-0.10672	6.54174	0.50259	-0.18185	-0.34653	0.43866
11	2.21199	-0.24575	8.02174	0.67859	0.17165	-0.38097	0.18823
12	1.22157	-0.11350	13.37291	0.61324	-0.11090	-0.20596	0.58780
		-0.18500		0.81729			

#	S1	S2	S3	S4	S5	C1	C2
1	0.066291	0.018355	0.005082	0.260908	0.073649	0.019272	0.005336
2	0.001378	-0.016304	0.192883	-0.088714	-1.232380	0.002735	-0.032355
3	0.010250	0.008563	0.007153	0.418505	0.355589	0.016507	0.013790
4	0.001531	-0.011688	0.089216	-0.046456	-0.326407	0.001885	-0.014388
5	-0.048036	0.111406	-0.258378	0.000000	0.599241	-0.017166	0.039813
6	-0.035860	-0.050068	-0.069907	-0.546719	-0.860945	-0.022969	-0.032070
7	0.000000	0.000000	0.000000	0.000000	0.000000	0.000000	0.000000
8	-0.001193	0.013151	-0.145019	0.026405	1.307957	-0.001533	0.016903
9	0.003513	0.009463	0.025490	0.071466	0.261166	0.004748	0.012791
10	0.060008	-0.075962	0.096157	0.177419	-0.346309	0.008113	-0.010270
11	-0.055105	0.027227	-0.013452	-0.255856	0.133062	-0.006002	0.002965
12	0.005747	-0.016401	0.046806	0.071829	-0.338569	0.001792	-0.005114
Sum	0.008525	0.017742	-0.023968	0.088786	-0.373949	0.007382	-0.002600

4.6 Wide-angle lens for rangefinder camera (Schneider Super-Angulon)

Another form of wide-angle lens uses a nearly symmetric negative-positive-negative construction. The general layout was invented by Roosinov in 1946, and has the advantage that the illumination at the edge of the field is much better than in the older designs. Bertele further developed it.

The lens shown in Fig. 4.18 was designed at Schneider for the Leica in about 1958. Note that the back focus is very small and it is clearly not suitable for use in

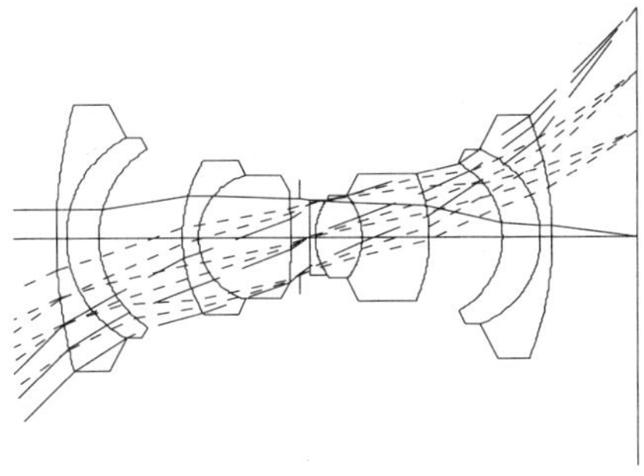

Figure 4.18 The Schneider Super-Angulon.

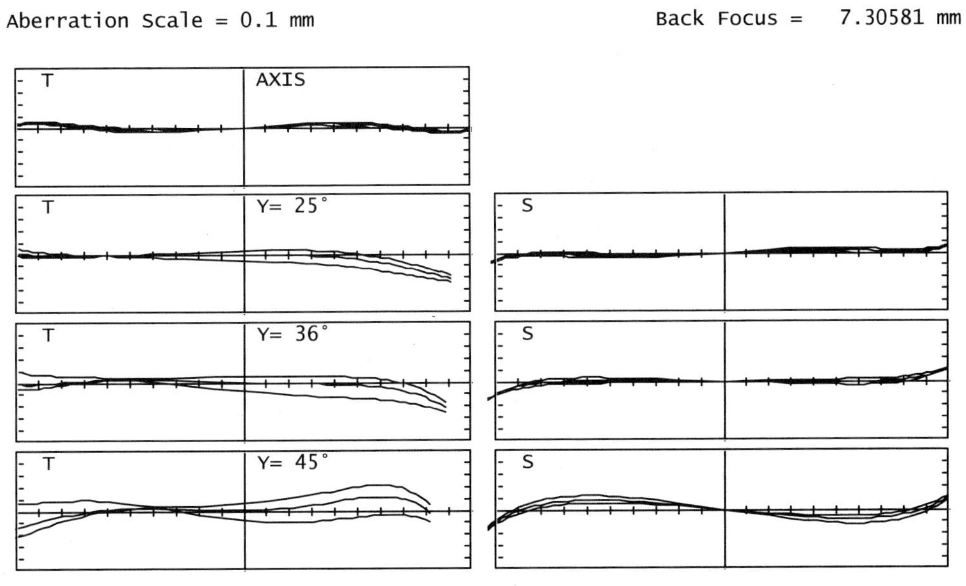

Figure 4.19 Transverse ray aberrations of the Schneider Super-Angulon.

a single-lens reflex camera. The focal length here is 20 mm, with a semifield angle of 45 deg.

EFL = 20.002
WAVELENGTHS [nm]
587.60 656.30 486.10
SURFACES

# SURF SPACE	RADIUS	SEPN	INDEX1	V	CLR RAD	GLASS
0		0.00000	1.000000			
1 S	44.16200	0.89400	1.658299	57.24	11.567	S-LAK11
2 S	10.00500	2.78200	1.464499	65.74	8.723	S-FK3
3 S	9.32600	7.14000	1.000000		7.746	
4 S	13.23400	1.38200	1.589038	52.99	6.648	A-BALF6
5 S	5.53000	7.93200	1.607287	49.38	5.343	A-BAF5
6 S	-52.67400	0.79000	1.000000		3.974	
7#S	Plane	0.79000	1.000000		3.301	
8 S	50.22100	0.51400	1.529438	51.61	3.233	S-KZFN2

9	S	5.75600	4.00000	1.617717	49.75	3.235	S-SSKN8
10	S	-5.70200	5.71000	1.784695	26.07	3.596	S-SF56A
11	S	-14.24400	6.24600	1.000000		5.488	
12	S	-7.32400	3.24000	1.533748	55.28	6.388	A-ZK5
13	S	-8.08200	1.00000	1.658438	50.86	7.538	S-SSKN5
14	S	-28.81800	7.50581	1.000000		10.513	
15	S	Plane				20.241	

LAGRANGE INVARIANT = -2.4997

#	H	U	HBAR	UBAR	D(U/N)	A	ABAR
1	2.49970	0.00000	-10.12373	1.00000	-0.01355	0.05660	0.77076
2	2.47961	-0.02247	-9.50327	0.69403	0.01857	0.37373	-0.42423
3	2.50007	0.00735	-7.66666	0.66018	0.13027	0.40337	-0.23710
4	3.46604	0.13529	-3.48994	0.58497	-0.14281	0.39719	0.32126
5	3.44953	-0.01195	-2.84609	0.46589	-0.00424	0.97224	-0.07751
6	3.29968	-0.01889	0.85371	0.46644	-0.05665	-0.13105	0.72365
7	3.24564	-0.06841	1.43820	0.73986	0.00000	-0.06841	0.73986
8	3.19160	-0.06841	2.02269	0.73986	0.02478	-0.00486	0.78014
9	3.15730	-0.06673	2.26417	0.46980	-0.01387	0.73688	1.32015
10	2.78523	-0.09302	3.95497	0.42270	0.03586	-0.94067	-0.43826
11	2.56474	-0.03861	6.51333	0.44805	-0.18857	-0.39026	-0.01646
12	1.25180	-0.21020	9.26664	0.44081	0.15963	-0.38112	-0.82443
13	1.00047	-0.07757	11.62445	0.72772	0.01293	-0.30884	-1.08988
14	0.93803	-0.06243	12.40559	0.78114	-0.08733	-0.15752	0.58155
		-0.12497		1.01203			

#	S1	S2	S3	S4	S5	C1	C2
1	0.000109	0.001478	0.020122	0.056168	1.038828	0.000981	0.013362
2	-0.006432	0.007301	-0.008288	-0.049838	0.065981	-0.001956	0.002221
3	-0.052989	0.031147	-0.018308	-0.212508	0.135675	-0.004865	0.002860
4	0.078089	0.063161	0.051087	0.175023	0.182885	0.009630	0.007789
5	0.013814	-0.001101	0.000088	0.008073	-0.000651	0.002199	-0.000175
6	0.003211	-0.017728	0.097895	0.044821	-0.788062	0.003308	-0.018269
7	0.000000	0.000000	0.000000	0.000000	0.000000	0.000000	0.000000
8	-0.000002	0.000300	-0.048133	0.043070	0.813278	-0.000104	0.016702
9	0.023781	0.042605	0.076329	0.038733	0.206138	0.002250	0.004032

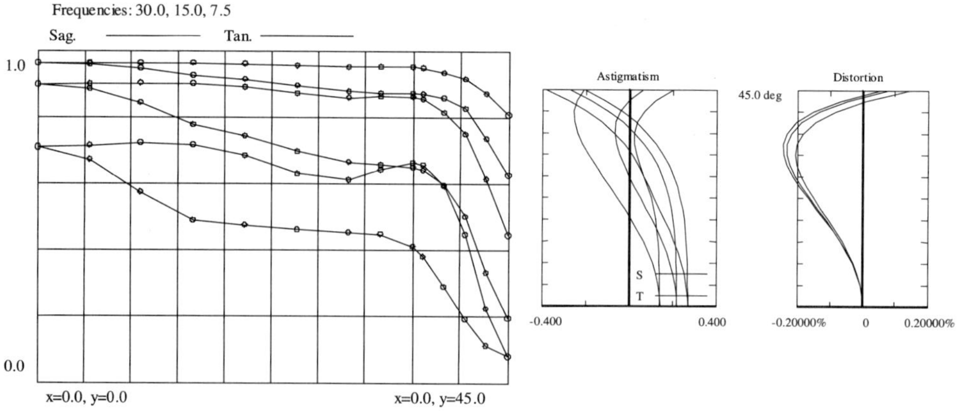

Figure 4.20 MTF as function of field, astigmatism, and distortion of the Schneider Super-Angulon.

10	-0.088388	-0.041179	-0.019185	-0.063378	-0.038466	-0.024086	-0.011221
11	0.073659	0.003106	0.000131	0.192877	0.008138	0.016884	0.000712
12	-0.029025	-0.062786	-0.135816	-0.296900	-0.936036	-0.003003	-0.006497
13	-0.001234	-0.004355	-0.015367	-0.037899	-0.187976	-0.000467	-0.001647
14	0.002033	-0.007504	0.027704	0.086085	-0.420089	0.001153	-0.004258
Sum	0.016625	0.014444	0.028258	-0.015675	0.079643	0.001926	0.005609

4.7 Wide-angle lenses for SLR cameras

In modern single-lens reflex cameras, a back focus of about 38 to 40 mm is required for the mirror mechanism. This means that earlier forms of wide-angle lens are not acceptable, and the inverted telephoto construction is necessary, as illustrated in Fig. 4.21. Although the lens is much larger than the Biogon (for example), correction of field curvature is very good because of the separated negative power at the front of the lens. Most aberrations are well corrected, as may be seen in Figs. 4.22 and 4.23.

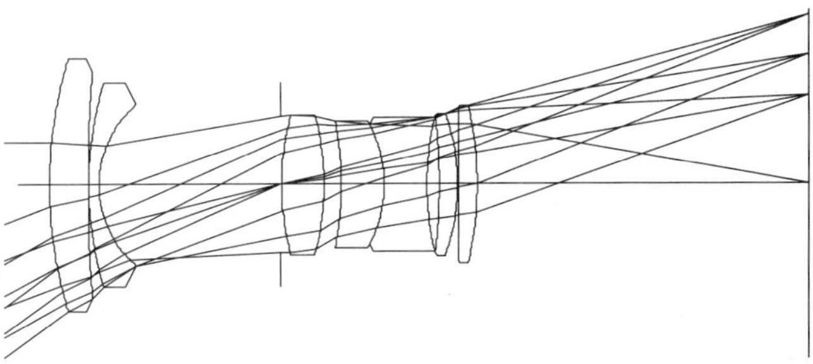

Figure 4.21 A wide-angle lens for SLR camera.

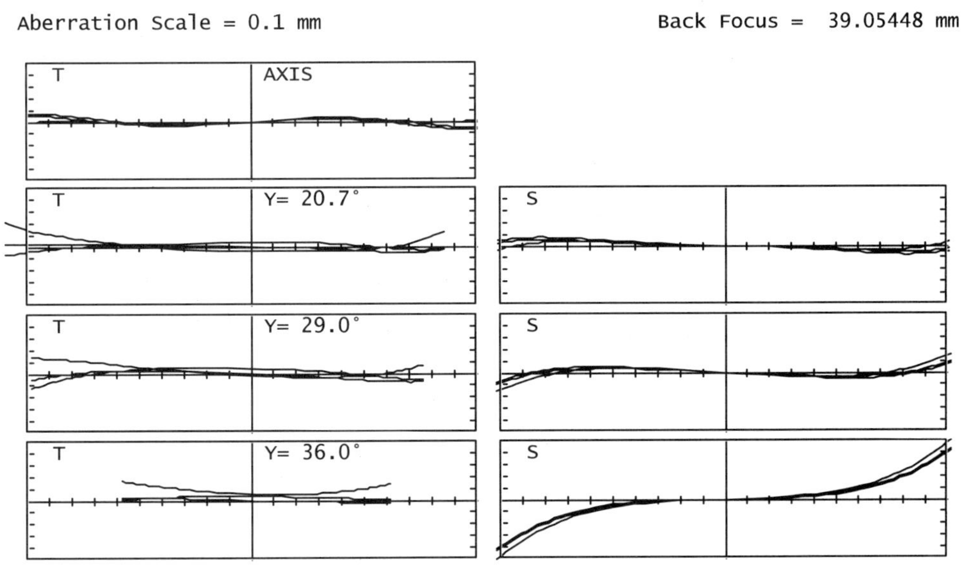

Figure 4.22 Transverse ray aberrations of a wide-angle lens for SLR camera.

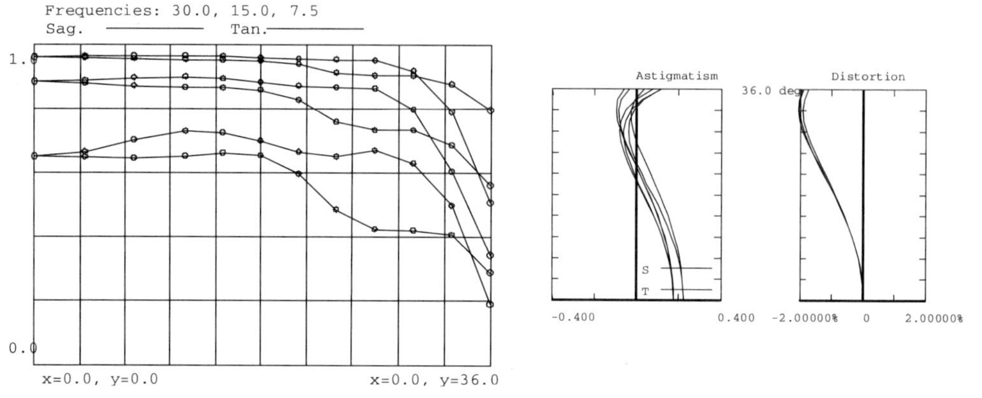

Figure 4.23 MTF as function of field, astigmatism, and distortion of a wide-angle lens for an SLR camera.

EFL = 28.499

WAVELENGTHS [nm]

587.60 656.30 486.10

SURFACES

# SURF SPACE	RADIUS	SEPN	INDEX1	V	CLR RAD	GLASS
0		0.00000	1.000000			
1 S	51.91800	4.50000	1.654457	33.70	15.019	H-FD9
2 S	166.15800	0.10000	1.000000		13.498	
3 S	42.85600	1.20000	1.719996	50.32	12.138	H-LAC10
4 S	12.64200	21.21800	1.000000		9.606	
5#S	Plane	0.10000	1.000000		8.105	
6 S	33.36500	5.00000	1.833996	37.33	8.299	H-NBFD10
7 S	-32.90000	1.98900	1.000000		8.217	
8 S	-41.15800	5.00000	1.516797	64.17	7.545	H-BSC7
9 S	-15.74600	5.00000	1.784694	26.05	7.148	H-FDS3
10 S	35.35500	1.42400	1.000000		8.012	
11 S	-77.33400	2.20000	1.804196	46.48	8.059	H-TAF3
12 S	-25.24200	0.10000	1.000000		8.456	
13 S	-1016.66400	2.20000	1.804196	46.48	9.101	H-TAF3
14 S	-41.16300	39.17448	1.000000		9.364	
15 S	Plane				20.512	

LAGRANGE INVARIANT = -3.6324

#	H	U	HBAR	UBAR	D(U/N)	A	ABAR
1	4.99959	0.00000	-13.74277	0.72654	-0.02302	0.09630	0.46184
2	4.82817	-0.03809	-11.29544	0.54385	-0.02098	-0.01495	0.78731
3	4.82377	-0.04401	-11.20991	0.85529	0.00174	0.06855	0.59372
4	4.73653	-0.07270	-10.48180	0.60676	0.18698	0.51938	-0.38247
5	7.80701	0.14471	-1.00474	0.44665	0.00000	0.14471	0.44665
6	7.82148	0.14471	-0.96007	0.44665	-0.15981	0.37913	0.41788
7	7.68300	-0.02770	0.32306	0.25663	-0.23045	-0.47908	0.45264
8	7.19459	-0.24556	1.24289	0.46246	0.17809	-0.42036	0.43226
9	6.68293	-0.10233	2.81880	0.31518	0.05443	-0.79898	0.20653
10	6.56662	-0.02326	4.29252	0.29474	0.11726	0.28996	0.74271
11	6.71504	0.10423	5.17724	0.62130	-0.05076	0.01740	0.55435
12	6.92728	0.09647	6.00049	0.37420	-0.10011	-0.32108	0.24624
13	6.92262	-0.04664	6.04888	0.48396	0.03400	-0.05345	0.47801

14	6.87242	-0.02282	6.64485	0.27089	-0.16278	-0.34239	0.19750
		-0.17543		0.35893			

#	S1	S2	S3	S4	S5	C1	C2
1	0.001067	0.005120	0.024553	0.100531	0.599898	0.005651	0.027101
2	0.000023	-0.001192	0.062793	-0.031412	-1.652802	0.000847	-0.044615
3	-0.000039	-0.000341	-0.002954	0.128879	1.090616	0.002751	0.023823
4	-0.238903	0.175930	-0.129556	-0.436896	0.417139	-0.020463	0.015069
5	0.000000	0.000000	0.000000	0.000000	0.000000	0.000000	0.000000
6	0.179674	0.198035	0.218272	0.179831	0.438787	0.036124	0.039815
7	0.406380	-0.383953	0.362763	0.182373	-0.515051	0.044839	-0.042364
8	-0.226405	0.232816	-0.239409	-0.109227	0.358508	-0.016057	0.016512
9	-0.232213	0.060027	-0.015517	-0.082927	0.025448	-0.061773	0.015968
10	-0.064742	-0.165829	-0.424754	-0.164088	-1.508260	-0.032138	-0.082318
11	0.000103	0.003287	0.104739	-0.076050	0.914173	0.001120	0.035695
12	0.071494	-0.054831	0.042052	0.232994	-0.210942	0.021328	-0.016357
13	-0.000672	0.006013	-0.053773	-0.005785	0.532631	-0.003548	0.031731
14	0.131147	-0.075650	0.043637	0.142877	-0.107587	0.022564	-0.013015
Sum	0.026914	-0.000569	-0.007154	0.061101	0.382558	0.001243	0.007046

4.8 Telephoto lens

When moderately long focal lengths are required, designs similar to normal 50-mm focal length systems are often used. The example shown in Fig. 4.24 has a focal length of 100 mm and an aperture of *f*/3.5, and gives very good performance, as may be seen in Figs. 4.25 and 4.26.

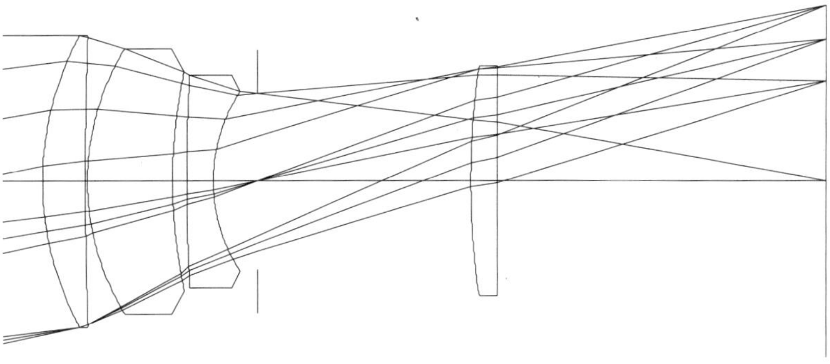

Figure 4.24 A telephoto lens.

EFL = 100.001
WAVELENGTHS [nm]
587.60 656.30 486.10
SURFACES

#	SURF SPACE	RADIUS	SEPN	INDEX1	V	CLR RAD	GLASS
0			0.00000	1.000000			
1	S	38.02200	5.20000	1.612718	58.61	17.867	S-SK4
2	S	527.02400	0.25000	1.000000		17.646	
3	S	31.46000	10.35000	1.620408	60.30	16.295	S-SK16
4	S	65.90300	1.69600	1.000000		13.356	
5	S	239.93500	3.20000	1.717355	29.50	12.979	S-SF1
6	S	20.51900	5.30800	1.000000		11.021	

7#S	Plane	25.74600	1.000000		8.561	
8 S	102.92600	3.25000	1.740765	27.59	13.838	S-SF13
9 S	-1468.63800	40.17036	1.000000		14.084	
10 S	Plane				21.836	

LAGRANGE INVARIANT = -3.0366

#	H	U	HBAR	UBAR	D(U/N)	A	ABAR
1	14.28584	0.00000	-7.72596	0.21256	-0.08851	0.37573	0.00936
2	13.54355	-0.14275	-6.63916	0.20900	-0.12595	-0.18877	0.31674
3	13.48993	-0.21447	-6.55682	0.32934	0.03147	0.21433	0.12092
4	10.42086	-0.29653	-3.62733	0.28304	-0.19940	-0.22427	0.36946
5	9.77232	-0.38240	-2.90738	0.42450	0.24283	-0.34167	0.41238
6	9.00535	-0.23968	-2.10021	0.25224	0.04278	0.34210	0.25741
7	8.49164	-0.09678	-0.19058	0.35976	0.00000	-0.09678	0.35976
8	5.99993	-0.09678	9.07193	0.35976	0.05059	-0.03849	0.44791
9	5.73862	-0.08040	9.62171	0.16916	-0.09667	-0.14676	0.28307
		-0.14286		0.28962			

#	S1	S2	S3	S4	S5	C1	C2
1	0.178510	0.004447	0.000111	0.092136	0.002298	0.034797	0.000867
2	0.060787	-0.101996	0.171142	-0.006647	-0.276011	0.016574	-0.027810
3	-0.019503	-0.011003	-0.006208	0.112216	0.059809	0.018358	0.010358
4	0.104514	-0.172172	0.283631	-0.053568	-0.378997	0.014839	-0.024446
5	-0.277019	0.334352	-0.403552	0.016052	0.467699	-0.047277	0.057061
6	-0.045088	-0.033926	-0.025528	-0.187706	-0.160447	-0.043621	-0.032823
7	0.000000	0.000000	0.000000	0.000000	0.000000	0.000000	0.000000
8	-0.000450	0.005233	-0.060898	0.038122	0.265063	-0.003562	0.041452
9	0.011949	-0.023047	0.044451	0.002672	-0.090886	0.012991	-0.025056
Sum	0.013700	0.001887	0.003149	0.013276	-0.111473	0.003100	-0.000397

Aberration Scale = 0.1 mm Back Focus = 40.03036 mm

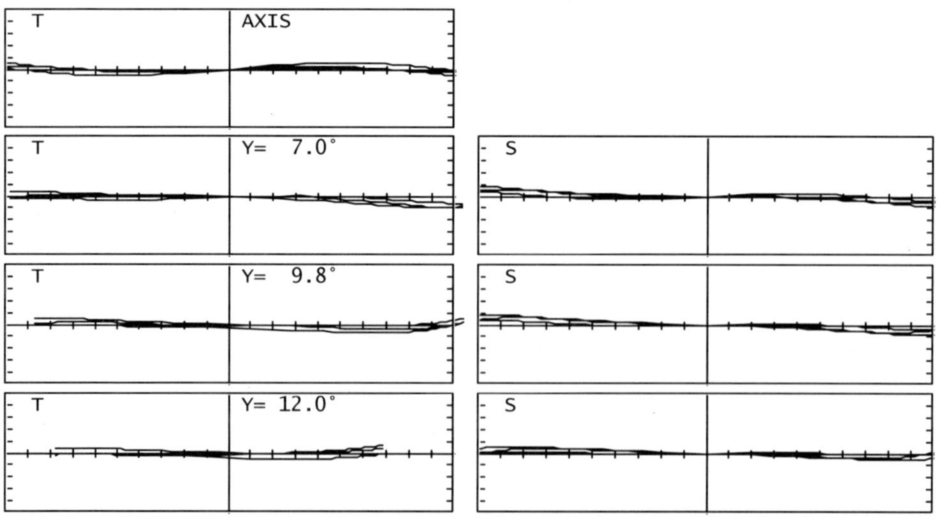

Figure 4.25 Transverse ray aberrations of a telephoto lens.

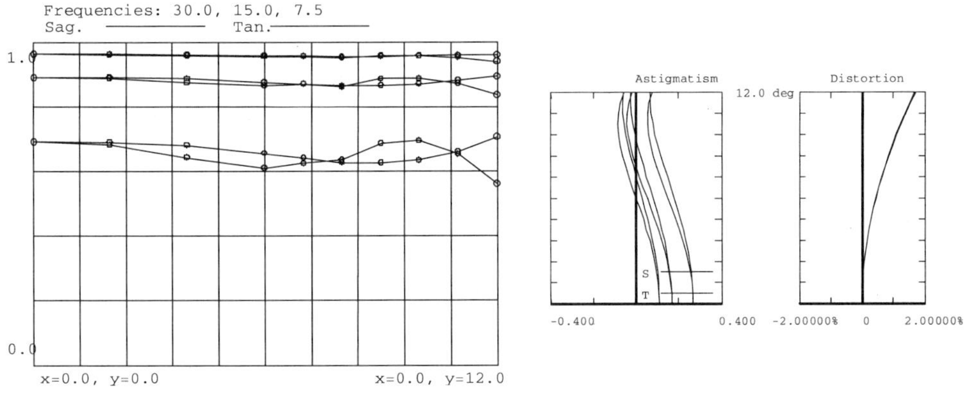

Figure 4.26 MTF as function of field, astigmatism, and distortion of a telephoto lens.

4.9 Long-focus telephoto lens

This example, Fig. 4.27, derived from U.S. Patent 4,709,998 by Yamanashi (Olympus), is typical of longer-focal-length telephoto lenses. Aberrations and MTF are shown in Figs. 4.28 and 4.29. When such a long focal length lens is required to focus at a close position, the movements can be quite large, and in this design the space before the rear triplet is used to focus.

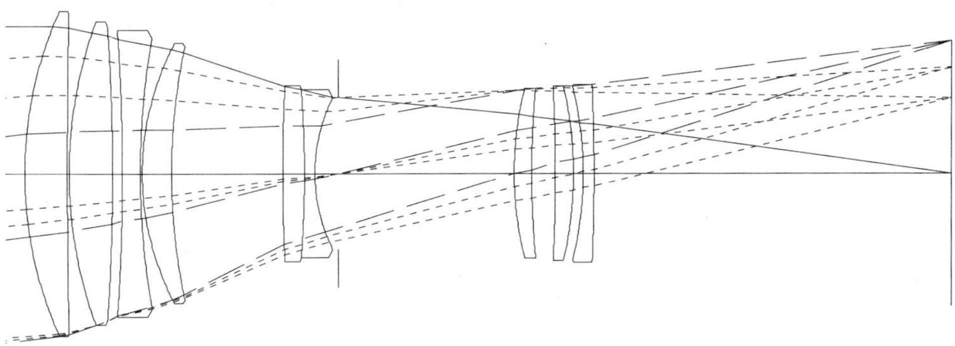

Figure 4.27 A long-focus telephoto lens.

EFL = 169.992
WAVELENGTHS [nm]
587.60 656.30 486.10
SURFACES

#	SURF SPACE	RADIUS	SEPN	INDEX1	V	CLR RAD	GLASS
0			0.00000	1.000000			
1	S	64.41600	6.67600	1.487488	70.41	25.878	H-FC5
2	S	1503.81800	0.29000	1.000000		25.552	
3	S	72.12600	6.77800	1.456499	90.23	24.151	H-FCD10
4	S	-334.88800	1.55000	1.000000		23.552	
5	S	-314.20400	2.81000	1.672696	32.16	22.825	H-E-FD5
6	S	131.35300	0.29000	1.000000		21.595	

7	S	45.15500	4.54700	1.487488	70.41	20.665	H-FC5
8	S	96.50900	17.32100	1.000000		20.098	
9	S	288.72100	3.38900	1.805175	25.45	13.940	H-FD60
10	S	-183.27200	1.65200	1.612496	44.86	13.377	H-ADF4
11	S	27.76100	3.63100	1.000000		12.136	
12#	S	Plane	27.47100	1.000000		12.044	
13	S	56.88500	2.70800	1.517283	69.65	13.415	H-PCS1
14	S	146.77100	3.67900	1.000000		13.474	
15	S	-276.54900	2.81000	1.664455	35.88	13.748	H-BAFD2
16	S	-62.01200	1.55000	1.000000		13.874	
17	S	-57.04600	1.65200	1.677897	50.53	13.846	H-LACL9
18	S	-351.52500	56.20849	1.000000		14.120	
19	S	Plane				21.276	

LAGRANGE INVARIANT = -2.8804

#	H	U	HBAR	UBAR	D(U/N)	A	ABAR
1	23.45883	0.00000	-10.14040	0.12278	-0.08024	0.36418	-0.03464
2	22.66205	-0.11935	-9.24491	0.13414	-0.08995	-0.15512	0.19038
3	22.61269	-0.17019	-9.18791	0.19653	0.02250	0.14333	0.06914
4	21.15468	-0.21511	-8.00273	0.17486	-0.19445	-0.40531	0.28949
5	20.62436	-0.34214	-7.59107	0.26559	0.23564	-0.40778	0.28975
6	20.12377	-0.17815	-7.17220	0.14906	-0.08842	-0.04172	0.15800
7	20.06724	-0.19493	-7.11054	0.21261	0.00892	0.24948	0.05514
8	18.80914	-0.27669	-6.22599	0.19454	-0.13055	-0.12167	0.19341
9	13.32598	-0.31656	-1.75852	0.25792	0.20801	-0.27041	0.25183
10	12.66190	-0.19595	-1.26510	0.14560	-0.03261	-0.47844	0.27529
11	12.28587	-0.22762	-0.99447	0.16382	0.04519	0.34659	0.20639
12	11.93740	-0.09597	-0.11499	0.24221	0.00000	-0.09597	0.24221
13	9.30098	-0.09597	6.53890	0.24221	0.01754	0.06753	0.35716
14	8.97874	-0.11900	6.86507	0.12045	-0.07048	-0.08773	0.25372
15	8.43092	-0.14890	7.62644	0.20695	0.10247	-0.17939	0.17937
16	8.21373	-0.07729	8.00675	0.13534	-0.17022	-0.34911	0.01036
17	7.87791	-0.21666	8.22295	0.13948	0.17295	-0.35475	-0.00467
18	7.75677	-0.07333	8.45648	0.14137	-0.09430	-0.16007	0.19683
		-0.13800		0.22089			

#	S1	S2	S3	S4	S5	C1	C2
1	0.249632	-0.023742	0.002258	0.042210	-0.004229	0.039763	-0.003782
2	0.049047	-0.060197	0.073883	-0.001808	-0.088461	0.016361	-0.020081
3	-0.010451	-0.005041	-0.002432	0.036053	0.016218	0.011258	0.005431
4	0.675773	-0.482657	0.344728	0.007765	-0.251761	0.029782	-0.021271
5	-0.808137	0.574220	-0.408010	-0.010619	0.297456	-0.105180	0.074736
6	0.003098	-0.011731	0.044424	-0.025402	-0.072035	0.010501	-0.039765
7	-0.011137	-0.002461	-0.000544	0.060215	0.013188	0.023301	0.005150
8	0.036349	-0.057783	0.091856	-0.028174	-0.101234	0.010651	-0.016932
9	-0.202685	0.188762	-0.175795	0.012817	0.151782	-0.063165	0.058826
10	0.094519	-0.054384	0.031292	0.002997	-0.019729	0.054901	-0.031589
11	-0.066691	-0.039714	-0.023650	-0.113519	-0.081684	-0.036051	-0.021468
12	0.000000	0.000000	0.000000	0.000000	0.000000	0.000000	0.000000
13	-0.000744	-0.003936	-0.020816	0.049724	0.152881	0.003075	0.016261
14	0.004870	-0.014086	0.040737	-0.019272	-0.062080	0.003856	-0.011151
15	-0.027801	0.027798	-0.027795	-0.011976	0.039767	-0.016829	0.016828
16	0.170405	-0.005059	0.000150	0.053409	-0.001590	0.031908	-0.000947
17	-0.171474	-0.002255	-0.000030	-0.058759	-0.000773	-0.022345	-0.000294
18	0.018740	-0.023045	0.028338	0.009535	-0.046573	0.009927	-0.012207
Sum	0.003313	0.00468	-0.001407	0.005196	-0.058856	0.001713	-0.002258

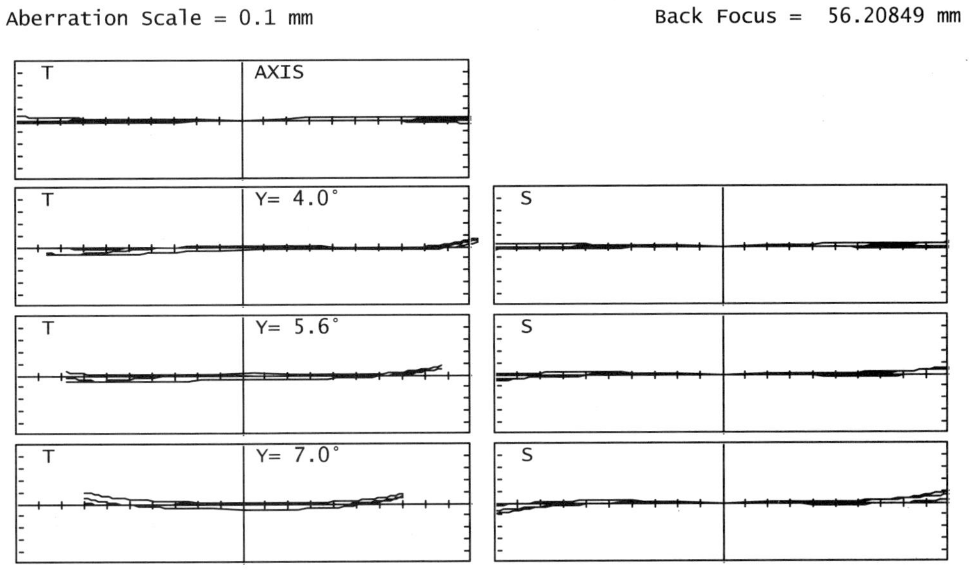

Figure 4.28 Transverse ray aberrations of a long-focus telephoto lens.

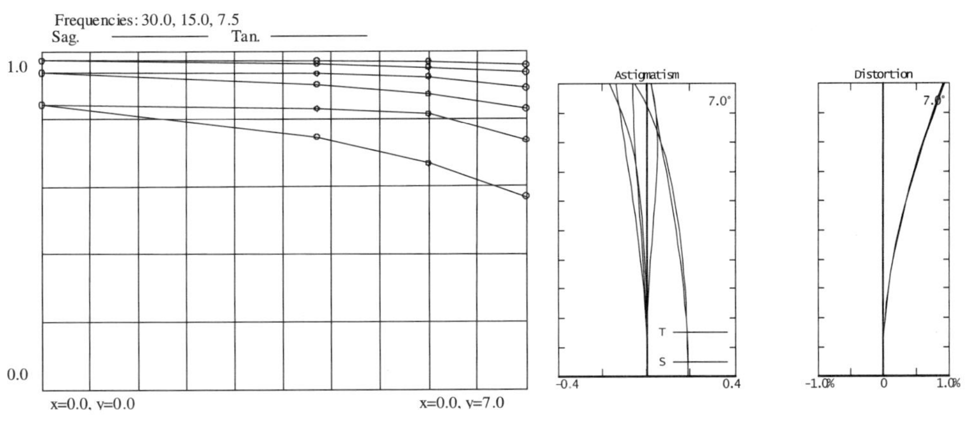

Figure 4.29 MTF as function of field, astigmatism, and distortion of a long-focus telephoto lens.

4.10 Lens for compact point-and-shoot camera

Modern compact 35-mm cameras use lenses with focal lengths of 35 or even 30 mm. This makes the camera quite compact, but forces the designer to use more complex constructions, if good performance is required. The example in Fig. 4.30 uses two aspheric surfaces to give optimum performance at an aperture of *f*/4.5. It is basically a triplet with a meniscus field flattener. This is quite a common arrangement. Performance is shown in Figs. 4.31 and 4.32.

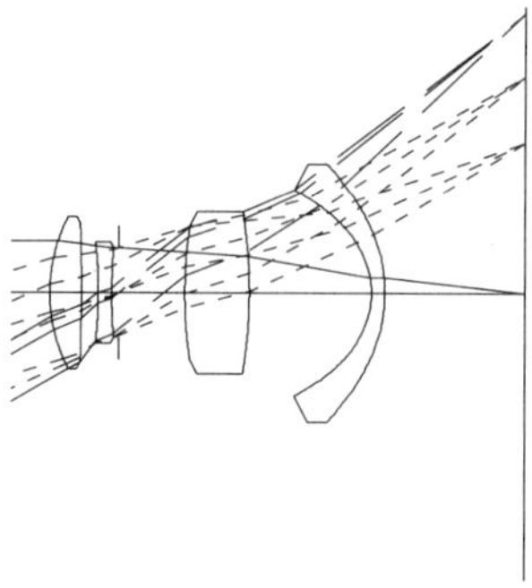

Figure 4.30 A compact point-and-shoot lens.

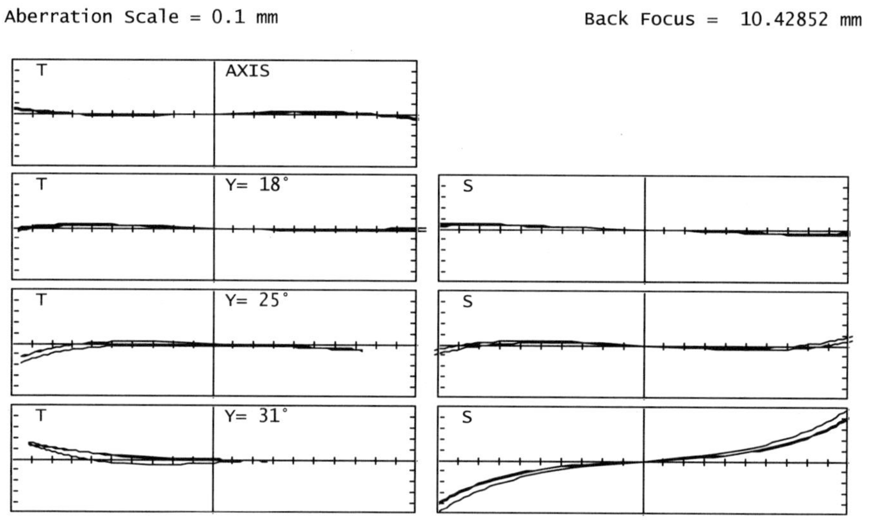

Figure 4.31 Transverse ray aberrations of a compact point-and-shoot lens.

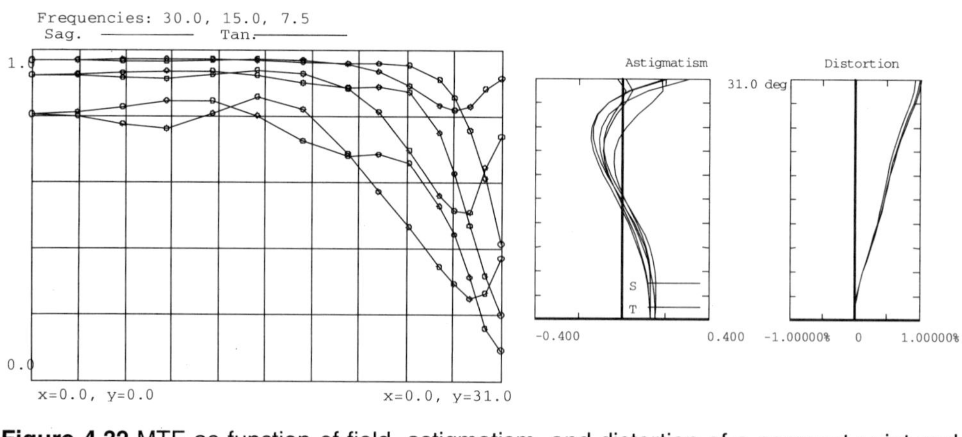

Figure 4.32 MTF as function of field, astigmatism, and distortion of a compact point-and-shoot lens.

```
EFL =     34.996
WAVELENGTHS  [nm]
      587.60        656.30        486.10
SURFACES
  # SURF SPACE     RADIUS        SEPN       INDEX1       V       CLR RAD  GLASS
  0                           0.00000    1.000000
  1 S            13.08300     2.29300    1.607288    59.44      5.745   S-SK7
  2 S          -152.03000     1.23100    1.000000               5.259
  3 S           -36.56400     0.98700    1.805176    25.42      3.824   S-SF6
  4 S            42.63500     0.57300    1.000000               3.435
  5#S             Plane       4.92700    1.000000               3.355
  6 S            34.48000     4.74700    1.664457    35.81      5.021   S-BASF2
  7 S           -37.52600     9.07000    1.000000               6.142
  8 A            -6.81200     0.97300    1.625879    35.69      7.770   S-F1
  9 A           -12.50900    10.52852    1.000000               9.747
 10 S             Plane                                        21.433
ASPHERIC SURFACE  8 CC =  -0.675998  (ELLIPSE)
 A4 = -0.000193056
ASPHERIC SURFACE  9 CC =  -0.134362  (ELLIPSE)
 A4 = -6.43256e-005   A6 = 9.33563e-007

LAGRANGE INVARIANT =      -2.3364
  #      H          U        HBAR       UBAR      D(U/N)        A         ABAR
  1   3.88845   0.00000   -2.96630   0.60086   -0.06987    0.29721    0.37413
  2   3.63095  -0.11230   -1.91267   0.45950   -0.12513   -0.21888    0.75877
  3   3.39091  -0.19500   -0.99411   0.74619    0.15807   -0.28774    0.77338
  4   3.32512  -0.06666   -0.59809   0.40123   -0.02061    0.02046    0.69898
  5   3.29215  -0.05753   -0.18954   0.71300    0.00000   -0.05753    0.71300
  6   3.00870  -0.05753    3.32344   0.71300    0.01584    0.02973    0.80939
  7   2.67926  -0.06940    5.17426   0.38989   -0.12126   -0.23435    0.41946
  8   1.20129  -0.16295   10.22934   0.55734    0.14306   -0.33930   -0.94432
  9   1.16982  -0.03234   11.12534   0.92086   -0.09122   -0.20463    0.05117
                -0.11111              0.94055

  #     S1         S2         S3         S4         S5         C1         C2
  1  0.023999   0.030210   0.038028   0.157650   0.246318   0.007347   0.009248
  2  0.021767  -0.075458   0.261582   0.013567  -0.953825   0.005052  -0.017514
  3 -0.044378   0.119279  -0.320597  -0.066591   1.040682  -0.017119   0.046013
  4  0.000029   0.000980   0.033475  -0.057109  -0.807442  -0.001194  -0.040780
  5  0.000000   0.000000   0.000000   0.000000   0.000000   0.000000   0.000000
  6 -0.000042  -0.001146  -0.031215   0.063202   0.870890   0.000997   0.027144
  7  0.017842  -0.031935   0.057161   0.058071  -0.206251   0.006999  -0.012527
  8 -0.019785  -0.055065  -0.153255  -0.308481  -1.285078  -0.004397  -0.012237
  9  0.004468  -0.001117   0.000279   0.167989  -0.042075   0.002582  -0.000646
     ASPHERIC SURFACE  8
     0.000774   0.006594   0.056150   0.000000   0.478134
     ASPHERIC SURFACE  9
     0.000523   0.004971   0.047277   0.000000   0.449620

Sum  0.005197  -0.002689  -0.011114   0.028297  -0.209027   0.000267  -0.001298
```

4.11 Single lens for disposable cameras

Using modern fast films (ASA 800), disposable cameras are very common. They use a single plastic lens, which can be aspheric, and a film that is curved around a cylindrical surface, as shown in Fig. 4.33. Although it is possible to correct coma,

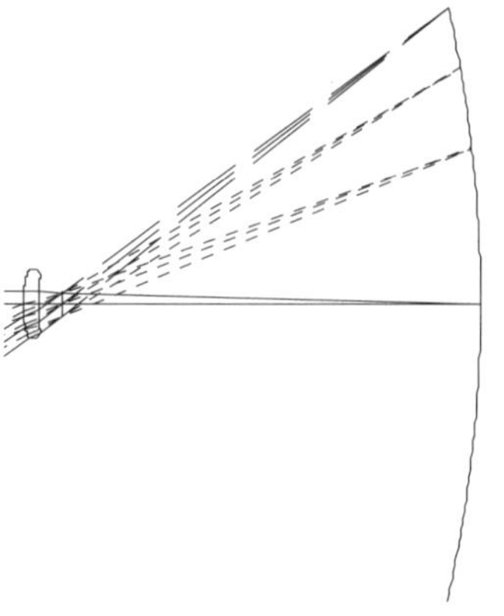

Figure 4.33 A single lens for disposable camera.

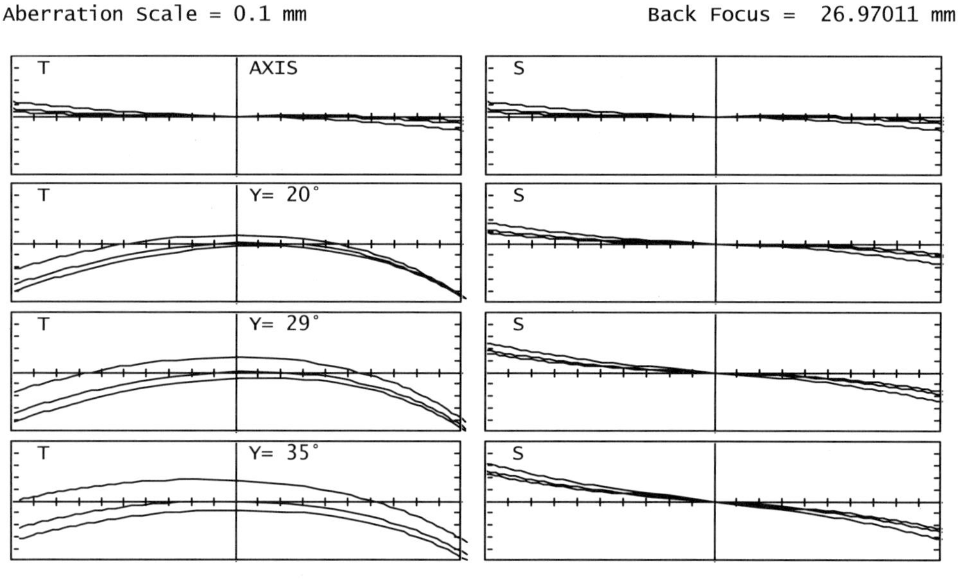

Figure 4.34 Transverse ray aberrations of a single lens for disposable camera.

the astigmatism is then quite large. It is probably a better compromise to balance coma against astigmatism. Clearly, field curvature, axial and lateral color are uncorrectable, but at an aperture of *f*/16 the performance shown in Figs. 4.34 and 4.35 is just acceptable.

```
EFL =     29.998
WAVELENGTHS   [nm]
      587.60       656.30       486.10
SURFACES
```

#	SURF SPACE	RADIUS	SEPN	INDEX1	V	CLR RAD	GLASS
0			0.00000	1.000000			
1	E	6.58900	1.00000	1.491754	57.42	2.415	G-PMM
2	E	11.31200	1.52700	1.000000		2.002	

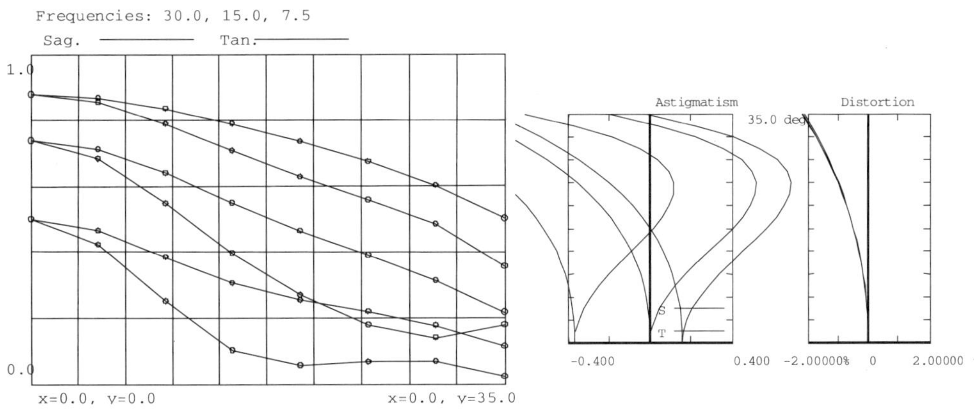

Figure 4.35 MTF as function of field, astigmatism, and distortion of a single lens for disposable camera.

```
  3#S             Plane      26.97011    1.000000              0.833
  4 C          -100.00000                                     20.753
CONIC    SURFACE    1 CC =   -4.498190
CONIC    SURFACE    2 CC =  -30.996000
CONIC   CYLINDER    4 CC =    0.000000   (SPHERE)

LAGRANGE INVARIANT =     -0.6462
  #      H          U          HBAR       UBAR      D(U/N)       A          ABAR
  1   0.92289    0.00000   -1.80092    0.70021   -0.03095    0.14007    0.42689
  2   0.87672   -0.04617   -1.24143    0.55949    0.00019    0.04674    0.67090
  3   0.82974   -0.03077   -0.04938    0.78065    0.00000   -0.03077    0.78065
                -0.03077               0.78065

  #     S1         S2         S3         S4         S5         C1         C2
  1  0.000560   0.001708   0.005205   0.020892   0.079539   0.000742   0.002262
  2 -0.000000  -0.000005  -0.000074  -0.012169  -0.175740  -0.000235  -0.003377
  3  0.000000   0.000000   0.000000   0.000000   0.000000   0.000000   0.000000
     ASPHERIC SURFACE  1
    -0.005610   0.010946  -0.021361   0.000000   0.041683
     ASPHERIC SURFACE  2
     0.006221  -0.008809   0.012474   0.000000  -0.017663

Sum  0.001172   0.003840  -0.003755   0.008723  -0.072181   0.000507  -0.001115
```

References

1. S.F. Ray, *Applied Photographic Optics*, Focal Press, London (1988).
2. R. Kingslake, *Lenses in Photography*, Garden City Books, (1951).
3. R. Kingslake, *A History of the Photographic Lens*, Academic Press, (1989).
4. R. Kingslake, *Optics in Photography*, SPIE Press, Bellingham, WA (1992).
5. R. Kingslake, "Development of the photographic objective," in *SPIE Proc.* Vol. 531, Bellingham, WA (1985).
6. W. Woltche, "Major landmarks from the century of anastigmatic lens design," *Opt. Eng.* **32**(8), 1740–1749 (1993).

7. J. Hoogland, "Systematics of photographic lens types," in *SPIE Proc*. Vol. 237, Bellingham, WA (1980).
8. W. Mandler, "Design of basic double-Gauss lenses," in *SPIE Proc*. Vol. 237, Bellingham, WA (1980).
9. W. Woltche, "Optical systems design with reference to the evolution of the double-Gauss lens," in *SPIE Proc*. Vol. 237, Bellingham, WA (1980).

CHAPTER 5
SECONDARY SPECTRUM AND APOCHROMATS

When longitudinal chromatic aberration has been corrected, we normally find that the focal positions for three wavelengths still do not coincide. This effect is known as secondary spectrum and is due to the nonlinearity of the curve of refractive index as a function of wavelength. More precisely, it stems from a difference between the nonlinearity of the positive (crown) element and the negative (flint) element of an achromat.[1]

5.1 Apochromatic doublets

The secondary change of power with wavelength is given by[1]

$$\delta K_{d-F} = K\frac{(p_1 - p_2)}{(V_1 - V_2)}. \tag{5.1}$$

A doublet is called apochromatic if this power change is zero, achieved by choosing two glasses that show a considerable difference in their V-values so as to minimize opposite crown and flint element powers, but with the same relative partial dispersion, p. Unfortunately, as Abbe was the first to point out,[2] in the visible spectrum the majority of normal glasses follow a linear relationship between V and p, so they cannot have the same p. An example of a 1000-mm *f*/8 achromatic doublet using normal glasses is shown in Fig. 5.1, with transverse ray aberrations in Fig. 5.2. Equation (5.1) tells us that this choice of glasses can be expected to give a secondary chromatic defocus error of about 0.5 mm, which corresponds to a transverse aberrations of $0.5 \times 0.0625 = 0.03$ mm. Although difficult to see in a black and white diagram, Fig. 5.2 shows approximately this difference between the extreme wavelengths and the center wavelength.

```
EFL =     999.978
WAVELENGTHS   [nm]
      587.60        656.30        486.10
SURFACES
  # SURF SPACE     RADIUS         SEPN        INDEX1       V       CLR RAD  GLASS
  0                              0.00000    1.000000
  1#S             646.32885   12.50000    1.563838    60.77     62.500  S-SK11
  2 S            -428.96362    5.00000    1.666796    32.99     62.130  S-SF19
  3 S           -1805.70603  991.10351    1.000000              61.679
  4 S               Plane                                         8.814
```

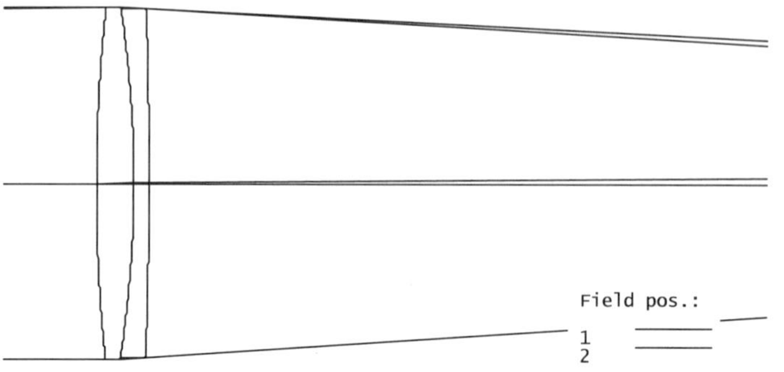

Figure 5.1 Achromatic doublet.

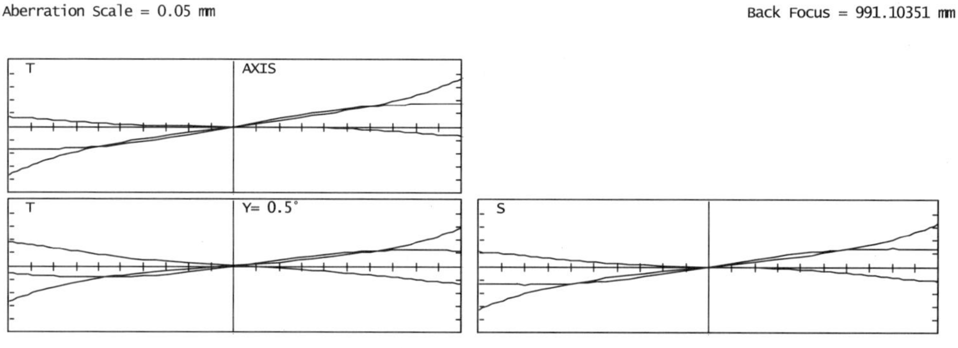

Figure 5.2 Transverse ray aberrations of achromatic doublet.

There are a few glasses, called "special" or "anomalous," that do not follow Abbe's normal line and may be selected for secondary spectrum correction.[3–8] These glasses tend to be more expensive and are difficult to polish (often with high chemical and thermal sensitivities), so their use should only be considered if the application demands a high quality of chromatic correction.

One possible pair of glasses with similar partial dispersions is FK51 and ZKN7. Unfortunately, the refractive indices of these glasses are too low for a doublet of high aperture, and the use of these materials is restricted to lenses with moderate apertures. Figure 5.3 shows such a doublet, again at an aperture of *f*/8. The secondary spectrum in Fig. 5.4 is fairly well corrected, since the three wavelengths all focus in approximately the same plane. However, there is considerable spherochromatism (a chromatic variation of spherical aberration) resulting from the higher elements powers that are needed because of a smaller difference in V-values than the normal glasses used in the achromatic doublet of Fig. 5.1.

```
EFL =    999.998
WAVELENGTHS   [nm]
      587.60        656.30        486.10
SURFACES
  # SURF SPACE    RADIUS        SEPN       INDEX1      V       CLR RAD   GLASS
  0                           0.00000    1.000000
```

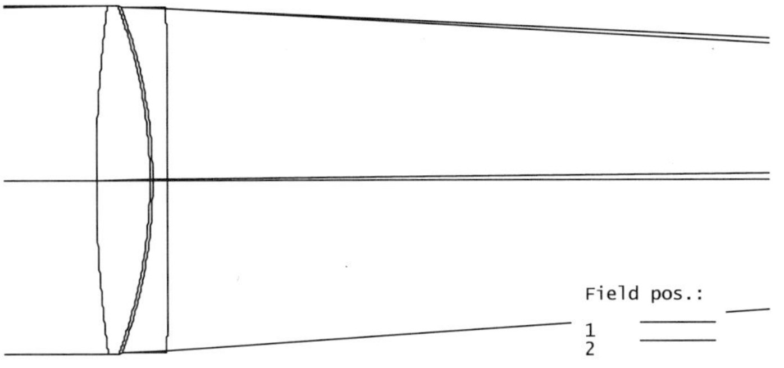

Figure 5.3 FK51/ZKN7 apochromatic doublet.

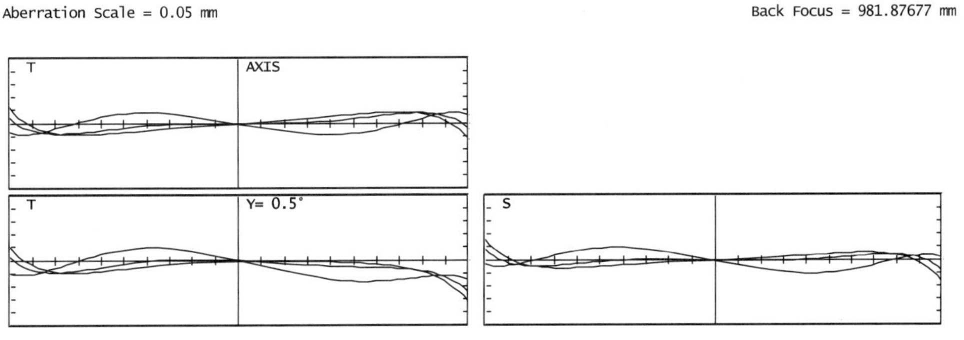

Figure 5.4 Transverse ray aberrations of FK51/ZKN7 apochromatic doublet.

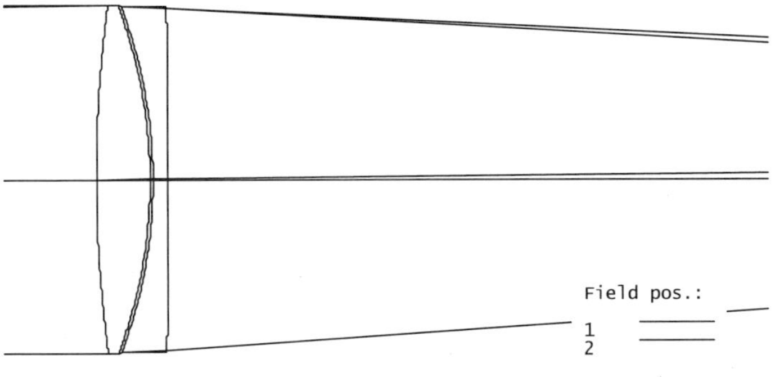

Figure 5.5 LGSK2/KZFSN4 apochromatic doublet.

1#S	501.69135	18.00000	1.486560	84.43	62.507	S-FK51
2 S	-188.48337	1.00000	1.000000		62.386	
3 S	-188.29362	5.00000	1.508468	61.16	62.130	S-ZKN7
4 S	-3612.32199	981.87924	1.000000		61.679	
5 S	Plane				8.740	

A second option is to use LgSK and/or KzFS glasses. In Fig. 5.5 we have an example with these glass types, in which the secondary spectrum is clearly corrected, as may be seen in Fig. 5.6.

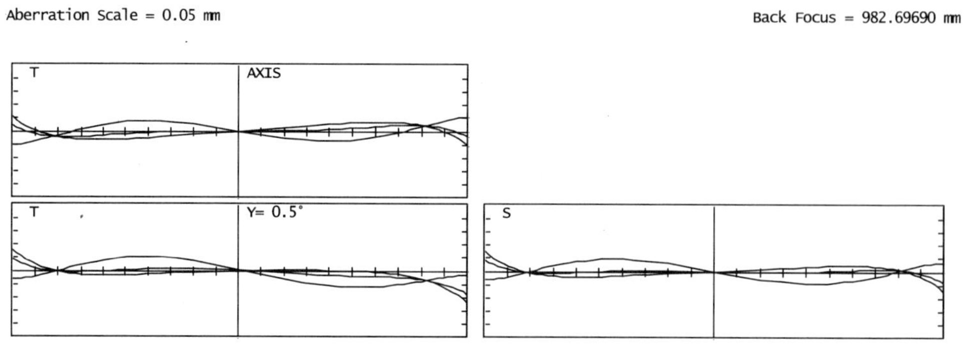

Figure 5.6 Transverse ray aberrations of LGSK2/KZFSN4 apochromatic doublet.

```
EFL =     999.998
WAVELENGTHS   [nm]
    587.60        656.30        486.10
SURFACES
# SURF SPACE      RADIUS          SEPN       INDEX1        V       CLR RAD   GLASS
0                              0.00000    1.000000
1#S               575.71849   17.00000    1.585988     61.02      62.503   User-LGSK2
2 S              -225.64378    1.00000    1.000000                62.336
3 S              -225.42647    5.00000    1.613398     44.28      62.085   S-KZFSN4
4 S             -6335.27723  982.69928    1.000000                61.686
5 S                Plane                                           8.743
```

5.2 Apochromatic triplets

An alternative method of correcting secondary spectrum is to use three glasses, which form a triangle in *V-p* space that is equivalent to two glasses on a line of constant *p*. This approach was used by H.D. Taylor[9] and is explained in detail by Kingslake.[10] A design of this type, following the example given by Kingslake, is shown in Fig. 5.7. The secondary spectrum and sphero-chromatism are very well corrected, as shown in Fig. 5.8. An equivalent description is given by Stephens.[11]

```
EFL =     999.999
WAVELENGTHS   [nm]
      587.60        656.30        486.10
SURFACES
  # SURF SPACE      RADIUS          SEPN       INDEX1        V       CLR RAD   GLASS
  0                              0.00000    1.000000
  1#S               551.85622   14.00000    1.436999     90.67      62.504   S-FK54
  2 S              -305.58492    1.00000    1.000000                62.390
  3 S              -316.56184    5.00000    1.613398     44.28      62.220   S-KZFSN4
  4 S             -1518.20507    5.00000    1.698947     30.06      62.166   S-SF15
  5 S              -905.11764  988.47149    1.000000                62.155
  6 S                Plane                                           8.738
```

Herzberger and McClure[12] have shown how to correct chromatic aberration at four wavelengths, giving what they call a superachromat. In many cases, however, sphero-chromatism and other aberrations are more important than the question of

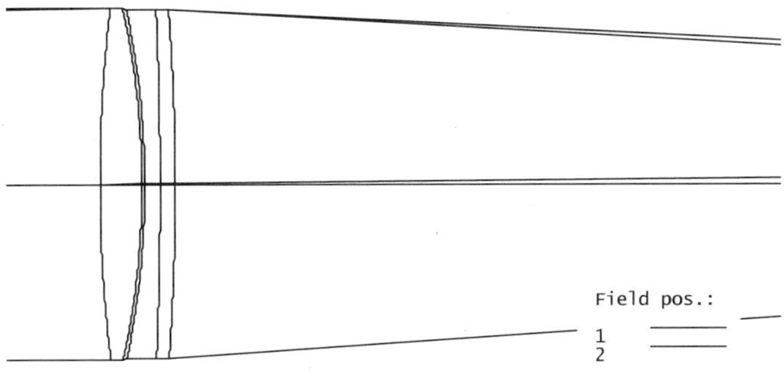

Figure 5.7 Apochromatic triplet.

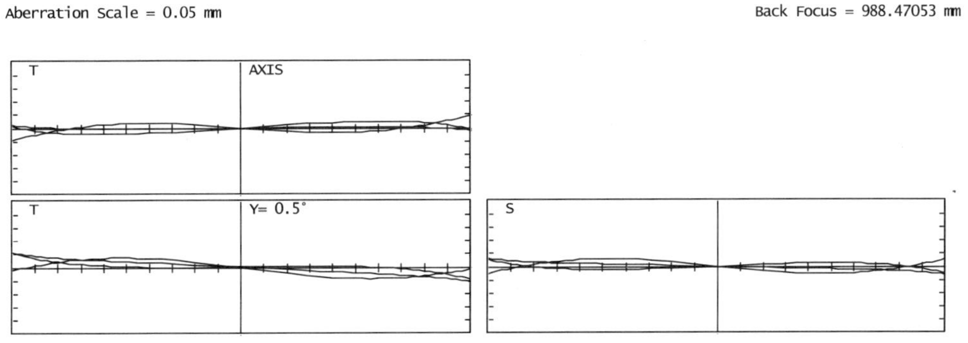

Figure 5.8 Transverse ray aberrations of apochromatic triplet.

whether we can correct the longitudinal chromatic aberration at three or four wavelengths.

5.3 Petzval lenses

Secondary spectrum is reduced in Petzval lenses because the second doublet operates at a lower aperture than the first; since the secondary spectrum for a doublet is proportional to h^2K, this reduces the total secondary spectrum.

Mercado has described several Petzval designs that use anomalous glasses.[13] One example, with a field flattener, is shown in Fig. 5.9. Transverse ray aberrations of this design are shown in Fig. 5.10.

```
EFL =       99.999
WAVELENGTHS   [nm]
 546.10        656.30        486.10
SURFACES
 # SURF SPACE     RADIUS          SEPN        INDEX1        V       CLR RAD   GLASS
 0                               0.00000    1.000000
 1 S             56.32056        5.16667    1.530190    77.17     14.598   User-PK51
 2 S            -83.51429        1.00000    1.000000              14.322
 3 S           -198.04688        3.33333    1.654260    58.68     13.755   User-HOYALAC7
```

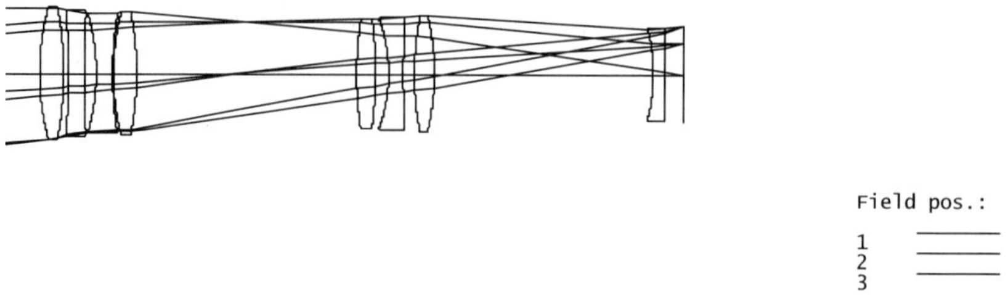

Figure 5.9 Apochromatic Petzval lens.

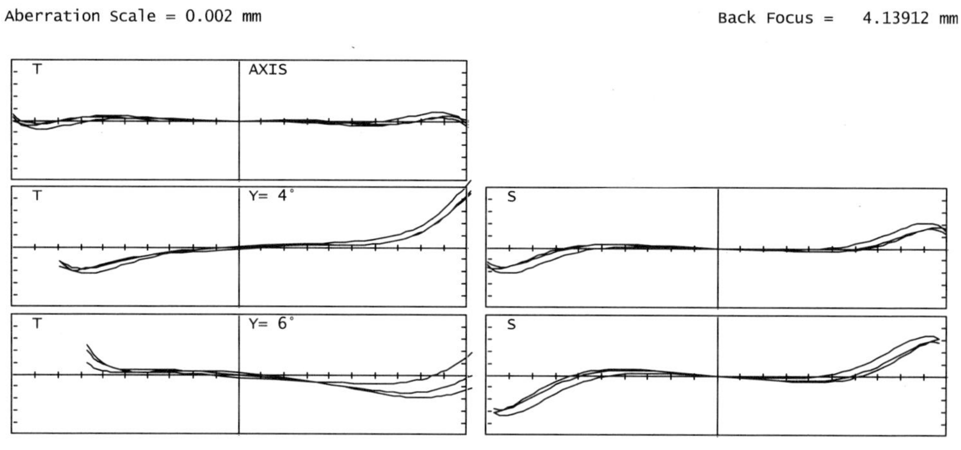

Figure 5.10 Transverse ray aberrations of apochromatic Petzval lens.

4 S	238.11136	2.50000	1.000000		13.213	
5 S	-50.17939	3.33333	1.654260	58.68	13.156	User-HOYALAC7
6 S	87.53147	0.16667	1.000000		13.273	
7 S	58.74077	4.66667	1.530190	77.17	13.366	User-PK51
8 S	-73.06602	22.83333	1.000000		13.376	
9#S	Plane	22.83333	1.000000		11.152	
10 S	86.58463	3.83333	1.530190	77.17	12.136	User-PK51
11 S	-96.31278	2.66667	1.000000		12.087	
12 S	-36.79518	3.00000	1.654260	58.68	11.940	User-HOYALAC7
13 S	238.25760	2.33333	1.000000		12.336	
14 S	68.46327	4.16667	1.530190	77.17	12.850	User-PK51
15 S	-57.67579	44.66667	1.000000		12.903	
16 S	-39.60591	2.16667	1.654260	58.68	9.834	User-HOYALAC7
17 S	419.74470	4.13912	1.000000		10.046	
18 S	Plane				10.512	

5.4 Double-Gauss lenses

When double-Gauss lenses of around 50-mm focal length are used as normal camera lenses secondary spectrum is not large enough to be a problem.[14] However, at longer focal lengths and smaller field angles, the secondary spectrum may need to be reduced. There appear to be two approaches described in the patent literature.

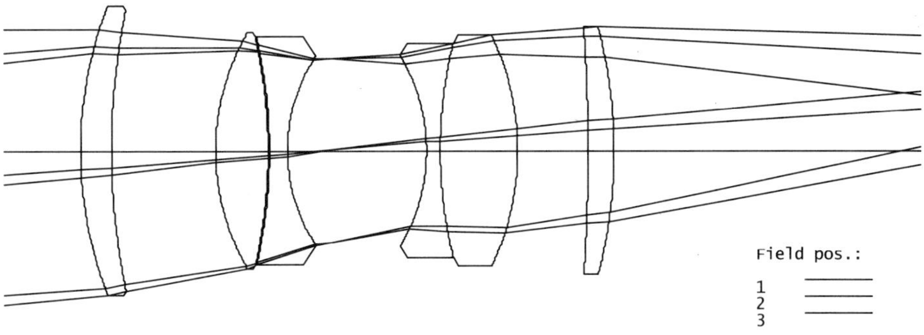

Figure 5.11 Apochromatic double Gauss.

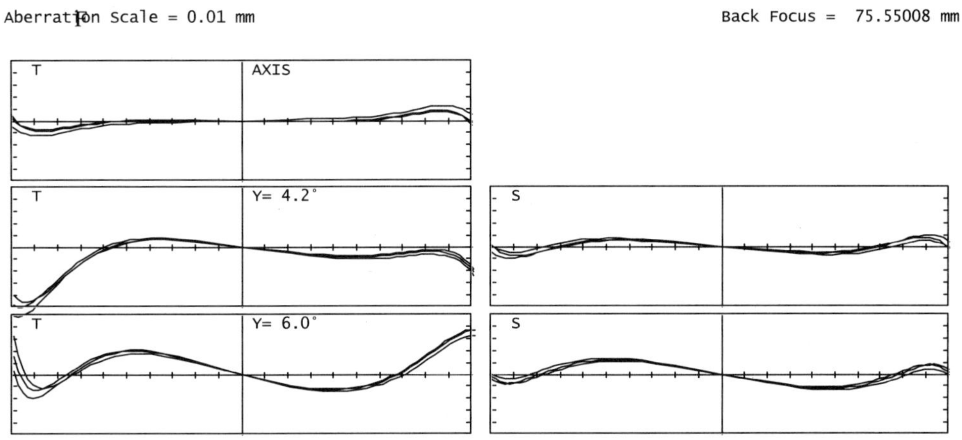

Figure 5.12 Transverse ray aberrations of apochromatic double Gauss.

One possibility is to use FK-51 or similar crown glasses in some or all of the positive elements. This is shown in the example in Fig. 5.11, which is derived from a patent by Mercado. Hoya ADC1[15] is the anomalous crown used in this design, which Fig. 5.12 indicates is well corrected for secondary spectrum.

EFL = 100.000
WAVELENGTHS [nm]
587.60 656.30 486.10
SURFACES

#	SURF SPACE	RADIUS	SEPN	INDEX1	V	CLR RAD	GLASS
0			0.00000	1.000000			
1	S	44.03850	3.00000	1.620000	62.19	14.899	User-ADC1
2	S	75.83000	10.05450	1.000000		14.537	
3	S	26.07100	5.04450	1.620000	62.19	12.166	User-ADC1
4	S	-56.70750	0.14050	1.000000		11.920	
5	S	-55.53050	1.75350	1.551150	49.52	11.794	User-SBF1
6	S	18.83500	4.26100	1.000000		9.785	
7#	S	Plane	9.17350	1.000000		9.451	
8	S	-20.37550	1.32900	1.551150	49.52	9.917	User-SBF1
9	S	49.25900	7.47750	1.620000	62.19	11.138	User-ADC1

10 S	-29.71250	6.74100	1.000000		11.929	
11 S	-197.84050	2.50000	1.620000	62.19	12.647	User-ADC1
12 S	-54.00505	75.55008	1.000000		12.750	
13 S	Plane				10.620	

An alternative approach is to use a relatively conventional double Gauss, with a secondary spectrum corrector in the central air space, as described by Dietzsch.[16] This technique seems to be limited to lenses of moderate aperture (*f*/4) by the steepness of the curves in the corrector.

5.5 Telephoto lenses

Because of the relatively high power of the first positive group, telephoto lenses would be expected to have worse secondary spectrum than a single doublet of the same focal length. This is unfortunate since these lenses are often expected to be used at long focal lengths and small fields, when secondary spectrum is most important. We have already seen in Fig. 4.28 an example of an apochromatic telephoto lens, which uses Hoya anomalous glasses.

5.6 Zoom lenses

For similar reasons, zoom lenses tend to suffer from relatively large amounts of secondary spectrum. High-performance zoom lenses for cinematography or television may justify the extra expense and higher thermal sensitivity resulting from the use of anomalous glasses.

5.7 Microscope objectives

In microscope objectives, correction of secondary spectrum is problematic because of the high numerical aperture and resolution. Historically, fluorite (calcium fluoride) has commonly been used to correct it. We will see in Chapter 7 some examples of modern microscope objective in which FK glasses are often used instead.

5.8 Secondary spectrum correction with normal glasses

We have seen that conventional lenses suffer from secondary spectrum, and this is usually corrected by using special glasses. There are, however, several ingenious ways of correcting secondary spectrum using normal glass types.

5.8.1 Liquids

Sigler[17] has noticed that liquids have very different partial dispersions from those of normal optical glasses. This suggests the possibility of correcting secondary spectrum with a combination of normal glasses and relatively low powered "liquid lenses" in the spaces between glass lenses. There are several obvious practical difficulties with such an approach, and Sigler has suggested possible solutions. However, this technique appears not to have been widely adopted.

5.8.2 Diffractive optics

Diffractive optical elements (DOEs) also have very different dispersion characteristics from normal glasses. In particular, since a DOE obeys the grating equation rather than Snell's law, its optical power is linearly proportional to wavelength. As a result, the Abbe V-value is always a small negative number, which allows a relatively weak positive-powered DOE to correct the chromatic aberration of a positive-powered refracting lens.

Such primary chromatic correction inevitably introduces a large amount of secondary spectrum. However, if a DOE is used in addition to normal primary chromatic correction with glasses of different dispersions, it is possible to arrange for the DOE to correct the secondary color.[18] A wide range of such "hybrid" optical designs have been investigated, both in the visible and infrared wavebands.[19] Although they can give significant improvements in chromatic correction, care has to be taken to ensure that this is not outweighed by the degradation of image quality caused by diffraction into unwanted orders, since the DOE is not 100% efficient.[20] Recently, Nakai et al. (Canon) have developed two-layer DOEs that have high diffraction efficiency across the visible spectrum.[21] These have been applied to high-performance, compact, telephoto lenses for photography.

5.8.3 McCarthy-Wynne principle

In a design by McCarthy, a zero-power doublet with positive C_1 is placed a long distance in front of a positive doublet with negative C_1.[22] McCarthy does not really explain how this system works, but the explanation is as follows:

Suppose the first lens has a power K_1 that is approximately proportional to wavelength, i.e. $K_1 = a\lambda$, where a is positive. The second lens has a power given by

$$K_2 = K_0 - b\lambda,$$

where b is also positive. Since the lenses have a separation d, the total power is given by

$$\begin{aligned} K &= K_1 + K_2 - dK_1K_2 \\ &= a\lambda + (K_0 - b\lambda) - da\lambda(K_0 - b\lambda). \end{aligned}$$

This gives a quadratic term proportional to $d\ a\ b$, where d, a, and b are all positive. This will be of the opposite sign to the variation of K due to secondary spectrum, and therefore it should be possible to correct the secondary spectrum as McCarthy claims.

Figure 5.13 McCarthy apochromat.

Figure 5.13 shows a design based on McCarthy's principle. It is clear from Fig. 5.14 that the secondary spectrum is considerably reduced, compared with the aberrations of a normal doublet of the same focal length.

EFL = 100.000
WAVELENGTHS [nm]
587.60 656.30 486.10
SURFACES

#	SURF SPACE	RADIUS	SEPN	INDEX1	V	CLR RAD	GLASS
0			0.00000	1.000000			
1	S	28.36302	4.47206	1.516798	64.14	10.000	S-BK7
2	S	-35.03766	1.00000	1.000000		9.844	
3	S	-32.49980	2.00000	1.620037	36.35	9.462	S-F2
4	S	35.23915	100.00000	1.000000		9.033	
5#	S	43.65083	5.00000	1.620037	36.35	10.523	S-F2
6	S	-1494.92528	1.00000	1.000000		10.205	
7	S	-176.26987	2.00000	1.516798	64.14	10.095	S-BK7
8	S	611.76492	92.99900	1.000000		9.917	
9	S	Plane				0.002	

LAGRANGE INVARIANT = 0.0000

#	H	U	HBAR	UBAR	D(U/N)	A	ABAR
1	10.00000	0.00000	0.00000	0.00000	-0.07920	0.35257	0.00000
2	9.46278	-0.12013	0.00000	0.00000	-0.24258	-0.59186	0.00000
3	9.14100	-0.32178	0.00000	0.00000	0.26562	-0.60305	0.00000
4	8.95904	-0.09098	0.00000	0.00000	0.06640	0.26448	0.00000
5	9.98368	0.01025	0.00000	0.00000	-0.06038	0.23896	0.00000
6	9.57762	-0.08121	0.00000	0.00000	-0.08541	-0.14195	0.00000
7	9.44208	-0.13554	0.00000	0.00000	0.08866	-0.18911	0.00000
8	9.29987	-0.07111	0.00000	0.00000	-0.05312	-0.08480	0.00000
		-0.10000		0.00000			

#	S1	S2	S3	S4	S5	C1	C2
1	0.098448	0.000000	0.000000	0.000000	0.000000	0.018728	0.000000
2	0.804113	0.000000	0.000000	0.000000	0.000000	0.029750	0.000000
3	-0.883004	0.000000	0.000000	0.000000	0.000000	-0.058038	0.000000
4	-0.041615	0.000000	0.000000	0.000000	0.000000	-0.024948	0.000000
5	0.034421	0.000000	0.000000	0.000000	0.000000	0.025118	0.000000
6	0.016482	0.000000	0.000000	0.000000	0.000000	0.014314	0.000000
7	-0.029936	0.000000	0.000000	0.000000	0.000000	-0.009485	0.000000
8	0.003552	0.000000	0.000000	0.000000	0.000000	0.004189	0.000000
Sum	0.002461	0.000000	0.000000	0.000000	0.000000	-0.000371	0.000000

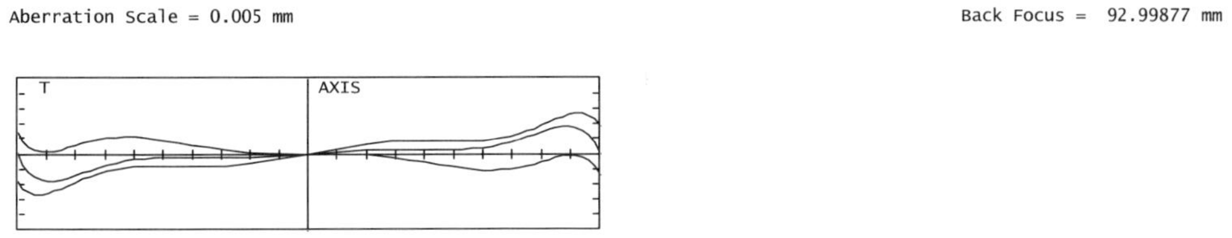

Figure 5.14 Transverse aberrations of McCarthy apochromat.

C.G. Wynne used what appears to be the same principle in 1977.[23,24] Wynne formed a unit magnification telescope with two different doublets, each with plane exterior surfaces and a curved interior surface, made of glass of the same mean refractive index but different dispersion. The power of these doublets is therefore approximately linearly dependent on wavelength, but one lens has a positive flint component and the other has a negative flint component. Therefore, at all wavelengths the powers of these doublets are equal but of opposite sign., i.e. $K_1 = -K_2$.

The combined power is therefore $+ d \cdot K_1^2$, and, as in the McCarthy example, this is of the correct sign to reduce the secondary spectrum. Wynne uses two of these pairs of doublets in order to form a unit magnification telescope with zero lateral color, and this is placed in front of a standard doublet. The design is shown in Fig. 5.15, with transverse ray aberrations in Fig. 5.16. Note that the triplet in Fig. 5.15 (surfaces 4–7) is actually two doublets in contact. Since the plane contact surface is between identical glass types, it does not need to be drawn or included in the prescription.

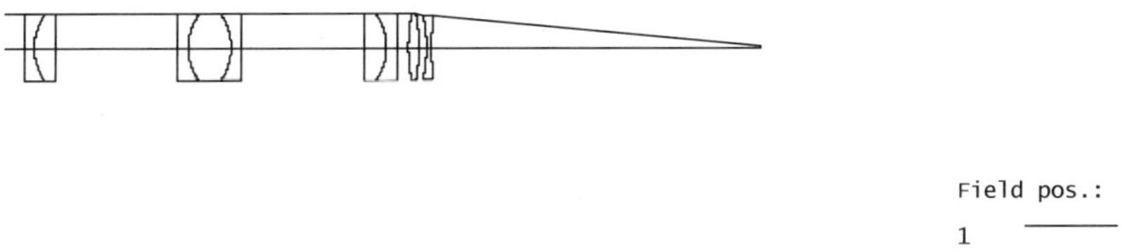

Figure 5.15 Wynne apochromat.

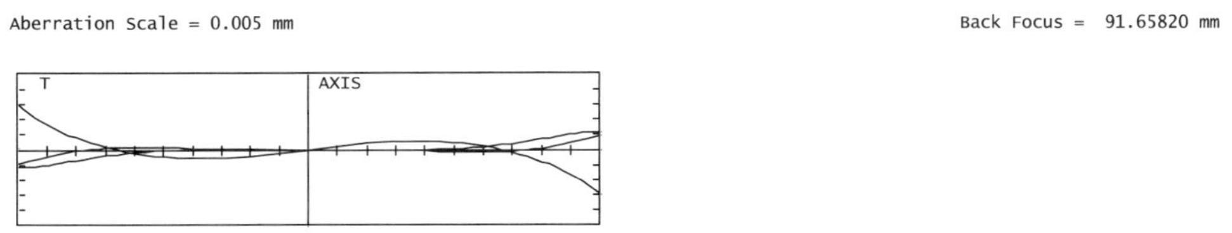

Figure 5.16 Transverse aberrations of Wynne apochromat.

EFL = 100.000
WAVELENGTHS [nm]
587.60 656.30 486.10
SURFACES

#	SURF SPACE	RADIUS	SEPN	INDEX1	V	CLR RAD	GLASS
0			0.00000	1.000000			
1	S	Plane	3.00000	1.691000	54.69	10.000	S-LAK9

2 S	18.14178	6.00000	1.688927	31.16	10.000	S-SF8
3 S	Plane	34.20000	1.000000		10.005	
4 S	Plane	3.00000	1.688927	31.16	10.111	S-SF8
5 S	18.14178	12.00000	1.691000	54.69	10.122	S-LAK9
6 S	-18.14178	3.00000	1.688927	31.16	10.122	S-SF8
7 S	Plane	34.20000	1.000000		10.110	
8 S	Plane	6.00000	1.688927	31.16	10.002	S-SF8
9 S	-18.14178	3.00000	1.691000	54.69	9.999	S-LAK9
10 S	Plane	3.00000	1.000000		9.999	
11#S	53.54849	3.00000	1.691000	54.69	9.999	S-LAK9
12 S	-55.63564	2.00000	1.000000		9.909	
13 S	-47.89385	1.50000	1.688927	31.16	9.378	S-SF8
14 S	370.21416	91.65809	1.000000		9.202	
15 S	Plane				0.003	

LAGRANGE INVARIANT = 0.0000

#	H	U	HBAR	UBAR	D(U/N)	A	ABAR
1	10.00000	0.00000	0.00000	0.00000	0.00000	0.00000	0.00000
2	10.00000	0.00000	0.00000	0.00000	0.00040	0.93210	0.00000
3	10.00406	0.00068	0.00000	0.00000	0.00074	0.00114	0.00000
4	10.04312	0.00114	0.00000	0.00000	-0.00074	0.00114	0.00000
5	10.04515	0.00068	0.00000	0.00000	-0.00040	0.93631	0.00000
6	10.04512	-0.00000	0.00000	0.00000	-0.00040	-0.93631	0.00000
7	10.04307	-0.00068	0.00000	0.00000	-0.00075	-0.00115	0.00000
8	10.00365	-0.00115	0.00000	0.00000	0.00075	-0.00115	0.00000
9	9.99956	-0.00068	0.00000	0.00000	0.00040	-0.93207	0.00000
10	9.99954	-0.00001	0.00000	0.00000	-0.00001	-0.00001	0.00000
11	9.99951	-0.00001	0.00000	0.00000	-0.04512	0.18673	0.00000
12	9.77057	-0.07631	0.00000	0.00000	-0.20527	-0.42601	0.00000
13	9.26977	-0.25040	0.00000	0.00000	0.20936	-0.44395	0.00000
14	9.16581	-0.06931	0.00000	0.00000	-0.05896	-0.07524	0.00000
		-0.10000		0.00000			

#	S1	S2	S3	S4	S5	C1	C2
1	0.000000	0.000000	0.000000	0.000000	0.000000	0.000000	0.000000
2	-0.003479	0.000000	0.000000	0.000000	0.000000	0.052358	0.000000
3	-0.000000	0.000000	0.000000	0.000000	0.000000	-0.000150	0.000000
4	0.000000	0.000000	0.000000	0.000000	0.000000	0.000150	0.000000
5	0.003542	0.000000	0.000000	0.000000	0.000000	-0.052831	0.000000
6	0.003542	0.000000	0.000000	0.000000	0.000000	-0.052832	0.000000
7	0.000000	0.000000	0.000000	0.000000	0.000000	0.000152	0.000000
8	-0.000000	0.000000	0.000000	0.000000	0.000000	-0.000151	0.000000
9	-0.003479	0.000000	0.000000	0.000000	0.000000	0.052354	0.000000
10	0.000000	0.000000	0.000000	0.000000	0.000000	0.000001	0.000000
11	0.015731	0.000000	0.000000	0.000000	0.000000	0.013952	0.000000
12	0.363990	0.000000	0.000000	0.000000	0.000000	0.031103	0.000000
13	-0.382492	0.000000	0.000000	0.000000	0.000000	-0.053867	0.000000
14	0.003060	0.000000	0.000000	0.000000	0.000000	0.009027	0.000000
Sum	0.000415	0.000000	0.000000	0.000000	0.000000	-0.000734	0.000000

5.8.4 Schupmann principle

In the Schupmann medial telescope,[25] the C_1 of a single positive refracting objective lens is corrected by a negative element of the same glass type, located next to,

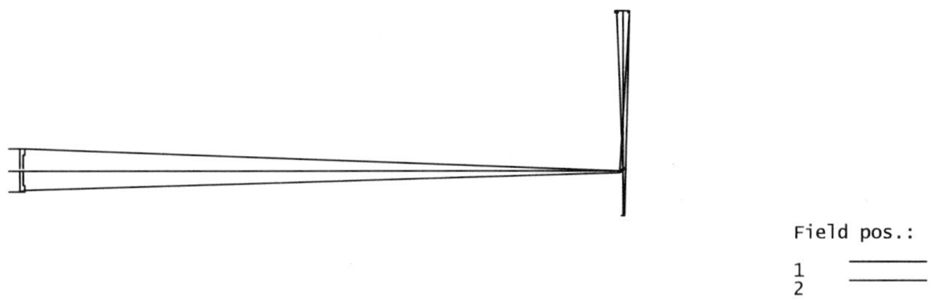

Figure 5.17 Schupmann medial telescope (Bingham).

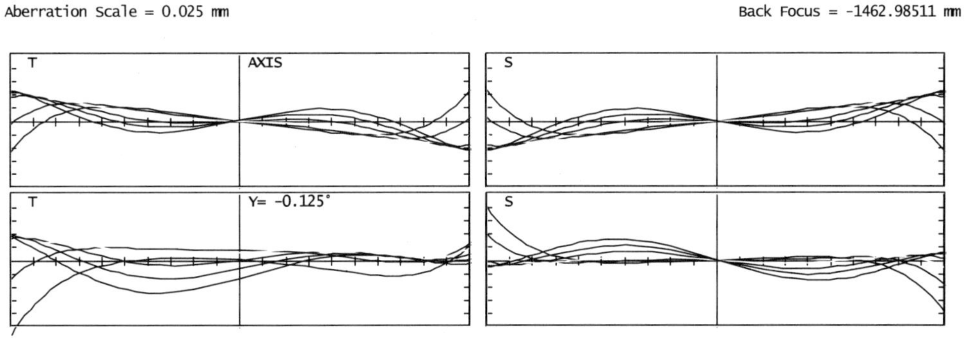

Figure 5.18 Transverse ray aberrations of Schupmann medial telescope (Bingham).

or in contact with, a positive-powered 1× relay mirror. Baker[26] developed several variants in the early 1950s, and Bingham[27] has recently described a design based on this principle. This is shown in Fig. 5.17. The system has to be used 0.125 deg off-axis in order for the beam to clear the fold prism. Figure 5.18 shows relatively small residual chromatic aberrations for a 5-m focal length, over a wide bandwidth from 450 nm to 700 nm.

As Baker pointed out, it is surprising that this design type has not been more widely used in astronomy. However, the general principle of achromatization with a single glass type has been used in many catadioptric designs.[1]

```
EFL = -5002.005
WAVELENGTHS  [nm]
     600.00       450.00       500.00       550.00       650.00      700.00
SURFACES
  # SURF SPACE    RADIUS         SEPN       INDEX1      V      CLR RAD  GLASS
  0                           0.00000    1.000000
  1#S          1142.02700   30.00000    1.516295   42.13   150.000  S-BK7
  2 S          2338.21700 4000.00000    1.000000           150.000
  3 S           307.29960   22.00000    1.516295   42.13    20.000  S-BK7
  4 SDZ            Plane     0.00000    1.516295   42.13    20.000  S-BK7
 ** *** ***  New Axis
  5 SM             Plane     0.00000    1.516295   42.13    20.000  S-BK7
  6 SDZ            Plane     8.00000    1.516295   42.13    20.000  S-BK7
 ** *** ***  New Axis
  7 S           952.29050 1134.14700    1.000000            20.000
```

8	S	-204.11860	7.99941	1.516295	42.13	48.739	S-BK7
9	SM	-455.34180	-7.99941	1.516295	42.13	50.011	S-BK7
10	S	-204.11860	0.00000	1.000000		48.941	
11	S	Plane	-1463.32931	1.000000		12.000	
12	SD	Plane				11.024	

DECENTERED SURFACES

No	Dx	Dy	Alpha	Beta	Gamma
4	0.0000	0.0000	-45.0000	0.0000	0.0000
**	New Axis	*.****	*.****	*.****	*.****
6	0.0000	0.0000	135.0000	0.0000	0.0000
**	New Axis	*.****	*.****	*.****	*.****
12	0.0000	0.0000			

LAGRANGE INVARIANT = -0.3274

#	H	U	HBAR	UBAR	D(U/N)	A	ABAR
1	-150.06015	0.00000	0.87267	-0.00218	0.02951	-0.13140	-0.00142
2	-148.71792	0.04474	0.82170	-0.00170	0.00550	-0.02860	-0.00204
3	-8.70922	0.03500	-8.75746	-0.00239	-0.01341	0.00666	-0.03089
4	-7.98907	0.03273	-8.57873	0.00812	0.00000	0.04963	0.01232
5	-7.98907	0.03273	-8.57873	0.00812	0.00000	0.04963	0.01232
6	-7.98907	-0.03273	-8.57873	-0.00812	0.00000	0.04963	0.01232
7	-7.72720	-0.03273	-8.51374	-0.00812	0.02386	0.03733	-0.00124
8	43.81431	-0.04545	0.22240	-0.00770	0.02252	-0.16921	0.00661
9	44.63872	-0.10306	0.26601	-0.00545	-0.12931	0.00762	0.00738
10	43.89472	-0.09301	0.30027	0.00428	0.03134	0.18505	0.00872
		-0.03000		0.00725			

#	S1	S2	S3	S4	S5	C1	C2
1	0.076447	0.000825	0.000009	0.000032	0.000000	0.159368	0.001719
2	0.000669	0.000048	0.000003	-0.000016	-0.000001	-0.034379	-0.002456
3	-0.000005	0.000024	-0.000111	0.000119	-0.000034	-0.000469	0.002175
4	0.000000	0.000000	0.000000	0.000000	0.000000	0.000000	0.000000
5	0.000000	0.000000	0.000000	0.000000	0.000000	0.000000	0.000000
6	0.000000	0.000000	0.000000	0.000000	0.000000	0.000000	0.000000
7	0.000257	-0.000009	0.000000	-0.000038	0.000001	0.002332	-0.000077
8	-0.028253	0.001104	-0.000043	-0.000179	0.000009	-0.059921	0.002342
9	0.000335	0.000325	0.000314	-0.000310	0.000004	0.000000	0.000000
10	-0.047103	-0.002221	-0.000105	-0.000179	-0.000013	-0.065650	-0.003095
Sum	0.002347	0.000096	0.000068	-0.000571	-0.000034	0.001281	0.000607

Offner[28] has also described why a positive single element placed at the intermediate image of the Schupmann medial telescope is capable of fully correcting secondary spectrum if its focal length is chosen such that it images the positive objective lens (and pupil) onto the negative achromatizing element, in effect removing the separation between them. The long separation between chromatic elements is used to good effect in the McCarthy-Wynne designs, but with a single glass type, which has zero secondary spectrum if the refracting elements are in contact, separation introduces unwanted secondary spectrum. Offner also pointed out that if the field lens is made a doublet with two different glass dispersions, the pupil imaging can be made achromatic, in which case all orders of chromatic aberration may be corrected.

As originally proposed by Schupmann, the field lens in Bingham's design is obtained by adding power to the fold prism refracting faces (surfaces 3 and 7 in the prescription). This is not exactly at the intermediate image, but close enough not to introduce large aberrations, other than some lateral color, C_{II}.

5.9 Transverse secondary spectrum

The variation of focal length with wavelength away from the pupil produces a variation of image height with wavelength, C_{II}. If the lens has secondary spectrum, this transverse aberration becomes a secondary spectrum effect. Correction is very difficult, particularly in wide-angle inverse telephoto lenses and zoom lenses, but glasses with anomalous relative partial dispersions should help.

References

1. M.J. Kidger, *Fundamental Optical Design*, SPIE Press, Bellingham, WA (2002).
2. J.L. Rayces and M. Rosete-Aguilar, Differential equation of the normal glass dispersion," in *SPIE Proc.* Vol. 3482, Bellingham, WA (1998).
3. B.H. Walker, "The 20% rule: selecting optical glass," *Optical Spectra*, 42–45 (Dec. 1978).
4. J. Hoogland, "The design of apochromatic lenses," *Perkin Elmer Symposium on Recent Development in Optical Design*, 6-1–6-8 (1968).
5. J. Hoogland, "High performance apochromatic refractor for long focal length reconnaissance," in *SPIE Proc.* Vol. 242, Bellingham, WA (1980).
6. P.N. Robb, "Selection of optical glasses, 1: two materials," *Appl. Opt.* **24** (1985).
7. P.N. Robb and R.I. Mercado, "Calculation of refractive indices using Buchdahl's chromatic coordinate," *Appl. Opt.* **22** (1983).
8. J.L. Rayces and M. Rosete-Aguilar, "Selection of glasses for achromatic doublets with reduced secondary color," in *SPIE Proc.* Vol. 4093, Bellingham, WA (2000).
9. H.D. Taylor, British Patent 17994 (1892).
10. R. Kingslake, *Lens Design Fundamentals*, Academic Press (1978).
11. R.E. Stephens, "Selection of glasses for three-color achromats," *JOSA* **49**, 398–401 (1959).
12. M. Herzberger, and N. McClure, "The design of superachromat lenses," *Appl. Opt.* **2**, 553–560 (1963).
13. R.I. Mercado and L. Ryzikov, "Designs of apochromats and superachromatic objectives," in *SPIE Proc.* Vol. 3482, Bellingham, WA (1998).
14. W. Mandler , "Design of basic double-Gauss systems," in *SPIE Proc.* Vol. 237, Bellingham, WA (1980).
15. Hoya Glass Works, *Hoya Abnormal Dispersion Glasses*, Technical Report HGW-0-11E.
16. E. Dietzsch, U.S. Patent 4,206,972 (1980).
17. R.D. Sigler, "Designing apochromatic telescope objectives with liquid lenses," in *SPIE Proc.* Vol. 1535, Bellingham, WA (1991).

18. A.I. Tudorovskii, "An objective with a phase plate," *Optics and Spectroscopy* **6** (2), (1959).
19. C.W. Chen, "Application of diffractive optical elements in visible and infrared optical systems," in *Lens Design,* Warren Smith, Ed., SPIE Press, Bellingham, WA (1992).
20. A. Woods et al., "Hybrid optics in dual waveband infrared systems," in *SPIE Proc*. Vol. 3482, Bellingham, WA (1998).
21. T. Nakai et al., "Research on multi-layer diffractive optical elements and their application to photographic lenses," in *Proc. ODF 2002*, Technical Digest AP022144, Optical Society of Japan (2002).
22. E.L. McCarthy, *Optical System with Corrected Secondary Spectrum*, U.S. Patent 2,698,555 ( ).
23. C.G. Wynne, Secondary spectrum correction with normal glasses, *Optics Communications* **21** (3), (1977).
24. M. Rosete-Aguilar, "Correction of secondary spectrum using normal glasses," in *SPIE Proc*. Vol. 2774, Bellingham, WA (1996).
25. L. Schupmann, *Die Medial Fernrohre. Eine neue Konstruktion fur grosse astronomische Instrumente*. Pub. B.G. Teubner, Leipzig (1899).
26. J.G. Baker, "The catadioptric refractor," *The Astronomical Journal* **59**, 74–84 (1954).
27. R.G. Bingham, "Aberrations in large lenses," in S*PIE Proc*. Vol. 4411, Bellingham, WA (2001).
28. A. Offner, "Field lenses and secondary axial aberration," *Appl. Opt.* **8,** 1735 (1969).

CHAPTER 6
LENSES FOR LASER APPLICATIONS

The aberration theory that we use in the design of "white-light" systems is applicable to the design of lenses that are intended for use in instruments involving lasers. However, for most designs the laser can be regarded as monochromatic, so that chromatic aberrations need not be considered. We also need to be aware of the intensity profile of the laser beam. This will affect lens apertures and the position of optimum focus, especially in conditions of very low numerical aperture, when the basic geometrical optics approximations break down.

6.1 Gaussian beams

A Gaussian beam is formed as the lowest-order transverse mode in a stable laser resonator with spherical mirrors, and in most of these cases it has a circular cross section. On the other hand, a Gaussian beam with an elliptical cross section often describes light emitted by laser diodes.

The energy of a Gaussian beam is concentrated into a narrow cone and its transverse amplitude distribution is a Gaussian, with its center on the beam axis. The width of a Gaussian distribution is minimal in the position of maximum contraction of the beam—in its center—and the width increases in both directions with distance from the center. The center of a Gaussian beam is known as the beam waist. In the vicinity of the waist, constant phase surfaces are nearly plane. As distance from the beam waist increases these phase surfaces become increasingly curved; for very large distances from the beam waist they become nearly spherical, with curvature centering on the beam waist center. The parameters of Gaussian beams are usually given at the $1/e$ fraction of maximum amplitude, which corresponds with $1/e^2$ level of maximum energy in the beam cross section.

When a Gaussian beam passes through a lens, it is important to ensure that all energy within the $1/e^2$ level passes through without obstruction. If this condition is not met, then the outgoing beam from the lens cannot be described as Gaussian, and Gaussian calculations will not apply. It should be noted that the $1/e^2$ requirement is an absolute minimum, and in many applications (e.g. when considering the coupling efficiency of a Gaussian beam into a single-mode optical fiber) a greater fraction of the energy must be transmitted without obstruction.

Several parameters characterize Gaussian beams:[1,2]

- Beam waist location, z, is the distance between the beam waist center and any plane perpendicular to the beam axis.
- Beam waist radius, w_0, is the radius of a Gaussian beam in the waist.
- Beam radius, w, is the radius of a Gaussian beam in a plane perpendicular to the beam axis, at the z-distance from the waist.
- Constant phase surface radius, R, is the radius of constant phase surface at the z-distance from the waist.
- Rayleigh range, z_R, is the distance from the waist, in which radius of the constant phase surface is $2z_R$. The value $2z_R$ is also called *confocal parameter* of the beam.
- Beam divergence, θ, is half the value of the full divergence. For this reason, Gaussian beams are often characterized by a full divergence value 2θ.

Collimated light usually cannot propagate through a medium without divergence, which follows from the wave nature of light. This means that Gaussian beam waist locations cannot be located at infinity.

6.2 Laser beam expanders

In many instruments, it is necessary to change the diameter of a "collimated" laser beam, and this is done using a simple type of Galilean telescope. To take a specific simple example, consider a system required to change the diameter of a laser beam from 5 to 20 mm. We will assume a wavelength of about 600 nm, and a refractive index of 1.52; since the light is monochromatic, we will not consider the effects of chromatic aberration. To make the system reasonably compact, we will use lenses with focal lengths of 80 and 20 mm, with a thin-lens separation of 60 mm.

6.2.1 Two-lens beam expanders

If we consider the two-lens beam expander shown in Fig. 6.1, LASER1, two design parameters (i.e. the shapes of the two lenses) are available once the powers have been chosen. Clearly we need to correct the spherical aberration, but it is also desirable to correct the coma; and with two shapes available, we might expect to be able to correct both S_1 and S_2. Although a laser beam expander is nominally used on-axis, if S_2 can be corrected, the system will be relatively insensitive to errors in alignment.

If we think about the spherical aberration of the two lenses, both lenses work at the same numerical aperture (NA) of 0.125 (the output beam radius is 10 mm and

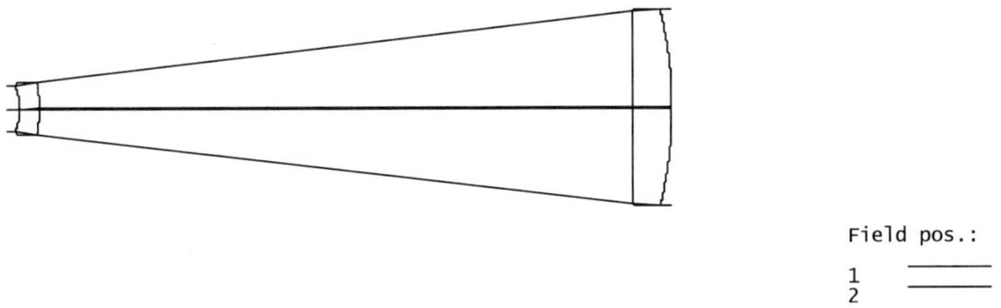

Figure 6.1 Two-lens beam expander, LASER1.

the focal length of the larger lens is 80 mm), and to a first approximation the minimum S_1 will depend on

$$h^4K^3 = hNA^3. \tag{6.1}$$

Since the magnification is 4×, the larger (positive) lens will intrinsically have four times the S_1 of the smaller negative lens. Even if we bend the positive lens for minimum spherical aberration, the negative lens must be bent well away from minimum spherical aberration, and there will be two solutions with very different shapes of the negative lens. LASER1 represents the first solution.

All designs in this section use the same fictitious refractive index (1.52), which is quite low. Use of a higher-index glass would reduce the (small) higher-order aberrations and would probably permit a more compact system.

```
WAVELENGTHS  [nm]
      600.00
SURFACES
  # SURF SPACE     RADIUS        SEPN       INDEX1      V     CLR RAD  GLASS
  0                             0.00000   1.000000
  1#S            -6.30940      2.00000   1.520000              2.500  User-USER
  2 S           -15.98075     60.00000   1.000000              2.828
  3 S           468.76682      4.00000   1.520000             10.509  User-USER
  4 S           -49.48315      0.00000   1.000000             10.714
  5 S             Plane                                       10.715

LAGRANGE INVARIANT =     -0.0044

  #      H         U         HBAR       UBAR      D(U/N)       A         ABAR
  1   2.50000   0.00000   0.00000   0.00175   0.08918  -0.39623   0.00175
  2   2.77111   0.13555   0.00230   0.00115   0.02669  -0.05753   0.00153
  3   9.72345   0.11587   0.10253   0.00167  -0.07039   0.13661   0.00189
  4   9.99999   0.06914   0.10663   0.00102  -0.04548  -0.20209  -0.00172
             -0.00000             0.00044

  #     S1         S2         S3         S4         S5         C1         C2
  1 -0.035004  0.000154 -0.000001 -0.000001  0.000000  0.000000  0.000000
  2 -0.000245  0.000006 -0.000000  0.000000 -0.000000  0.000000  0.000000
  3  0.012774  0.000177  0.000002  0.000000  0.000000  0.000000  0.000000
  4  0.018576  0.000158  0.000001  0.000000  0.000000  0.000000  0.000000

Sum -0.003899  0.000495  0.000003 -0.000000  0.000000  0.000000  0.000000
```

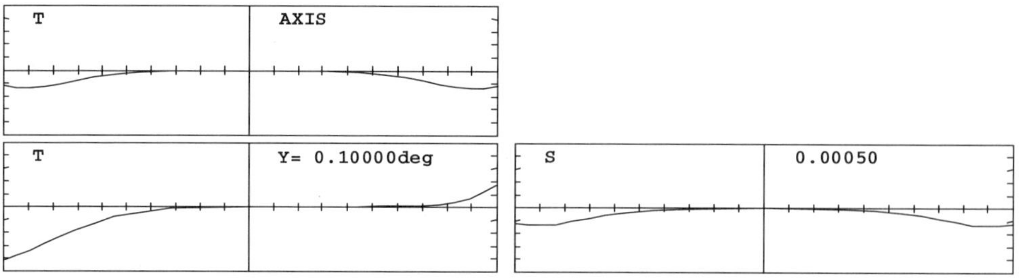

Figure 6.2 Wavefront aberrations for LASER1.

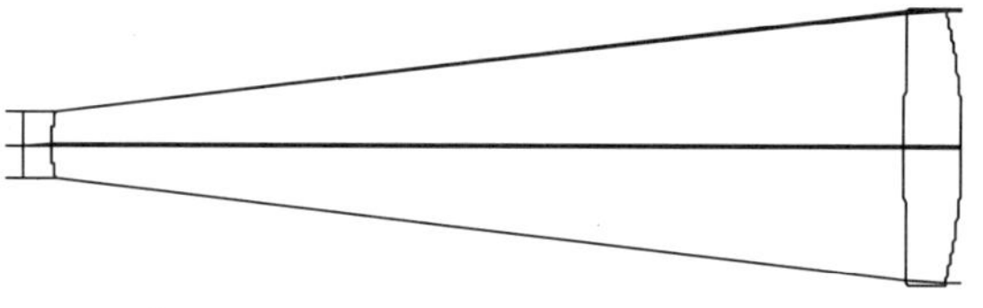

Figure 6.3 Two-lens beam expander, LASER2.

Although it has been optimized, this design has significant coma, as may be seen in the off-axis tangential section optical path difference (OPD) curves in Fig. 6.2. This implies that it needs an alignment accuracy of about 0.04 deg, if the wavefront aberration is to remain less than Rayleigh's criterion of $\lambda/4$.

The second solution is represented by LASER2, shown in Fig. 6.3.

WAVELENGTHS [nm]
600.00

SURFACES

# SURF SPACE	RADIUS	SEPN	INDEX1	V	CLR RAD	GLASS
0		0.00000	1.000000			
1#S	-1445.05940	2.00000	1.520000		2.500	User-USER
2 S	10.91288	60.00000	1.000000		2.501	
3 S	601.89291	4.00000	1.520000		10.120	User-USER
4 S	-46.55499	0.00000	1.000000		10.334	
5 S	Plane				10.335	

LAGRANGE INVARIANT = -0.0044

#	H	U	HBAR	UBAR	D(U/N)	A	ABAR
1	2.50000	0.00000	0.00000	0.00175	0.00039	-0.00173	0.00175
2	2.50118	0.00059	0.00230	0.00115	0.11969	0.34928	0.00207
3	9.70606	0.12008	0.11358	0.00185	-0.07174	0.13621	0.00204
4	10.00000	0.07348	0.11820	0.00116	-0.04834	-0.21480	-0.00210
		0.00000		0.00044			

#	S1	S2	S3	S4	S5	C1	C2
1	-0.000000	0.000000	-0.000000	-0.000000	0.000000	0.000000	0.000000
2	-0.036522	-0.000216	-0.000001	-0.000001	-0.000000	0.000000	0.000000

3	0.012918	0.000194	0.000003	0.000000	0.000000	0.000000	0.000000
4	0.022306	0.000218	0.000002	0.000000	0.000000	0.000000	0.000000
Sum	-0.001298	0.000196	0.000004	-0.000000	0.000000	0.000000	0.000000

We can see from Fig. 6.4 that the maximum wavefront aberration of LASER2 at 0.1 deg off-axis is about 0.00015 mm, i.e., about $\lambda/4$, so we can assume that LASER2 needs to be aligned with a precision of at least 0.1 deg.

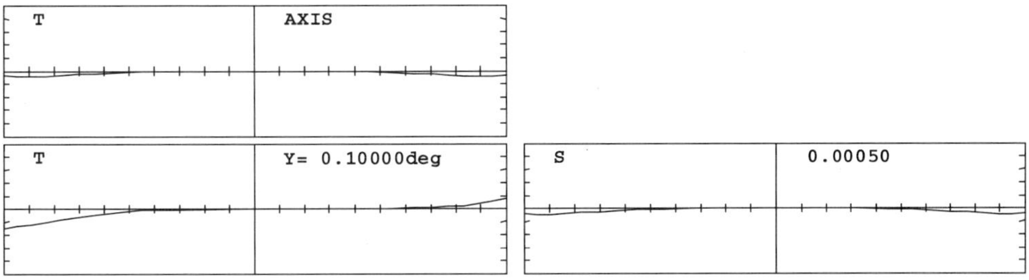

Figure 6.4 Wavefront aberrations for LASER2.

Both of these designs are not corrected for coma, although in both optimizations we can effectively vary the shapes of the two lenses; simplistically, one would think that with two shapes available as variables it should be possible to control both spherical aberration and coma. But despite many attempts to control spherical aberration and coma, no solution was found. Why? The sine condition explains. Remember, an afocal system is free from coma if

$$y \text{ (exit pupil)} = y \text{ (entrance pupil)} \times \text{expansion factor}, \tag{6.2}$$

where the expansion factor is the ratio of the paraxial ray heights in the exit and entrance pupils (in this case, 4).

Therefore, we would expect that if coma is corrected for, then the height of a marginal ray in the output space should be very close to 10 mm (2.5 mm times 4), but a ray trace will show that this is not the case. In fact, y at surface 4 is 10.67 mm in LASER1 and 10.26 mm in LASER2. This is consistent with the fact that both designs have nonzero coma, but LASER2 is better than LASER1.

If we remember that the lenses are thin, and that lens 1 (the negative lens) will have negative S_1, it is clear that the marginal ray will be aberrated when it hits lens 2, and will therefore hit lens 2 at a height greater than 10 mm, as we see from the ray trace. The key to correcting the coma is therefore to correct the spherical aberration of lens 1, which implies that we must also correct the spherical aberration of lens 2, so that the total spherical aberration is corrected. Note that if we ray-traced from the other end of the beam expander, we would ultimately come to the same conclusion.

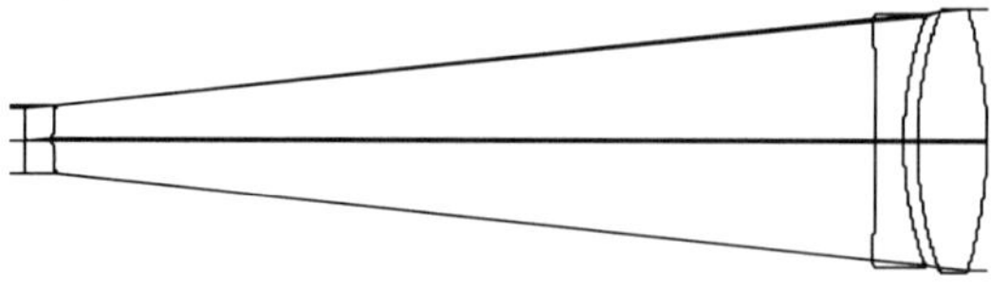

Figure 6.5 Three-element beam expander, LASER3.

6.2.2 Three-lens beam expanders

If we design a beam expander with a plus-plus positive group, we find that the coma is reduced, but we can do better if the positive group is made from a plus-minus combination, in which the spherical aberration is of course correctable. Figure 6.5 shows such a design, LASER3.

WAVELENGTHS [nm]
600.00
SURFACES

#	SURF SPACE	RADIUS	SEPN	INDEX1	V	CLR RAD	GLASS
0			0.00000	1.000000			
1	#S	-47.59613	2.00000	1.520000		2.500	User-USER
2	S	15.82977	59.00000	1.000000		2.541	
3	S	-211.03789	2.00000	1.520000		9.269	User-USER
4	S	28.89451	1.00000	1.000000		9.618	
5	S	31.33044	5.00000	1.520000		9.972	User-USER
6	S	-36.90386	0.00000	1.000000		10.162	
7	S	Plane				10.264	

LAGRANGE INVARIANT = -0.0044

#	H	U	HBAR	UBAR	D(U/N)	A	ABAR
1	2.50000	0.00000	0.00000	0.00175	0.01182	-0.05253	0.00175
2	2.53594	0.01797	0.00230	0.00115	0.09880	0.27082	0.00197
3	9.06237	0.11062	0.10972	0.00182	-0.05307	0.06768	0.00130
4	9.23730	0.08747	0.11247	0.00138	0.24164	0.61888	0.00801
5	9.53649	0.29919	0.11659	0.00412	-0.23820	0.60357	0.00784
6	9.99999	0.09270	0.12376	0.00143	-0.06099	-0.27097	-0.00292
		0.00000		0.00044			

#	S1	S2	S3	S4	S5	C1	C2
1	-0.000082	0.000003	-0.000000	-0.000000	0.000000	0.000000	0.000000
2	-0.018375	-0.000133	-0.000001	-0.000000	-0.000000	0.000000	0.000000
3	0.002203	0.000042	0.000001	-0.000000	0.000000	0.000000	0.000000
4	-0.854925	-0.011062	-0.000143	-0.000000	-0.000002	0.000000	0.000000
5	0.827528	0.010744	0.000139	0.000000	0.000002	0.000000	0.000000
6	0.044782	0.000482	0.000005	0.000000	0.000000	0.000000	0.000000
Sum	0.001131	0.000076	0.000001	-0.000000	0.000000	0.000000	0.000000

Figure 6.6 shows that the coma is much smaller—note that the scale has been reduced to 0.0001 mm (about $\lambda/5$). The marginal ray height is 10.15 mm at the output surface, which is consistent with the results given above.

If we want to reduce the coma still further, we can see that the marginal ray height must be even closer to 10 mm. How can we achieve this? One way would be make the negative lens a doublet, so that its spherical aberration was zero, or to make it aspheric, but there is another approach. Suppose that we use a plus-plus positive group. As we know, the negative lens will have negative S_1, and the marginal ray height will be greater than 10 mm, if the positive lenses are in contact. If we split the two positive lenses, with a significant separation between them, it might be possible for the positive S_1 of the middle lens to correct the marginal ray height at the final lens; this is exactly what happens, as can be seen in Fig. 6.7 and Fig. 6.8. For LASER4, the marginal ray height is 10.04 mm, while the wavefront aberration at 0.1 deg is only 0.00002 mm, i.e., about λ/30.

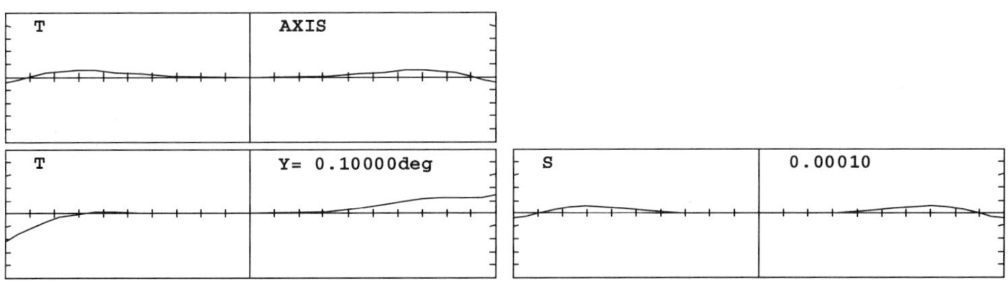

Figure 6.6 Wavefront aberrations for LASER3.

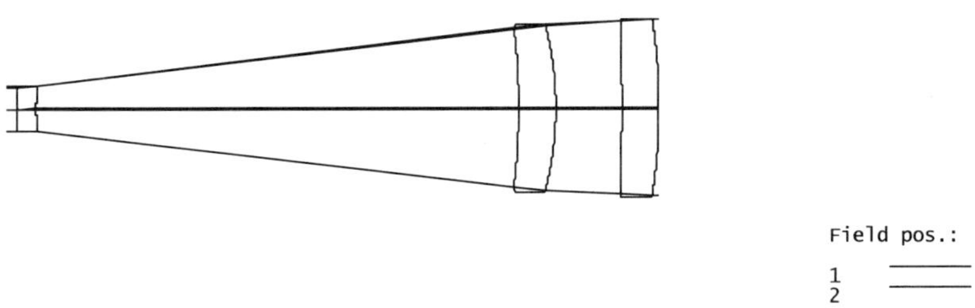

Figure 6.7 Three-element beam expander, LASER4.

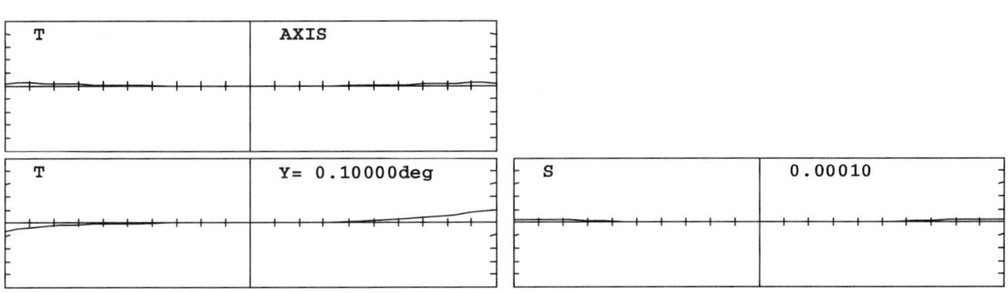

Figure 6.8 Wavefront aberrations for LASER4.

WAVELENGTHS [nm]
600.00

SURFACES

#	SURF SPACE	RADIUS	SEPN	INDEX1	V	CLR RAD	GLASS
0			0.00000	1.000000			
1	#S	-18.86067	2.00000	1.520000		2.500	User-USER
2	S	25.28875	51.89371	1.000000		2.605	
3	S	-80.71737	4.00000	1.520000		9.055	User-USER
4	S	-39.56230	7.10629	1.000000		9.463	
5	S	-276.54995	4.00000	1.520000		9.929	User-USER
6	S	-69.74596	0.00000	1.000000		10.102	
7	S	Plane				10.102	

LAGRANGE INVARIANT = -0.0044

#	H	U	HBAR	UBAR	D(U/N)	A	ABAR
1	2.50000	0.00000	0.00000	0.00175	0.02983	-0.13255	0.00175
2	2.59069	0.04535	0.00230	0.00115	0.09236	0.22464	0.00188
3	8.93198	0.12220	0.09532	0.00179	-0.04440	0.01154	0.00061
4	9.40498	0.11825	0.10165	0.00158	-0.02167	-0.18160	-0.00150
5	9.80380	0.05612	0.10926	0.00107	-0.02385	0.02067	0.00068
6	10.00000	0.04905	0.11262	0.00084	-0.03227	-0.14338	-0.00118
		-0.00000		0.00044			

#	S1	S2	S3	S4	S5	C1	C2
1	-0.001310	0.000017	-0.000000	-0.000000	0.000000	0.000000	0.000000
2	-0.012075	-0.000101	-0.000001	-0.000000	-0.000000	0.000000	0.000000
3	0.000053	0.000003	0.000000	-0.000000	0.000000	0.000000	0.000000
4	0.006723	0.000055	0.000000	0.000000	0.000000	0.000000	0.000000
5	0.000100	0.000003	0.000000	-0.000000	0.000000	0.000000	0.000000
6	0.006634	0.000055	0.000000	0.000000	0.000000	0.000000	0.000000
Sum	0.000123	0.000032	0.000000	-0.000000	0.000000	0.000000	0.000000

6.3 F-Theta lenses

Laser-printing systems generally require lenses with the property that the image size is proportional to the angle of the input beam, rather than to the tangent of this angle (which is the common case when designing photographic lenses). These lenses are known as F-theta lenses; they always have an external pupil (the scanning mirror). Sometimes F-theta lenses are required to be telecentric (exit pupil at infinity), so that the size of the image is not affected by the longitudinal position of the image plane, but this does not seem to be a general requirement. In so far as the pupil is external, an F-theta lens is similar to an eyepiece, but eyepieces are necessarily over-telecentric in general, and are also usually corrected for chromatic aberration.

Consider the following specification:

Focal length = 250 mm,

f/20,

30 deg semi-field,

$\lambda = 0.0006$ mm,

$n = 1.6$.

Since we clearly expect to correct all of the monochromatic Seidel aberrations, the simplest possible solution worth considering is a doublet; F-theta lenses often resemble one half of a double-Gauss lens, so if we are to attempt a doublet design we might expect a Gauss-type doublet to result. We should remember that there are other solutions for doublets, however. In practice, a doublet does not provide enough design parameters for an adequate F-theta lens, but a triplet as shown in Fig. 6.9, FTHETA3, is typical of simple F-theta lenses. Design data are given below.

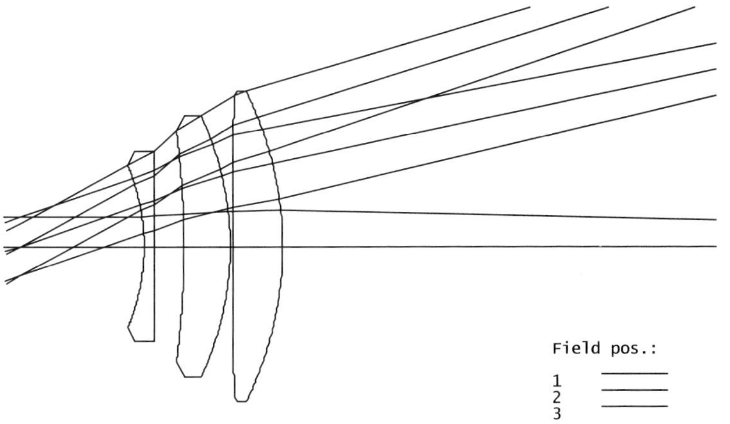

Figure 6.9 F-Theta lens, FTHETA3.

```
EFL =    250.000
WAVELENGTHS   [nm]
      600.00
SURFACES
```

# SURF	SPACE	RADIUS	SEPN	INDEX1	V	CLR RAD	GLASS
0			0.00000	1.000000			
1#S		Plane	24.74367	1.000000		6.250	
2 S		-46.98943	2.00000	1.600000		17.274	User-USER
3 S		-5.32113e+004	5.86374	1.000000		19.948	
4 S		-204.07491	9.18119	1.600000		24.525	User-USER
5 S		-69.59270	0.50000	1.000000		27.448	
6 S		1556.98954	9.83472	1.600000		31.887	User-USER
7 S		-78.46052	304.48018	1.000000		32.750	
8 S		Plane				130.871	

LAGRANGE INVARIANT =   -3.6084

#	H	U	HBAR	UBAR	D(U/N)	A	ABAR
1	6.25000	0.00000	0.00000	0.57735	0.00000	0.00000	0.57735
2	6.25000	0.00000	14.28577	0.57735	0.03117	-0.13301	0.27333
3	6.34976	0.04988	15.23547	0.47485	0.04856	0.07961	0.75930
4	6.81729	0.07973	19.68951	0.75959	-0.04076	0.04633	0.66311
5	7.38984	0.06236	24.38042	0.51093	-0.00291	-0.07012	0.25695

6	7.40787	0.03606	24.68406	0.60728	-0.02309	0.04082	0.62314
7	7.61200	0.02076	28.35837	0.37361	-0.03797	-0.12202	0.01947
		-0.02500		0.38091			

#	S1	S2	S3	S4	S5	C1	C2
1	0.000000	0.000000	0.000000	0.000000	0.000000	0.000000	0.000000
2	-0.003447	0.007083	-0.014556	-0.103913	0.243451	0.000000	0.000000
3	-0.001954	-0.018640	-0.177773	0.000092	-1.694595	0.000000	0.000000
4	0.000596	0.008536	0.122179	-0.023927	1.406330	0.000000	0.000000
5	0.000106	-0.000388	0.001420	0.070163	-0.262304	0.000000	0.000000
6	0.000285	0.004352	0.066424	0.003136	1.061791	0.000000	0.000000
7	0.004303	-0.000687	0.000110	0.062233	-0.009950	0.000000	0.000000
	-0.000111	0.000257	-0.002196	0.007784	0.744722	0.000000	0.000000

Note from Fig. 6.10 that this lens is diffraction limited, in the sense that its wavefront aberrations are less than $\lambda/4$. It should be noted that the patent literature contains examples of F-theta lenses that are simpler than this example, having only one or two glasses. These lenses are intended for low-resolution applications, and they are not capable of the resolution that is attainable with three-lens systems.

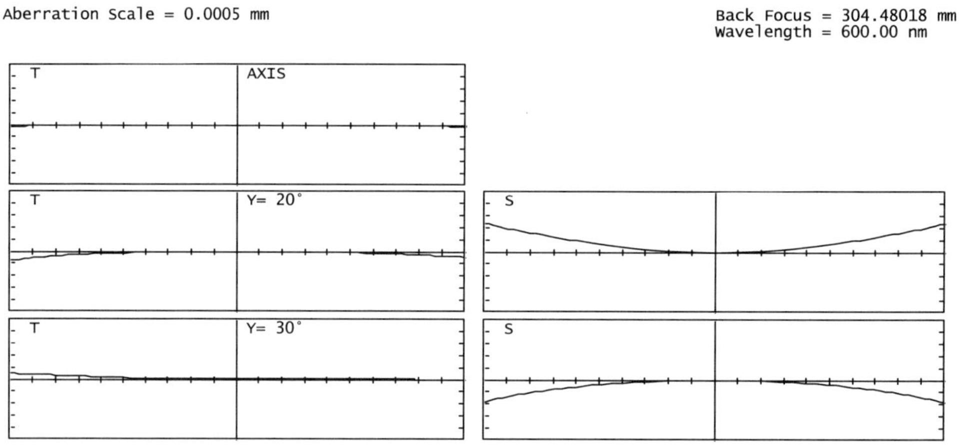

Figure 6.10 Wavefront aberrations for FTHETA3.

Note also the similarity between this design and half of a double-Gauss objective. In fact, if we take two of these F-theta lenses "front to front" and reoptimize, we obtain a very good monochromatic double-Gauss design.

6.4 Lenses for optical disks

In optical disk players, a high-resolution lens is needed in order to read the data encoded on the disk. This lens must accurately follow the track that is recorded on the disk; it does so by using a servo system, and this in turn requires that the lens is very lightweight. Optical disk players are now made in very large quantities, and although many published optical disk lenses are essentially similar to lenses made

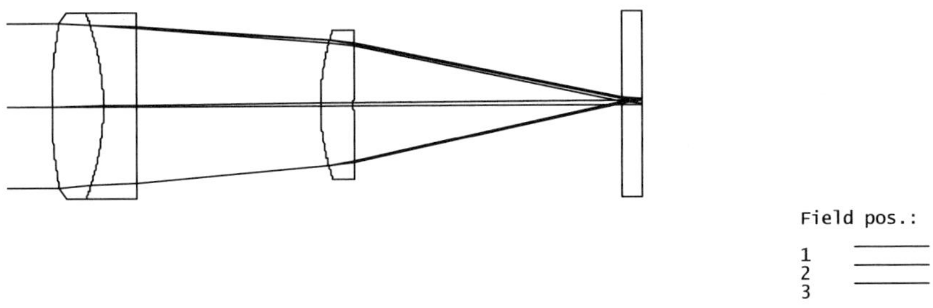

Figure 6.11 Sugiyama optical disk lens.

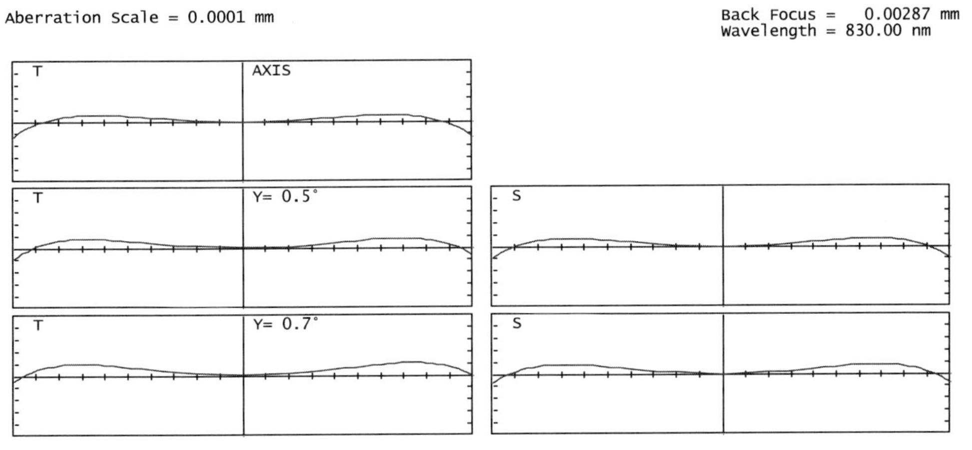

Figure 6.12 Wavefront aberrations of Sugiyama optical disk lens.

for collimating laser diodes, as described below, a lighter design can be achieved by the use of a single lens, sometimes with both surfaces aspheric. If the lens were thin there would be little point in using two aspherics, but in practice the lenses are quite thick, and a useful improvement is obtained by the use of two aspherics.

The three-glass design in Fig. 6.11 by Sugiyama (U.S. Patent 4,753,524) has a numerical aperture of 0.22. The patent claims that the lens is corrected for chromatic aberration, so that it can work over a range of wavelengths, and also that the effects of temperature changes have been allowed for. Wavefront aberrations in Fig. 6.12 show that the lens is well within the diffraction limit of $\lambda/4$.

```
EFL =       3.990
WAVELENGTHS  [nm]
      830.00
SURFACES
  # SURF SPACE   RADIUS       SEPN      INDEX1     V    CLR RAD  GLASS
  0                         0.00000   1.000000
  1#S           5.10520     0.52160   1.677900            0.878  User-USER
  2 S          -2.86920     0.32600   1.784720            1.000  User-USER-2
  3 S           Plane       1.85520   1.000000            1.000
  4 S           2.69400     0.32600   1.618000            0.800  User-USER-3
```

5 S	20.78560	2.68400	1.000000	0.800	
6 S	Plane	0.20000	1.510000	1.000	User-USER-4
7 S	Plane	0.00287	1.000000	1.000	
8 S	Plane			1.000	

LAGRANGE INVARIANT = -0.0107

#	H	U	HBAR	UBAR	D(U/N)	A	ABAR
1	0.87776	0.00000	0.00000	0.01222	-0.04140	0.17193	0.01222
2	0.84153	-0.06946	0.00380	0.00728	0.01464	-0.60868	0.01000
3	0.82596	-0.04775	0.00606	0.00693	-0.05847	-0.08522	0.01236
4	0.66785	-0.08522	0.02898	0.01236	-0.00585	0.16268	0.02312
5	0.61981	-0.14736	0.03014	0.00353	-0.12892	-0.19018	0.00806
6	0.02933	-0.22000	0.04787	0.00661	0.12351	-0.22000	0.00661
7	0.00019	-0.14570	0.04874	0.00437	-0.12351	-0.22000	0.00661
		-0.22000		0.00660			

#	S1	S2	S3	S4	S5	C1	C2
1	0.001074	0.000076	0.000005	0.000009	0.000001	0.000000	0.000000
2	-0.004565	0.000075	-0.000001	-0.000001	0.000000	0.000000	0.000000
3	0.000351	-0.000051	0.000007	0.000000	-0.000001	0.000000	0.000000
4	0.000103	0.000015	0.000002	0.000016	0.000003	0.000000	0.000000
5	0.002890	-0.000122	0.000005	-0.000002	-0.000000	0.000000	0.000000
6	-0.000175	0.000005	-0.000000	0.000000	0.000000	0.000000	0.000000
7	0.000001	-0.000000	0.000000	0.000000	-0.000000	0.000000	0.000000
Sum	-0.000321	-0.000002	0.000019	0.000022	0.000002	0.000000	0.000000

The single aspheric lens in Fig. 6.13 by Arai et al. (U.S. Patent 4,449,792) has aspherics on both surfaces and a numerical aperture of 0.50. Wavefront aberrations are shown in Fig. 6.14.

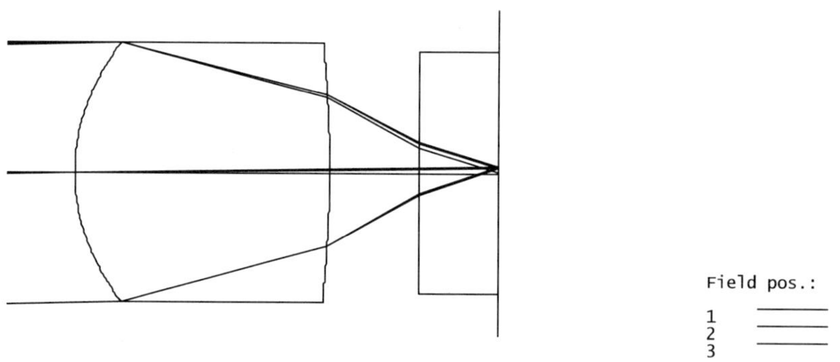

Figure 6.13 Arai optical disk lens.

EFL = 4.300

WAVELENGTHS [nm]

780.00

SURFACES

# SURF SPACE	RADIUS	SEPN	INDEX1	V	CLR RAD	GLASS
0		0.00000	1.000000			
1#A	3.37103	4.00000	1.702140		2.150	User-USER
2 A	-14.76876	1.40096	1.000000		2.150	
3 S	Plane	1.20000	1.510720		2.000	User-USER-2

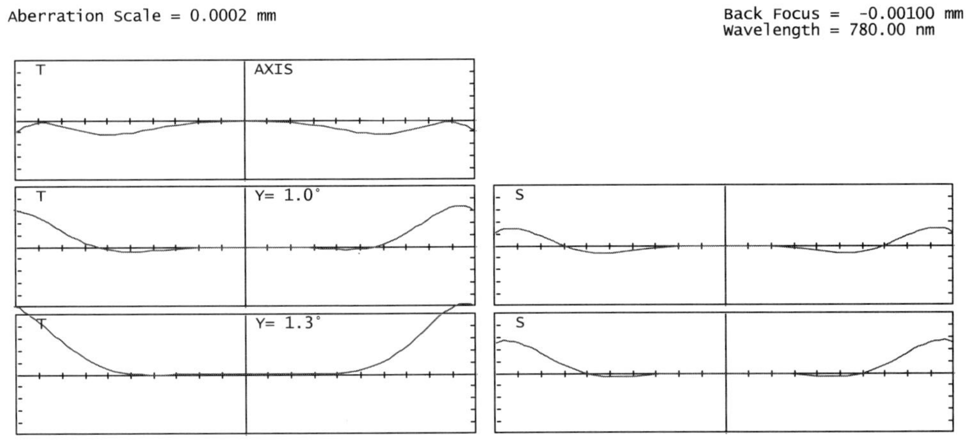

Figure 6.14 Wavefront aberrations of Arai optical disk lens.

4 S	Plane	0.00000	1.000000	2.000
5 S	Plane			2.688

ASPHERIC SURFACE 1 CC = -2.416880
A4 = 0.0062875 A6 = -0.00021838 A8 = 6.7164e-006
ASPHERIC SURFACE 2 CC = -149.999000

LAGRANGE INVARIANT = -0.0488

#	H	U	HBAR	UBAR	D(U/N)	A	ABAR
1	2.15000	0.00000	0.00000	0.02269	-0.15456	0.63779	0.02269
2	1.09764	-0.26309	0.05333	0.01333	-0.34544	-0.57432	0.01655
3	0.39716	-0.50000	0.08157	0.02016	0.28092	-0.50000	0.02016
4	-0.00000	-0.33097	0.09758	0.01334	-0.28092	-0.50000	0.02016
		-0.50000		0.02016			

#	S1	S2	S3	S4	S5	C1	C2
1	0.135176	0.004810	0.000171	0.000291	0.000016	0.000000	0.000000
2	0.125066	-0.003603	0.000104	0.000066	-0.000005	0.000000	0.000000
3	-0.027893	0.001125	-0.000045	0.000000	0.000002	0.000000	0.000000
4	-0.000000	0.000000	-0.000000	0.000000	0.000000	0.000000	0.000000
	ASPHERIC SURFACE 1						
	-0.191901	0.000000	0.000000	0.000000	0.000000		
	ASPHERIC SURFACE 2						
	-0.047459	-0.002306	-0.000112	0.000000	-0.000005		
Sum	-0.007012	0.000025	0.000118	0.000358	0.000008	0.000000	0.000000

6.5 Laser diode collimators

Since laser diodes produce a diverging beam, it is almost always necessary for the beam to be collimated. Collimators for laser diodes need to operate at high numerical apertures (about 0.60) to be compatible with the laser diode itself. Naturally, spherical aberration must be very well corrected, and it is also desirable for coma to be corrected so that alignment of the collimator is facilitated. The basic lens

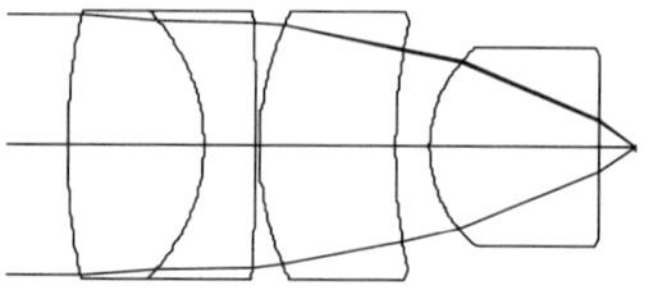

Figure 6.15 Laser diode collimator.

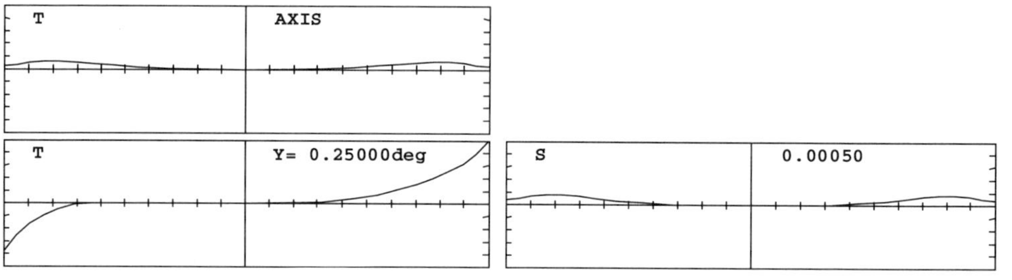

Figure 6.16 Wavefront aberrations of laser diode collimator.

EFL = 6.500
WAVELENGTH 780 nm

#	SURF SPACE	RADIUS	SEPN	INDEX1	CLR RAD	GLASS
0			0.000	1.000000		
1	S	19.234	4.000	1.582525	4.000	S-SK5
2	S	-5.660	1.500	1.659490	4.000	S-SF5
3	S	-80.675	0.100	1.000000	4.000	
4	S	9.355	4.000	1.582525	3.750	S-SK5
5	S	18.107	1.000	1.000000	3.000	
6	S	3.930	5.000	1.582525	2.750	S-SK5
7	S	Plane	1.000	1.000000	0.100	
8	S	Plane			0.100	

design is essentially straightforward, and one example is shown in Fig. 6.15, with wavefront aberrations in Fig. 6.16.

In addition to the basic lens design, it is necessary to correct for the unequal divergence of the laser beam in the planes parallel and perpendicular to the emitting junction; typically the beam widths in the two perpendicular directions are in the ratio of about 2:1, and often this has to be corrected so that the output beam from the collimator has a circular cross section rather than an elliptical cross section. This is commonly achieved by means of a pair of prisms arranged so that the beam diameter is increased in one direction.

References

1. H. Kogelnik and T. Li, "Laser beams and resonators," *Appl. Opt.* **5**, 1550–67 (1966); and *Bell. Syst. Tech. J.* **44**, 455 (1965).

2. R. Herloski, S. Marshall, and R. Antos, "Gaussian beam ray-equivalent modeling and optical design," *Appl. Opt.* **22**, 1168 (1983).

CHAPTER 7
MICROSCOPE OBJECTIVES

The function of a microscope objective is to resolve fine detail in an object and to magnify it. The resolution limit of two adjacent Airy disks (the images of point objects) is given by Rayleigh's formula:[1]

$$\text{Resolution} = 0.61\frac{\lambda}{n\sin U}, \tag{7.1}$$

where λ is the wavelength of light, n is the refractive index of the medium in which the object is immersed, and $\sin U$ is the sine of the marginal ray angle at the object. The product $n\sin U$ is the numerical aperture (NA). In order to produce good images close to the optical axis the objective must be aplanatic, so we can assume that it obeys the sine condition. In this case $\sin U$ is equal to the paraxial marginal ray angle u, even when this approaches the upper limit of 1.0.

Equation (7.1) indicates that increasing the NA and reducing λ can improve the resolution. This assumes that the objective design is diffraction limited when, by Rayleigh's criterion, its wavefront aberrations are less than $\lambda/4$. In fact, most modern designs have much smaller aberrations than this.

7.1 Classical microscope objectives

It is usual to design microscope objectives backwards, from image to object. As may be expected for a high-aperture small-field application, the classical design form is the Petzval lens, which in microscopy is more commonly referred to as a Lister objective. To achieve a higher NA, it is common to add one or more positive lenses to the basic Lister form. An example is shown in Fig. 7.1, where the last lens surface is concave, to

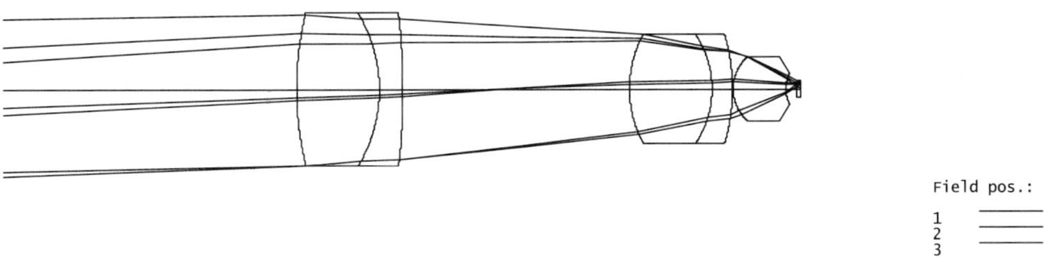

Figure 7.1 Classical Lister objective with aplanatic element.

reduce Petzval sum, and the previous surface operates close to the aplanatic condition, when spherical aberration, coma, and astigmatism are zero, if $\delta(u/n) = 0$. This example has an NA of 0.65 and operates at finite conjugates at a magnification of 40×. The design prescription includes a cover glass for the specimen, which introduces significant spherical aberration that must be corrected by the objective.

```
EFL =      4.716
OBJECT DISTANCE = 180, CURVATURE = 0, TILT ANGLE = 0
WAVELENGTHS   [nm]
      587.60       656.30       486.10
SURFACES
```

#	SURF SPACE	RADIUS	SEPN	INDEX1	V	CLR RAD	GLASS
0			0.00000	1.000000			
1	S	13.37510	3.07300	1.548138	45.73	3.048	S-LLF1
2	S	-6.16238	0.86600	1.761814	26.52	3.048	S-SF14
3	S	-28.39196	4.19500	1.000000		3.048	
4#	S	Plane	4.19500	1.000000		2.481	
5	S	4.70110	3.06100	1.539958	59.68	2.144	S-BAK2
6	S	-4.02330	0.76200	1.761814	26.52	2.144	S-SF14
7	S	-7.94052	0.02500	1.000000		2.144	
8	S	1.79285	1.92400	1.607379	56.63	1.272	S-SK2
9	S	1.36064	0.39020	1.000000		0.600	
10	S	Plane	0.18000	1.516798	64.14	0.314	S-BK7
11	S	Plane	-0.00007	1.000000		0.314	
12	S	Plane				0.248	

LAGRANGE INVARIANT =   -0.1625

#	H	U	HBAR	UBAR	D(U/N)	A	ABAR
1	2.92500	0.01625	-0.45828	0.05301	-0.05948	0.23494	0.01875
2	2.71931	-0.06693	-0.31578	0.04637	0.04023	-0.78678	0.15112
3	2.71473	-0.00530	-0.28587	0.03453	-0.07917	-0.17779	0.07858
4	2.37001	-0.08217	0.00153	0.06851	0.00000	-0.08217	0.06851
5	2.02529	-0.08217	0.28894	0.06851	-0.05057	0.34864	0.12997
6	1.39957	-0.20442	0.35916	0.02294	0.05619	-0.85049	-0.10215
7	1.29680	-0.13487	0.38300	0.03129	-0.28548	-0.52535	-0.02985
8	1.28775	-0.36203	0.38346	0.01838	0.05306	0.35624	0.23227
9	0.33221	-0.49664	0.24997	-0.06938	-0.34102	-0.40585	0.18377
10	0.07858	-0.65000	0.24999	0.00006	0.36747	-0.65000	0.00006
11	0.00144	-0.42853	0.25000	0.00004	-0.36747	-0.65000	0.00006
		-0.65000		0.00006			

#	S1	S2	S3	S4	S5	C1	C2
1	0.009604	0.000766	0.000061	0.000699	0.000061	0.005320	0.000425
2	-0.067717	0.013007	-0.002498	-0.000336	0.000544	-0.018325	0.003520
3	0.006793	-0.003003	0.001327	0.000402	-0.000764	0.007871	-0.003479
4	0.000000	0.000000	0.000000	0.000000	0.000000	0.000000	0.000000
5	0.012449	0.004641	0.001730	0.001970	0.001379	0.004148	0.001546
6	-0.056884	-0.006832	-0.000821	-0.000537	-0.000163	-0.012418	-0.001491
7	0.102174	0.005805	0.000330	0.001438	0.000100	0.011110	0.000631
8	-0.008670	-0.005653	-0.003686	0.005565	0.001226	0.003061	0.001996
9	0.018660	-0.008450	0.003826	-0.007333	0.001588	0.000900	-0.000407
10	-0.012200	0.000001	-0.000000	0.000000	0.000000	-0.000271	0.000000
11	0.000224	-0.000000	0.000000	0.000000	-0.000000	0.000005	-0.000000
Sum	0.004432	0.000283	0.000270	0.001868	0.003971	0.001400	0.002740

The aberrations curves in Figs. 7.2 and 7.3 show some residual spherical aberration, sphero-chromatism, and secondary spectrum over the visible waveband, and an absence of coma over the field. There is a noticeable amount of lateral color, which is difficult to correct within such a simple objective. Normally it is compensated elsewhere, either in the eyepiece for visual use, or relay optics if a camera is used. Figure 7.4 shows that neither the Petzval sum nor the distortion is fully corrected, but these were acceptable in older microscopes, for direct viewing through an eyepiece.

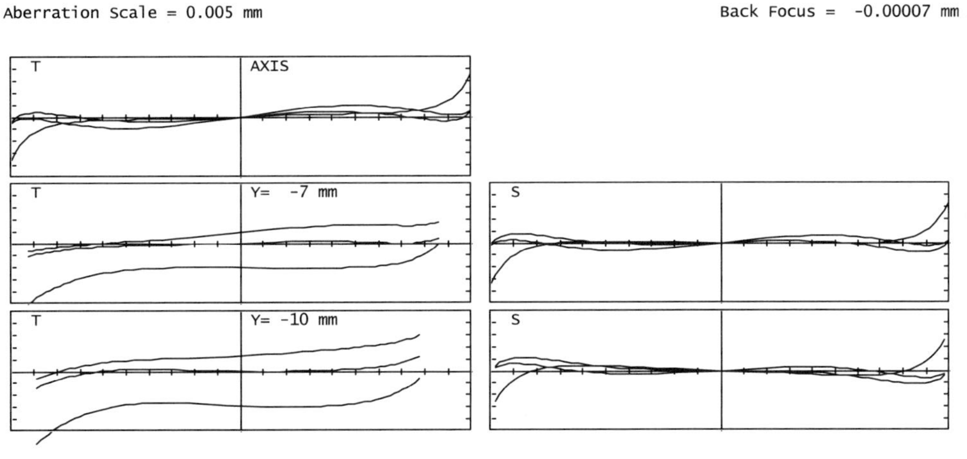

Figure 7.2 Transverse ray aberrations of Lister objective with aplanatic element.

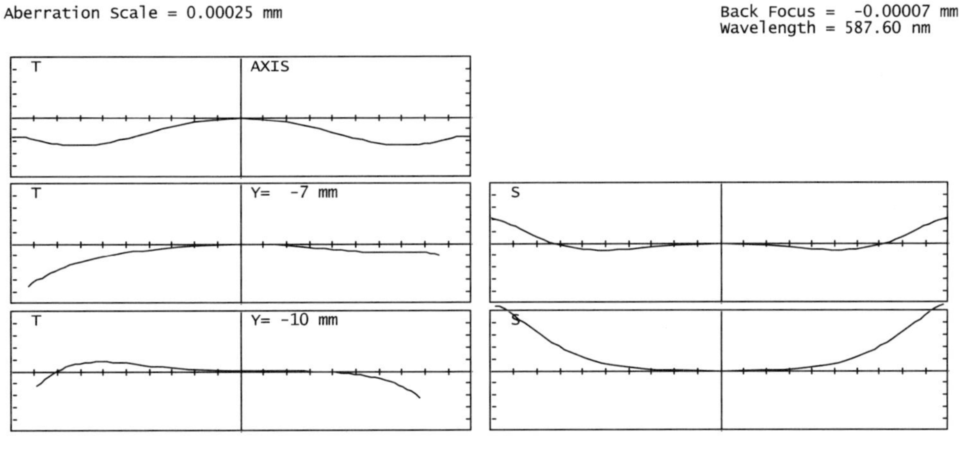

Figure 7.3 Wavefront aberrations of Lister objective with aplanatic element.

7.2 Flat-field microscope objectives

As they are more often used with a relay and camera, modern microscope objectives are required to give a flatter field, and therefore some stronger method of Petzval sum reduction is needed. Figure 7.5 shows an example of one type of apo-

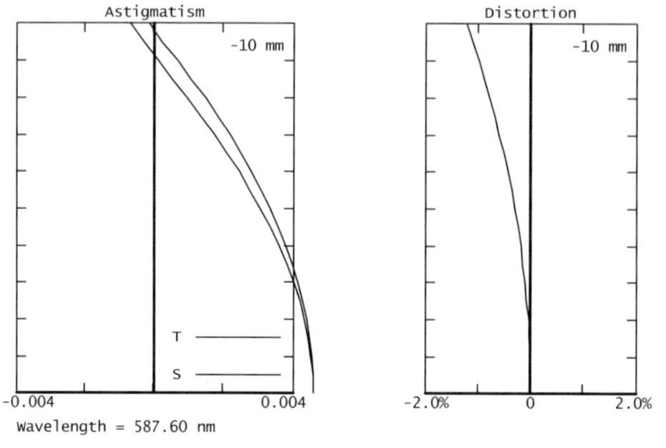

Figure 7.4 Astigmatism and distortion of Lister objective with aplanatic element.

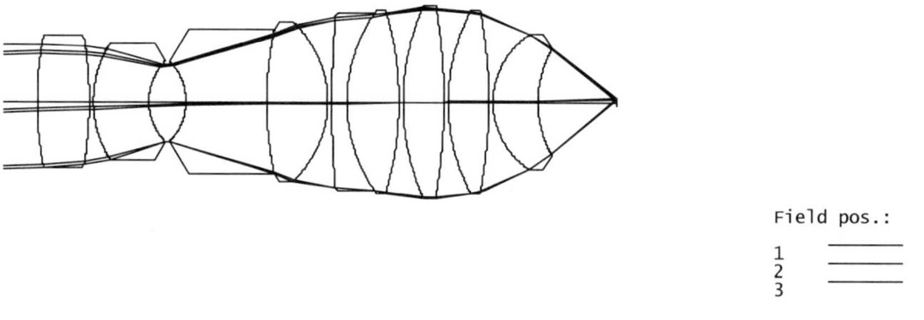

Figure 7.5 Double-Gauss-type flat-field microscope objective.

chromatic microscope objective, in which the field-flattening is obtained by the use of a bi-convex "air lens"—the double-Gauss form. Note that the back focus, or working distance, is relatively long. This particular design is for use without a cover glass, and with one infinite conjugate. This is typical of more modern objectives, as it gives more flexibility in the use of a variety of relay optics, such as a beamsplitter to divert light to both a camera and eyepieces. It does not use high-index glasses, but still gives diffraction-limited performance at a numerical aperture of 0.65, as may be seen in Figs. 7.6 and 7.7. The secondary spectrum correction is obtained mainly by the use of FK51 and FK54 crown and KZFSN4 flint glasses. Again the lateral color is not corrected, as may be seen in Fig. 7.6. The flatter field and lower distortion are apparent in Fig. 7.8.

```
EFL =      4.500
WAVELENGTHS   [nm]
      546.10         656.30         486.10
SURFACES
```

#	SURF SPACE	RADIUS	SEPN	INDEX1	V	CLR RAD	GLASS
0			1.30641	1.000000			
1	S	17.35664	2.60396	1.487936	84.67	3.399	S-FK51
2	S	-21.74533	0.20000	1.000000		3.219	
3	S	5.97609	2.63782	1.550985	45.97	2.961	S-LLF1

4	S	3.06343	1.83096	1.000000		2.077	
5	S	-3.01507	3.90374	1.677636	32.43	2.067	S-SF5
6	S	18.74531	2.93552	1.487936	84.67	3.699	S-FK51
7	S	-5.87789	0.20000	1.000000		4.072	
8	S	59.18729	0.70000	1.616690	44.51	4.430	S-KZFSN4
9	S	8.38842	2.52456	1.438150	90.90	4.581	S-FK54
10#	S	-17.32874	0.20000	1.000000		4.676	
11	S	12.83435	1.95758	1.438150	90.90	4.930	S-FK54
12	S	-30.45877	0.20000	1.000000		4.911	
13	S	10.76567	1.95157	1.487936	84.67	4.706	S-FK51
14	S	-26.59881	0.20000	1.000000		4.595	
15	S	4.31068	2.15003	1.566050	61.01	3.461	S-SK11
16	S	6.00286	3.80501	1.000000		2.724	
17	S	Plane				0.157	

LAGRANGE INVARIANT = -0.1021

#	H	U	HBAR	UBAR	D(U/N)	A	ABAR
1	2.92500	0.00000	-0.52550	0.03492	-0.03714	0.16852	0.00464
2	2.78110	-0.05526	-0.43854	0.03340	-0.10749	-0.27253	0.07970
3	2.75217	-0.14463	-0.42663	0.05953	-0.02097	0.31590	-0.01186
4	2.07463	-0.25685	-0.25848	0.06375	0.14037	0.65199	-0.03200
5	2.02843	-0.02524	-0.16258	0.05238	0.17825	-0.69800	0.10630
6	3.03053	0.25670	-0.12572	0.00944	0.05536	0.70187	0.00459
7	3.94066	0.31004	-0.09698	0.00979	-0.07417	-0.53622	0.03912
8	3.96750	0.13420	-0.09246	0.02262	-0.09867	0.20123	0.02106
9	4.00770	0.05744	-0.08225	0.01459	0.05061	0.86526	0.00773
10	4.32045	0.12388	-0.04392	0.01518	-0.01722	-0.18040	0.02548
11	4.33423	0.06892	-0.03934	0.02294	-0.10714	0.40663	0.01988
12	4.22664	-0.05496	-0.00628	0.01689	-0.10163	-0.27861	0.02458
13	4.19867	-0.13985	-0.00141	0.02437	-0.00927	0.25016	0.02424
14	3.76565	-0.22188	0.03065	0.01642	-0.25010	-0.54079	0.02272
15	3.68581	-0.39922	0.03542	0.02388	0.03909	0.45582	0.03209
16	2.47324	-0.56398	0.06182	0.01228	-0.28987	-0.23799	0.03535
		-0.65000		0.02505			

#	S1	S2	S3	S4	S5	C1	C2
1	0.003085	0.000085	0.000002	0.000197	0.000005	0.001909	0.000053
2	0.022203	-0.006493	0.001899	0.000157	-0.000601	0.002935	-0.000858
3	0.005761	-0.000216	0.000008	0.000620	-0.000024	0.006719	-0.000252
4	-0.123795	0.006075	-0.000298	-0.001210	0.000074	-0.010453	0.000513
5	-0.176157	0.026827	-0.004086	-0.001398	0.000835	-0.017633	0.002685
6	-0.082641	-0.000540	-0.000004	-0.000042	-0.000000	-0.018253	-0.000119
7	0.084041	-0.006131	0.000447	0.000582	-0.000075	0.008184	-0.000597
8	0.015852	0.001659	0.000174	0.000067	0.000025	0.006842	0.000716
9	-0.151857	-0.001357	-0.000012	-0.000096	-0.000001	-0.018094	-0.000162
10	0.002421	-0.000342	0.000048	0.000183	-0.000033	0.002612	-0.000369
11	0.076780	0.003753	0.000183	0.000248	0.000021	0.005907	0.000289
12	0.033343	-0.002942	0.000260	0.000104	-0.000032	0.003947	-0.000348
13	0.002437	0.000236	0.000023	0.000318	0.000033	0.004068	0.000394
14	0.275437	-0.011574	0.000486	0.000129	-0.000026	0.007887	-0.000331
15	-0.029939	-0.002108	-0.000148	0.000875	0.000051	0.009953	0.000701
16	0.040606	-0.006032	0.000896	-0.000628	-0.000040	0.003487	-0.000518
Sum	-0.002423	0.000902	-0.000121	0.000107	0.000213	0.000016	0.001796

The example in Fig. 7.9 shows a different method of Petzval sum correction. This system can be thought of as an inverted telephoto lens, with a negative lens

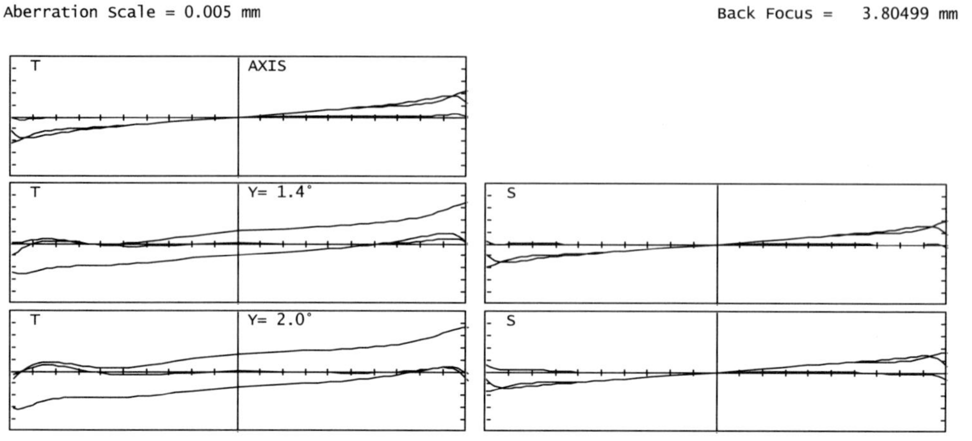

Figure 7.6 Transverse ray aberrations of double-Gauss-type flat-field microscope objective.

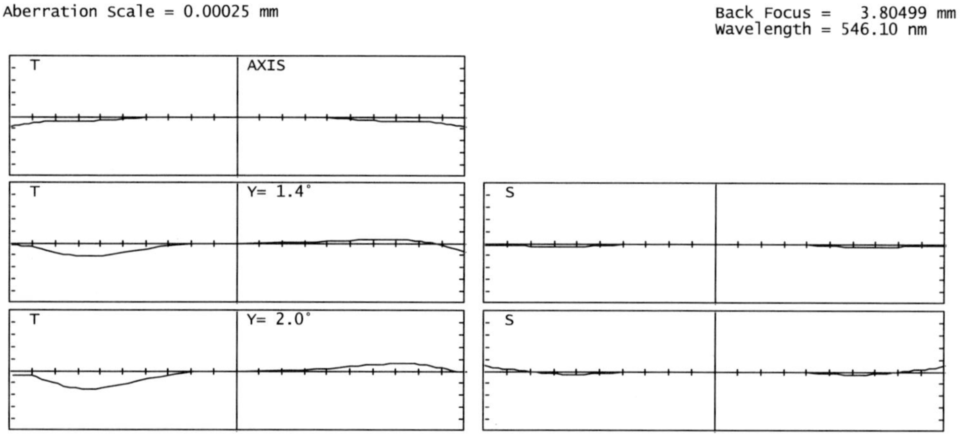

Figure 7.7 Wavefront aberrations of double-Gauss-type flat field microscope objective.

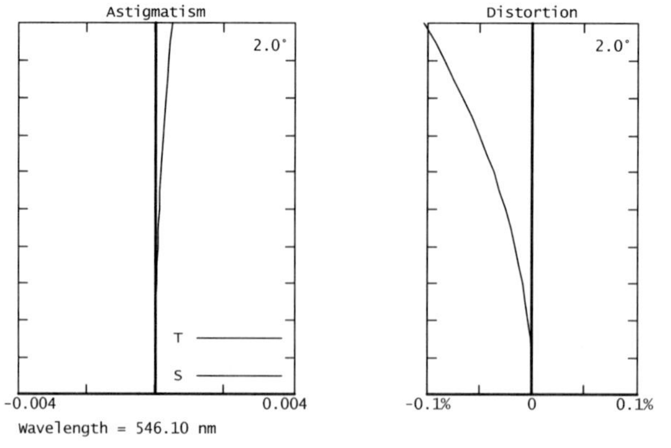

Figure 7.8 Astigmatism and distortion of double-Gauss-type flat-field microscope objective.

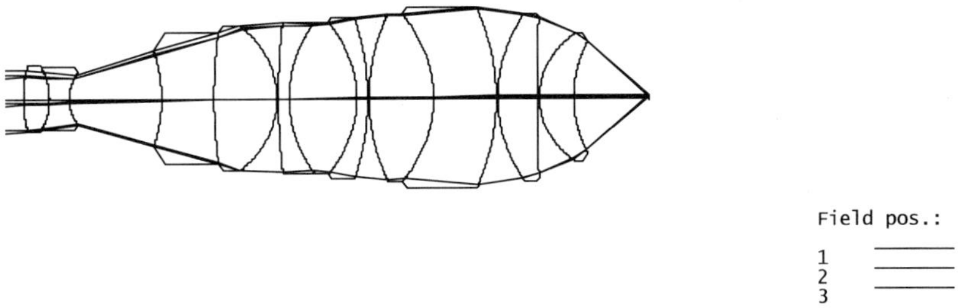

Figure 7.9 Inverse-telephoto-type flat-field microscope objective.

that is separated from the positive components. There are several elements of fluorite (CaF_2), or FK52 (a fluor-crown glass), to give the secondary spectrum correction.

EFL = 3.000

WAVELENGTHS [nm]

546.10 656.30 486.10

SURFACES

#	SURF SPACE	RADIUS	SEPN	INDEX1	V	CLR RAD	GLASS
0			-0.44023	1.000000			
1	S	9.99908	1.83246	1.746197	28.42	2.455	S-SF3
2	S	-5.33855	1.39770	1.736881	51.86	2.318	User-LAKN16
3	S	3.45242	6.42522	1.000000		1.961	
4	S	-12.26840	3.61285	1.720551	48.20	3.520	User-LAFN3
5	S	40.01262	4.34115	1.487470	82.01	4.859	S-FK52
6	S	-6.89792	0.10000	1.000000		5.413	
7	S	52.13524	0.75000	1.625087	53.37	5.560	S-SSK2
8	S	9.89282	4.64809	1.434963	95.64	5.535	User-CAF2
9#	S	-9.02826	0.75000	1.754579	35.17	5.614	S-LAFN7
10	S	-20.44716	0.10000	1.000000		5.976	
11	S	15.34850	4.50543	1.434963	95.64	6.259	User-CAF2
12	S	-9.94823	4.31758	1.768585	26.75	6.206	S-SF14
13	S	-17.32619	0.10000	1.000000		6.742	
14	S	11.31451	2.78035	1.487470	82.01	6.266	S-FK52
15	S	254.12859	0.10000	1.000000		5.977	
16	S	5.75727	2.36706	1.625087	53.37	4.788	S-SSK2
17	S	9.34330	5.25103	1.000000		4.180	
18	S	Plane				0.159	

LAGRANGE INVARIANT = -0.1101

#	H	U	HBAR	UBAR	D(U/N)	A	ABAR
1	2.10000	0.00000	-0.45673	0.05241	-0.05140	0.21002	0.00673
2	1.93554	-0.08975	-0.36596	0.04953	-0.00167	-0.78982	0.20620
3	1.80671	-0.09217	-0.29585	0.05016	0.27860	0.74885	-0.06171
4	3.25579	0.22553	-0.14174	0.02398	-0.08475	-0.03985	0.03554
5	4.13090	0.24222	-0.10886	0.00910	0.05845	0.59438	0.01098
6	5.41740	0.29635	-0.06501	0.01010	-0.14126	-0.72740	0.02904
7	5.42320	0.05797	-0.06305	0.01962	-0.06064	0.16199	0.01841
8	5.41994	-0.00434	-0.05364	0.01254	0.04983	0.88328	0.01156
9	5.73450	0.06767	0.00902	0.01348	0.05033	-0.81434	0.01791
10	5.86279	0.17105	0.01742	0.01121	-0.01373	-0.20297	0.01817

11	5.87117	0.08376	0.01933	0.01902	-0.12389	0.46629	0.02028
12	5.61176	-0.05758	0.07733	0.01287	0.07388	-0.89208	0.00732
13	5.86949	0.05969	0.12876	0.01191	-0.18855	-0.49356	0.00792
14	5.85401	-0.15480	0.13029	0.01535	-0.02916	0.36259	0.02687
15	5.09324	-0.27362	0.14850	0.00655	-0.21328	-0.37720	0.01061
16	5.05351	-0.39724	0.14950	0.01003	0.03906	0.48052	0.03599
17	3.67572	-0.58207	0.14046	-0.00382	-0.34182	-0.30659	0.01823
		-0.70000		0.00319			

#	S1	S2	S3	S4	S5	C1	C2
1	0.004761	0.000153	0.000005	0.000518	0.000017	0.006631	0.000213
2	0.002019	-0.000527	0.000138	0.000007	-0.000038	0.010480	-0.002736
3	-0.282264	0.023260	-0.001917	-0.001488	0.000281	-0.011067	0.000912
4	0.000438	-0.000391	0.000348	-0.000413	0.000058	-0.001127	0.001005
5	-0.085304	-0.001576	-0.000029	-0.000028	-0.000001	-0.011522	-0.000213
6	0.404905	-0.016167	0.000646	0.000575	-0.000049	0.015747	-0.000629
7	0.008630	0.000981	0.000111	0.000089	0.000023	0.006332	0.000720
8	-0.210714	-0.002759	-0.000036	-0.000100	-0.000002	-0.019330	-0.000253
9	-0.191384	0.004209	-0.000093	-0.000170	0.000006	-0.042299	0.000930
10	0.003315	-0.000297	0.000027	0.000255	-0.000025	0.014550	-0.001302
11	0.158145	0.006878	0.000299	0.000239	0.000023	0.008676	0.000377
12	-0.329923	0.002707	-0.000022	-0.000160	0.000001	-0.065459	0.000537
13	0.269588	-0.004328	0.000069	0.000304	-0.000006	0.047061	-0.000756
14	0.022441	0.001663	0.000123	0.000351	0.000035	0.008482	0.000629
15	0.154556	-0.004348	0.000122	-0.000016	-0.000003	0.007677	-0.000216
16	-0.045577	-0.003414	-0.000256	0.000809	0.000041	0.017501	0.001311
17	0.118106	-0.007021	0.000417	-0.000499	0.000005	0.008122	-0.000483
Sum	0.001738	-0.000976	-0.000046	0.000273	0.000367	0.000454	0.000046

The long separation between negative and positive lens groups in this form also allows better correction of lateral color, as may be seen in Fig. 7.10. The diffraction-limited performance, flat field, and low distortion are again apparent in Figs. 7.11 and 7.12.

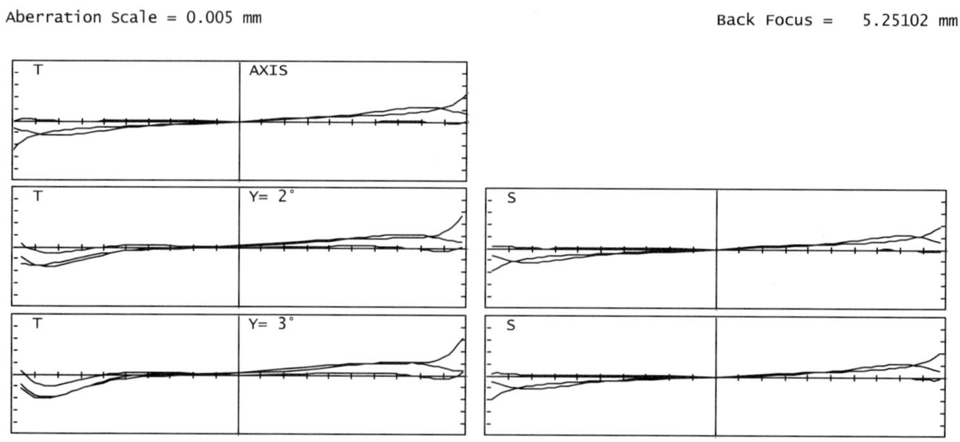

Figure 7.10 Transverse ray aberrations of inverse-telephoto-type flat-field microscope objective.

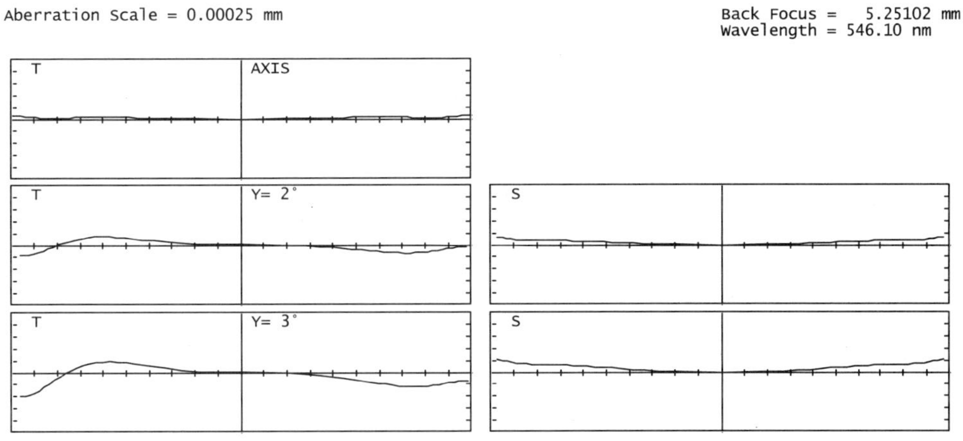

Figure 7.11 Wavefront aberrations of inverse-telephoto-type flat-field microscope objective.

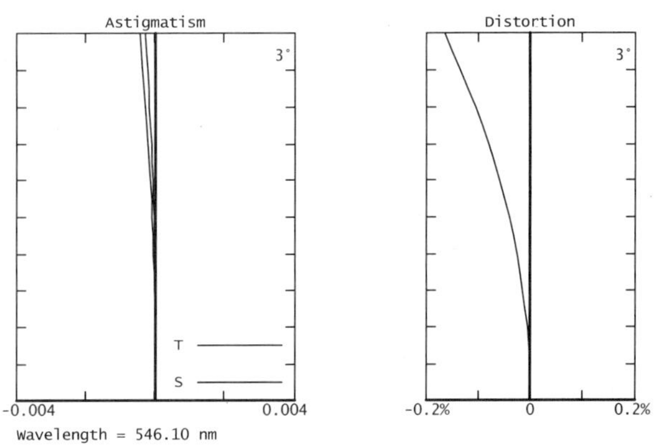

Figure 7.12 Astigmatism and distortion of inverse-telephoto-type flat-field microscope objective.

7.3 Oil-immersion objectives

If the object in immersed in a medium of refractive index n, NAs significantly greater than 1.0 are possible. Figure 7.13 shows a design based on a classical Lister objective with two aplantic elements, which has a $\sin U$ of 0.815 corresponding to $U = 55$ deg, and an NA of 1.25. The design prescription shows a BaK2 glass immersion material of $n = 1.54$, but in practice this would be an oil that matches this refractive index and dispersion as closely as possible, to minimize reflection losses and aberrations at the interface.[2] Note that the second aplanatic element is almost hemispherical. In some designs it may even be a hyper-hemisphere. Usually it is manufactured as a complete sphere, and then the flat face ground and polished afterwards.

Even though relatively normal glasses are used, Figs. 7.14 and 7.15 show an excellent level of spherical aberration and axial chromatic correction. Lateral color

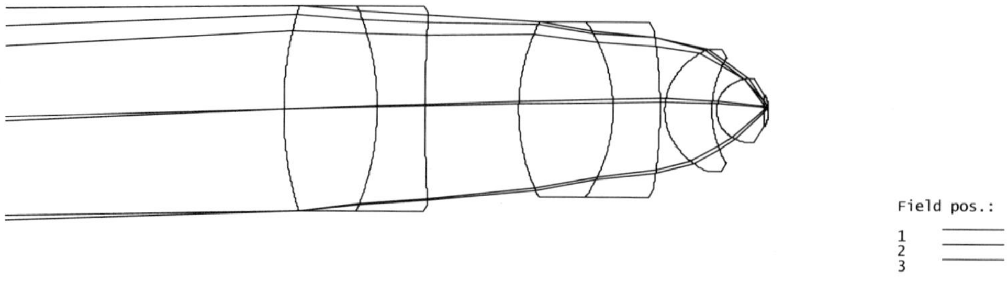

Figure 7.13 Oil-immersion objective; NA 1.25.

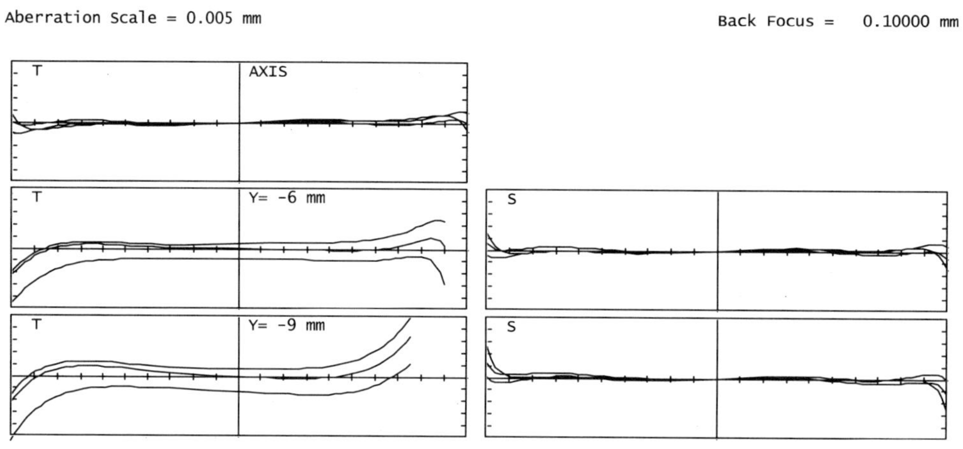

Figure 7.14 Transverse ray aberrations of oil-immersion objective.

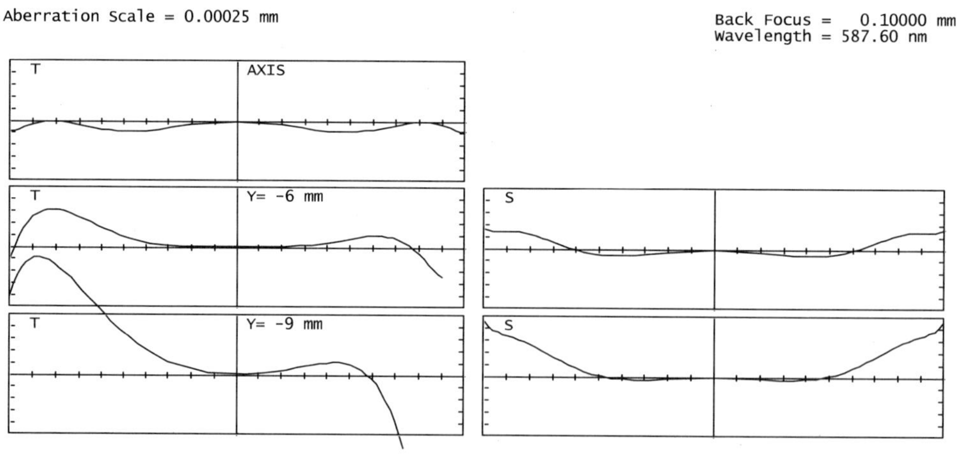

Figure 7.15 Wavefront aberrations of oil-immersion objective.

is increased by the dispersion of the immersion material. This would again have to be corrected by a compensating eyepiece or relay. The aberrations of Figs. 7.14, 7.15, and 7.16 are shown relative to the curved field in the prescription (surface

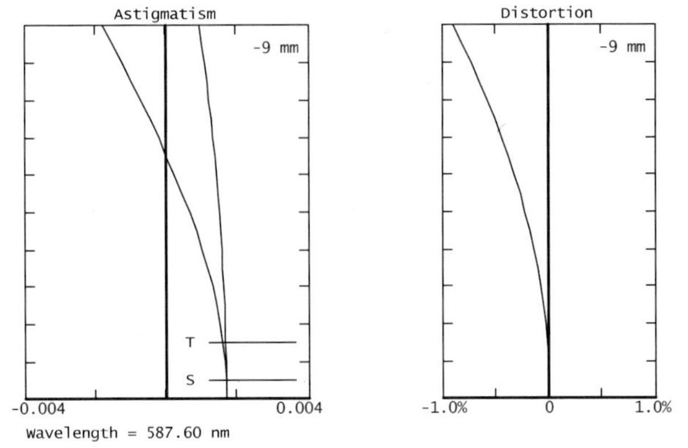

Figure 7.16 Astigmatism and distortion of oil-immersion objective.

11), as this design form is unable to correct Petzval sum. This is because the last lens surface next to immersion medium has to be plano. If a better field correction is required, oil immersion can also be used with variants of the more complex flat-field objectives decribed in the previous section.[3]

EFL = 1.947
OBJECT DISTANCE = 180, CURVATURE = 0, TILT ANGLE = 0
WAVELENGTHS [nm]
587.60 656.30 486.10
SURFACES

# SURF SPACE	RADIUS	SEPN	INDEX1	V	CLR RAD	GLASS
0		0.00000	1.000000			
1#S	8.53045	2.00000	1.607379	56.63	2.344	S-SK2
2 S	-6.50077	1.00000	1.755196	27.57	2.344	S-SF4
3 S	35.52938	2.00000	1.000000		2.200	
4 S	4.74007	2.00000	1.516798	64.14	1.995	S-BK7
5 S	-3.71742	1.00000	1.717355	29.50	1.995	S-SF1
6 S	-11.79206	0.10000	1.000000		1.800	
7 S	1.50685	1.00000	1.516798	64.14	1.381	S-BK7
8 S	2.50035	0.10000	1.000000		1.200	
9 S	0.72680	1.00000	1.539958	59.68	0.725	S-BAK2
10 S	Plane	0.10000	1.539958	59.68	0.350	S-BAK2
11 S	-0.85499				0.350	

LAGRANGE INVARIANT = -0.1170

#	H	U	HBAR	UBAR	D(U/N)	A	ABAR
1	2.34000	0.01300	0.00000	0.05000	-0.07245	0.28731	0.05000
2	2.14887	-0.09557	0.06221	0.03111	0.02545	-0.68494	0.03462
3	2.08919	-0.05968	0.09151	0.02929	-0.02634	-0.00154	0.05594
4	1.96850	-0.06034	0.19823	0.05336	-0.05917	0.35495	0.09518
5	1.60595	-0.18128	0.24009	0.02093	0.05566	-0.93023	-0.06621
6	1.49629	-0.10966	0.26612	0.02603	-0.21549	-0.40624	0.00594
7	1.46835	-0.27935	0.26897	0.02851	-0.06096	0.69511	0.20701
8	0.95217	-0.51618	0.22695	-0.04202	-0.24583	-0.20532	0.07394
9	0.89356	-0.58614	0.22526	-0.01683	0.05904	0.64331	0.29311
10	0.08186	-0.81170	0.10566	-0.11960	0.00000	-1.24999	-0.18418
		-0.81170		-0.11960			

#	S1	S2	S3	S4	S5	C1	C2
1	0.013995	0.002436	0.000424	0.000606	0.000179	0.004486	0.000781
2	-0.025660	0.001297	-0.000066	-0.000110	0.000009	-0.013150	0.000665
3	0.000000	-0.000005	0.000172	-0.000166	-0.000232	0.000050	-0.001824
4	0.014675	0.003935	0.001055	0.000984	0.000547	0.003712	0.000995
5	-0.077350	-0.005506	-0.000392	-0.000284	-0.000048	-0.013217	-0.000941
6	0.053212	-0.000779	0.000011	0.000485	-0.000007	0.008607	-0.000126
7	0.043251	0.012881	0.003836	0.003095	0.002064	0.005422	0.001615
8	0.009868	-0.003554	0.001280	-0.001865	0.000211	0.001038	-0.000374
9	-0.021834	-0.009948	-0.004533	0.006604	0.000944	0.003377	0.001539
10	0.000000	0.000000	0.000000	0.000000	0.000000	0.000000	0.000000
Sum	0.010157	0.000757	0.001788	0.009350	0.003666	0.000325	0.002329

References

1. W.T. Welford, *Geometrical Optics, Optical Instrumentation*, North-Holland Publishing (1962).
2. R.P. Cargille, "Microscope immersion oils," Data Sheet IO-1260, Cargille Labs, Cedar Grove, NJ (1989).
3. B.G. Broome, "Microscope objectives and their evolution to optical disc objectives," in *Lens Design,* Warren Smith, Ed., SPIE Press, Bellingham, WA (1992).

CHAPTER 8
MICROLITHOGRAPHIC PROJECTION OPTICS

Microlithographic projection optics image a pattern on a reticle, or mask, onto the surface of a silicon wafer. This pattern selectively exposes a photoresist on the wafer, which is subsequently etched to form one of many layers that build the structure of a microchip. The image on the wafer may be either the same size as the reticle, or a reduced copy of it, depending on the magnification of the projection lens.

The minimum feature size that can be resolved is given by a modified form of Rayleigh's equation:

$$Linewidth = k_1 \frac{\lambda}{\mathrm{NA}}. \tag{8.1}$$

This differs from the formula for microscopy because microlithography is primarily concerned with the imaging of extended patterns of lines, rather than adjacent points. The empirical variable k_1 depends on the details of the lithographic process, such as photoresist, illumination conditions, periodicity of the lines, and whether the imaged features include phase objects. In the early decades of microlithography, it had a lower limit of 0.8.[1] However, lithographic technology in the 1990s has been able to utilize lower image contrasts, allowing a reduction of $\underline{k}_1$ toward 0.25.[2]

8.1 Unit-magnification zero-power monocentric systems

The designs considered in this section are monocentric zero-power (and therefore telecentric) unit-magnification relays. Before considering these designs, we need to discuss monocentric systems in general. It is not difficult to see that a monocentric system has no Seidel coma, astigmatism, or transverse chromatic aberration. It can have spherical aberration and longitudinal chromatic aberration, and it will have a curved field, which will be concentric with all of the refracting and reflecting surfaces. Distortion will be zero in the sense that the image size, measured along the curved image surface will be proportional to the field angle.

If a monocentric system has zero power with an object at infinity, the image will be at infinity and will therefore be flat, since it must be monocentric. However, S_4 does not depend on the object position, so S_4 will be zero at all conjugates. If, therefore, we have a monocentric system of this type, because it has zero power, it will have zero S_4 and it will be telecentric. Since the object is at the common center of curvature of the system, there is no spherical aberration or longitudinal chromatic aberration at all, and there is no coma, distortion, or lateral color because the system is symmetric before and after the stop.

Several fifth-order aberrations are also automatically zero, in particular coma, distortion, sagittal astigmatism, and sagittal oblique spherical aberration (SOBSA); but fifth-order tangential astigmatism is not corrected and is a limiting factor in almost all cases. Tangential oblique spherical aberration (TOBSA) can also be significant at higher NAs.

One practical problem is that the image and the object are coincident and these systems can therefore only be used if the object and image occupy the opposite sides of the axis, or a beamsplitter is used. While this is a serious restriction, it does not prevent the use of these designs for microlithographic reproduction. Several different systems use this concept. All of them use reflecting surfaces, although, as we will see, several also use refracting components.

8.1.1 Dyson 1× relay

This is the first of the monocentric zero-power unit magnification designs (1959) and is shown in Fig. 8.1.[3] Note that all of the Seidel aberrations are zero. Since it uses a refracting surface (twice), there will inevitably be chromatic variation of S_4, and this can restrict the waveband that can be used. However, for microlithographic applications, this restriction is not prohibitive, since the light source can have a relatively narrow spectral bandwidth. This is typically one or more spectral lines of a mercury arc lamp, or an excimer laser.

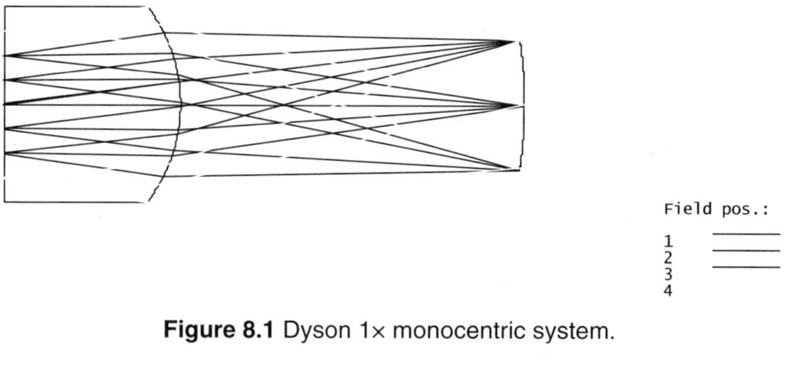

Figure 8.1 Dyson 1× monocentric system.

```
EFL = -4.08e+008
OBJECT DISTANCE = 1e-005, CURVATURE = 0, TILT ANGLE = 0
WAVELENGTHS  [nm]
     400.00
SURFACES
```

# SURF SPACE	RADIUS	SEPN	INDEX1	V	CLR RAD	GLASS
0		0.00000	1.000000			
1 S	Plane	34.34012	1.523000		20.000	User-USER
2 S	-34.34012	65.65988	1.000000		20.000	
3#SM	-100.00000	-65.65988	1.000000		13.132	

4	S	-34.34012	-34.34012	1.523000	20.000	User-USER
5	S	Plane	0.00001	1.000000	20.000	
6	S	Plane			15.324	

LAGRANGE INVARIANT = 3.0000

#	H	U	HBAR	UBAR	D(U/N)	A	ABAR
1	0.00000	0.20000	15.00000	0.06888	-0.11378	0.20000	0.06888
2	4.50954	0.13132	16.55302	0.04522	0.04510	-0.00000	-0.66526
3	13.13198	0.13132	4.52246	-0.18323	-0.26264	-0.00000	-0.22845
4	4.50953	0.13132	-13.44698	0.27367	0.04510	-0.00000	-0.66526
5	-0.00000	0.13132	-15.00000	0.04522	-0.11378	-0.20000	-0.06888
		0.20000		0.06888			

#	S1	S2	S3	S4	S5	C1	C2
1	0.000000	0.000000	0.000000	0.000000	0.000000	0.000000	0.000000
2	-0.000000	-0.000000	-0.090000	0.090000	0.094579	0.000000	0.000000
3	0.000000	0.000000	0.180000	-0.180000	0.123979	0.000000	0.000000
4	-0.000000	-0.000000	-0.090000	0.090000	-0.218557	0.000000	0.000000
5	-0.000000	-0.000000	-0.000000	0.000000	-0.000000	0.000000	0.000000
Sum	-0.000000	-0.000000	0.000000	-0.000000	0.000000	0.000000	0.000000

We can see from the aberration curves of Figs. 8.2 and 8.3 that at NA 0.2 there is no significant spherical aberration, coma, or sagittal astigmatism, and that the tangential astigmatism is predominantly fifth-order. Note that the maximum transverse ray aberration at 15 mm off-axis is about 0.16 mm, while at 10 mm it is about 0.03 mm. If the aberration is fifth-order it will be proportional to (object size)^4; the ratio of the object sizes is 0.66 and $0.66^4 \times 0.16$ mm = 0.03 mm, which is correct.

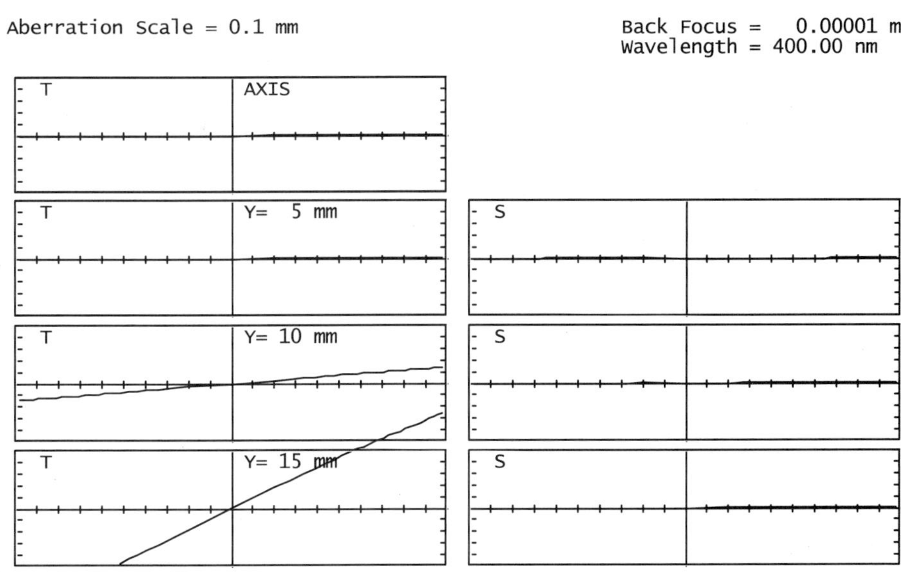

Figure 8.2 Transverse ray aberrations of Dyson system.

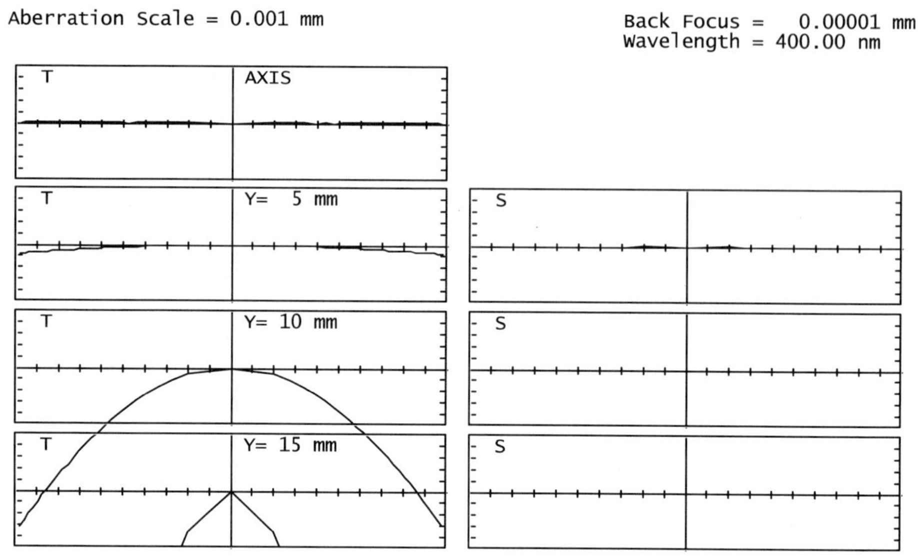

Figure 8.3 Wavefront aberrations of Dyson system.

8.1.2 Offner 1× relay

The Offner 1× relay is the only purely reflecting, or catoptric, monocentric 1× relay, and is shown in Fig. 8.4.[4,5] It has the same general properties of all monocentric designs, in particular zero Seidel aberrations, but has the advantage of a large air space next to the object/image plane, as well as a complete absence of chromatic aberrations.

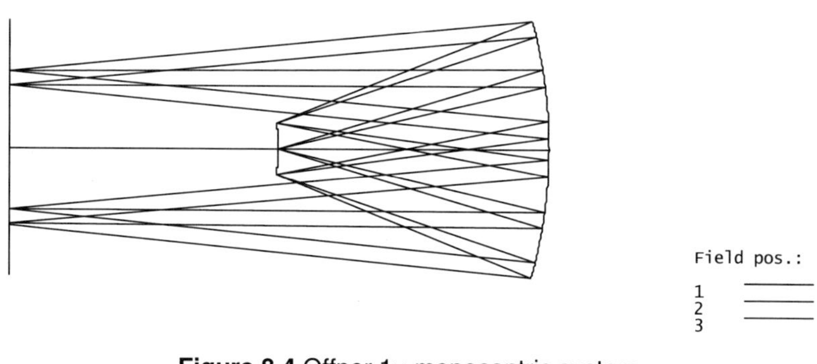

Figure 8.4 Offner 1× monocentric system.

```
EFL undefined
OBJECT DISTANCE = 100, CURVATURE = 0, TILT ANGLE = 0
WAVELENGTHS   [nm]
      400.00
SURFACES
  # SURF SPACE    RADIUS        SEPN       INDEX1     V     CLR RAD  GLASS
  0                           0.00000    1.000000
  1 SM          -100.00000  -50.00000    1.000000            25.000
```

2#SM	-50.00000	50.00000	1.000000	5.000
3 SM	-100.00000	-100.00000	1.000000	25.000
4 S	Plane			25.000

LAGRANGE INVARIANT = 1.5000

#	H	U	HBAR	UBAR	D(U/N)	A	ABAR
1	10.00000	0.10000	15.29284	0.00293	-0.20000	0.00000	-0.15000
2	5.00000	0.10000	0.14642	0.30293	0.20000	0.00000	-0.30000
3	10.00000	0.10000	-14.70716	-0.29707	-0.20000	0.00000	-0.15000
		0.10000		0.00293			

#	S1	S2	S3	S4	S5	C1	C2
1	0.000000	0.000000	0.045000	-0.045000	0.137636	0.000000	0.000000
2	0.000000	0.000000	-0.090000	0.090000	-0.005271	0.000000	0.000000
3	0.000000	0.000000	0.045000	-0.045000	-0.132365	0.000000	0.000000
Sum	0.000000	0.000000	-0.000000	0.000000	0.000000	0.000000	0.000000

Figures 8.5 and 8.6 show that the Offner design at NA 0.1 also has the characteristic of residual fifth-order tangential astigmatism.

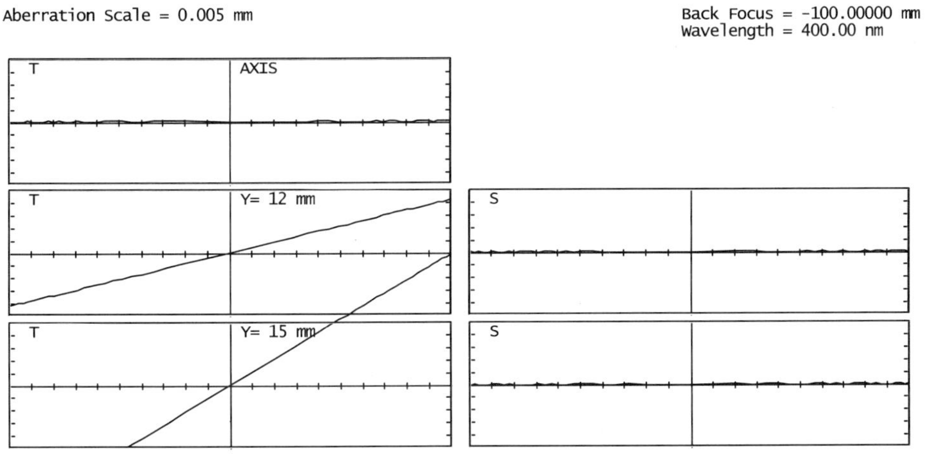

Figure 8.5 Transverse ray aberrations of Offner 1× relay.

8.2 Wynne-Dyson 1× relay

Wynne and Offner have both pointed out that the fifth-order tangential astigmatism of the Dyson and Offner systems is directly related to the spherical aberration of the chief ray, and the fact that the system is therefore not exactly telecentric.[6,7,8] Wynne proposed a means of reducing this pupil spherical aberration by introducing cemented interfaces between glasses of different refractive indices, in place of the thick single lens of the Dyson concept.[7] This introduced some third-order aberrations, which were balanced against the smaller residual fifth-order aberrations, including the spherical aberration of the marginal ray introduced by the need for a

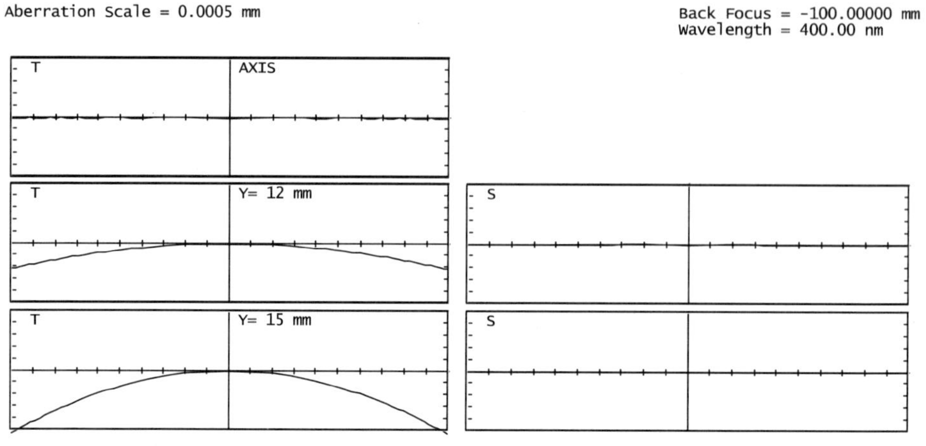

Figure 8.6 Wavefront aberrations of Offner 1× relay.

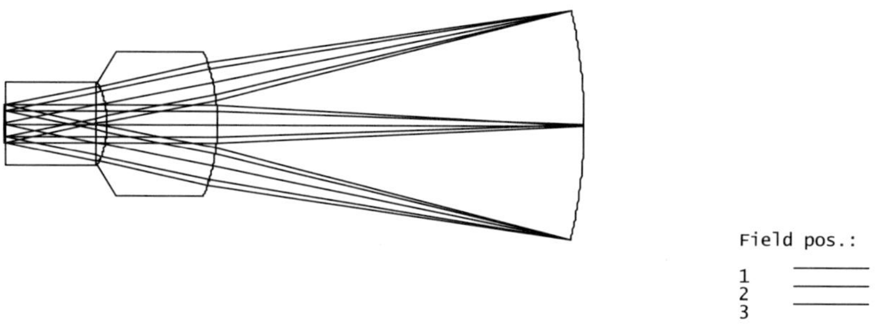

Figure 8.7 Wynne-Dyson 1× relay.

finite working distance at the reticle and wafer. The glasses also had different dispersions to optimize the residual chromatic aberration correction. Such a design is shown in Fig. 8.7.

```
EFL = -23473.838
OBJECT DISTANCE = 0.33, CURVATURE = 0, TILT ANGLE = 0
WAVELENGTHS  [nm]
      435.80       486.00       405.00
SURFACES
```

#	SURF SPACE	RADIUS	SEPN	INDEX1	V	CLR RAD	GLASS
0			0.00000	1.000000			
1	S	Plane	83.80000	1.683282		40.000	S-LAK23
2	S	Plane	10.16000	1.672564		40.000	S-LAK11
3	S	-94.00000	101.00000	1.761634		40.000	S-SF53
4	S	-194.80000	330.00000	1.000000		70.000	
5	#SM	-525.00000	-330.00000	1.000000		109.808	
6	S	-194.80000	-101.00000	1.761634		70.000	S-SF53
7	S	-94.00000	-10.16000	1.672564		40.000	S-LAK11
8	S	Plane	-83.80000	1.683282		40.000	S-LAK23
9	S	Plane	-0.24692	1.000000		40.000	
10	S	Plane				17.978	

LAGRANGE INVARIANT = 6.2300

#	H	U	HBAR	UBAR	D(U/N)	A	ABAR
1	0.11550	0.35000	17.80004	0.00011	-0.22648	0.35000	0.00011
2	17.53979	0.20793	17.80568	0.00007	0.00159	0.35000	0.00011
3	19.66587	0.20926	17.80637	0.00007	-0.00633	0.00008	-0.31672
4	40.80084	0.20926	18.78022	0.00964	0.09032	-0.00034	-0.15285
5	109.80724	0.20911	0.15450	-0.05644	-0.41831	-0.00005	-0.05674
6	40.77025	0.20920	-18.66545	0.05703	0.09043	0.00009	-0.15285
7	19.63683	0.20924	-17.75107	-0.00905	-0.00634	-0.00060	-0.31672
8	17.51074	0.20926	-17.75636	0.00052	0.00159	-0.35000	-0.00087
9	0.08642	0.20793	-17.79975	0.00052	-0.22648	-0.35000	-0.00087
		0.35000		0.00087			

#	S1	S2	S3	S4	S5	C1	C2
1	0.003204	0.000001	0.000000	0.000000	0.000000	0.000277	0.000000
2	-0.003412	-0.000001	-0.000000	0.000000	-0.000000	-0.000342	-0.000000
3	0.000000	-0.000003	0.012482	-0.012482	0.000582	0.000015	-0.058780
4	-0.000000	-0.000191	-0.086100	0.086143	0.019328	0.000225	0.101276
5	0.000000	0.000121	0.147859	-0.147859	0.000416	0.000000	0.000000
6	-0.000000	0.000051	-0.086131	0.086143	-0.019768	0.000060	-0.101200
7	0.000000	0.000024	0.012481	-0.012482	-0.000527	0.000111	0.058694
8	-0.003407	-0.000008	-0.000000	0.000000	-0.000000	-0.000342	-0.000001
9	0.002398	0.000006	0.000000	0.000000	0.000000	0.000207	0.000001
Sum	-0.001218	-0.000002	0.000591	-0.000537	0.000030	0.000211	-0.000011

Residual aberrations of this design are shown at NA 0.35 in Figs. 8.8 and 8.9. Note the small sagittal astigmatism and oblique spherical aberration (SOBSA), while residual fifth-order tangential astigmatism and oblique spherical aberration (TOBSA) are much smaller than the original Dyson design, giving a more uniform image quality across the field. The chromatic variation of tangential astigmatism is also apparent in Fig. 8.8.

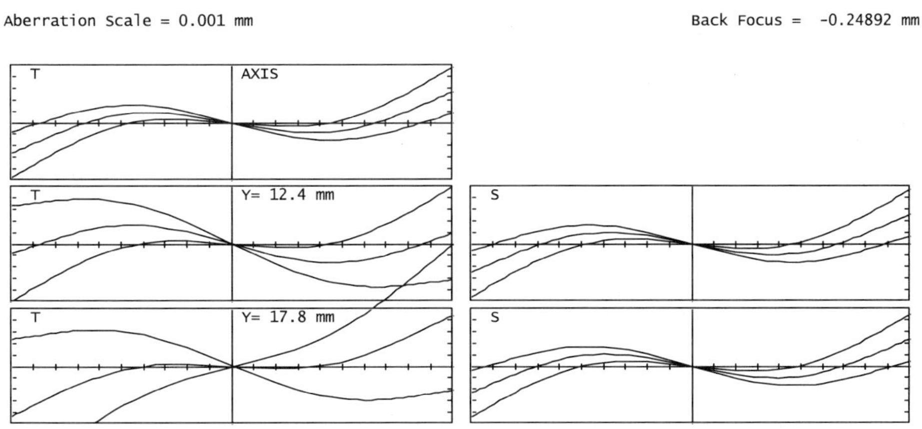

Figure 8.8 Transverse ray aberrations of Wynne-Dyson 1× relay.

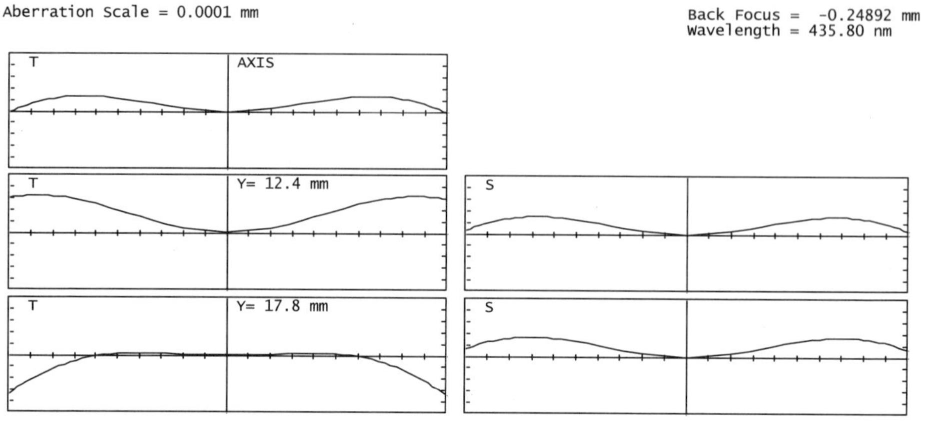

Figure 8.9 Wavefront aberrations of Wynne-Dyson 1× relay.

8.3 Wynne-Offner 1× relay

Offner applied similar reasoning to his original two-mirror monocentric design by departing from exact concentricity and adding one or two low-power meniscus elements to reduce chief ray spherical aberration, in a similar way to that used for spherical aberration of the marginal ray in Maksutov-Bowers catadioptric designs.[6] Again, residual third-order astigmatism was introduced, balanced against the remaining fifth-order astigmatism, which gave acceptable image quality over a narrow off-axis annular field, past which the reticle and wafer were scanned in synchronism to expose the whole wafer.

Two variations on this theme have also been described by Wynne.[9] They modify the basic Offner design by adding refracting surfaces, while maintaining the monocentric and zero-power conditions. In one case the glass is used to support the large mirror, while in the other case it supports the small mirror. The second case is shown in Fig. 8.10.

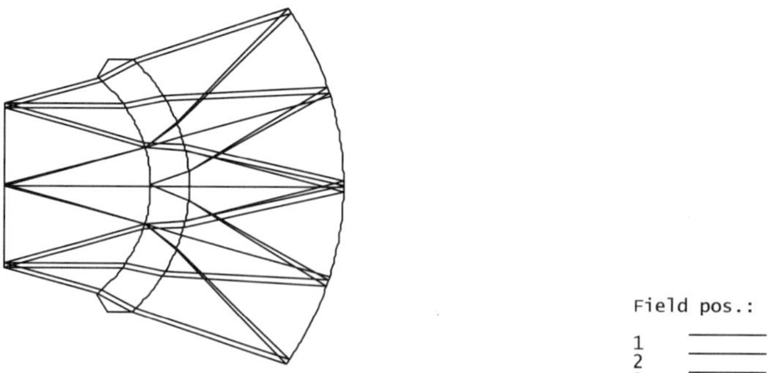

Figure 8.10 Wynne-Offner 1× relay.

EFL = -12114.924
OBJECT DISTANCE = 100, CURVATURE = 0, TILT ANGLE = 0
WAVELENGTHS [nm]
262.80 263.30 262.30
SURFACES

#	SURF SPACE	RADIUS	SEPN	INDEX1	V	CLR RAD	GLASS
0			0.00000	1.000000			
1	S	-100.00000	26.22000	1.501071		78.000	G-SILICA
2	S	-126.22000	104.87600	1.000000		91.000	
3	SM	-231.09600	-104.87600	1.000000		128.000	
4	S	-126.22000	-26.22000	1.501071		91.000	G-SILICA
5#	SM	-100.00000	26.22000	1.501071		78.000	G-SILICA
6	S	-126.22000	104.87600	1.000000		91.000	
7	SM	-231.09600	-104.87600	1.000000		128.000	
8	S	-126.22000	-26.22000	1.501071		91.000	G-SILICA
9	S	-100.00000	-100.00000	1.000000		78.000	
10	S	Plane				59.590	

LAGRANGE INVARIANT = 16.5200

#	H	U	HBAR	UBAR	D(U/N)	A	ABAR
1	28.00000	0.28000	58.99648	-0.00004	-0.09347	0.00000	-0.59000
2	35.34160	0.28000	64.15952	0.19691	0.09347	-0.00000	-0.46744
3	64.70688	0.28000	68.44655	0.04088	-0.56000	0.00000	-0.25531
4	35.34160	0.28000	10.60875	0.55149	0.09347	0.00000	-0.46744
5	28.00000	0.28000	0.23998	0.39545	0.37307	0.00000	-0.59000
6	35.34160	0.28000	-10.00294	-0.39065	0.09347	0.00000	-0.46744
7	64.70688	0.28000	-67.33738	-0.54669	-0.56000	0.00000	-0.25531
8	35.34160	0.28000	-63.55371	-0.03608	0.09347	0.00000	-0.46744
9	28.00000	0.28000	-58.51652	-0.19211	-0.09347	0.00000	-0.59000
		0.28000		0.00484			

#	S1	S2	S3	S4	S5	C1	C2
1	0.000000	0.000000	0.910999	-0.910999	0.640553	0.000000	-0.004958
2	-0.000000	-0.000000	-0.721755	0.721755	-0.612949	0.000000	0.004958
3	0.000000	-0.000000	2.361879	-2.361879	4.996764	0.000000	0.000000
4	-0.000000	0.000000	-0.721755	0.721755	-2.440925	0.000000	-0.004958
5	-0.000000	0.000000	-3.636210	3.636210	-0.062330	0.000000	0.000000
6	-0.000000	0.000000	-0.721755	0.721755	2.416181	-0.000000	0.004958
7	0.000000	-0.000000	2.361879	-2.361879	-4.915791	0.000000	0.000000
8	-0.000000	0.000000	-0.721755	0.721755	0.588205	0.000000	-0.004958
9	0.000000	-0.000000	0.910999	-0.910999	-0.609321	-0.000000	0.004958
Sum	0.000000	-0.000000	0.022527	-0.022527	0.000386	0.000000	0.000000

The aberration plots in Figs. 8.11 and 8.12 again show, at NA 0.28, some residual tangential astigmatism and oblique spherical aberration, as well as chromatic variation of tangential astigmatism over a spectral band of 1 nm in the deep ultraviolet, where fused silica is one of the few available transmitting materials.

8.4 Reduction lenses

For feature sizes less than about 1 μm, reticle manufacturers prefer the objective lens to reduce the size of the features projected from the reticle to wafer. This

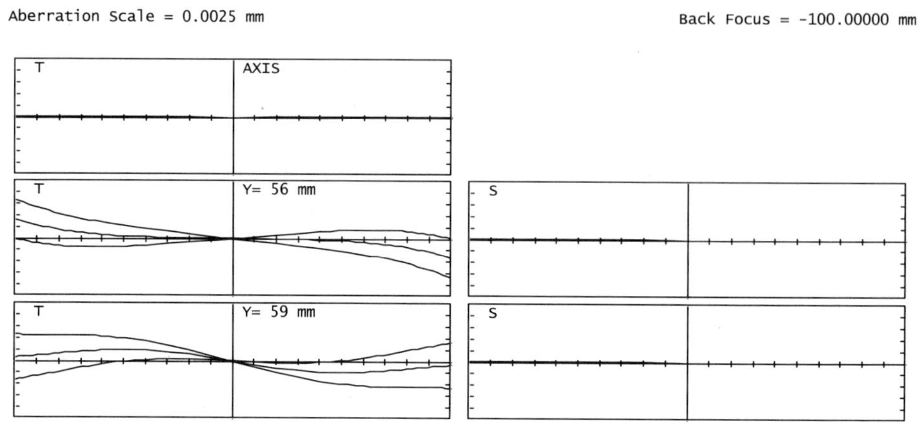

Figure 8.11 Transverse ray aberrations of Wynne-Offner 1× relay.

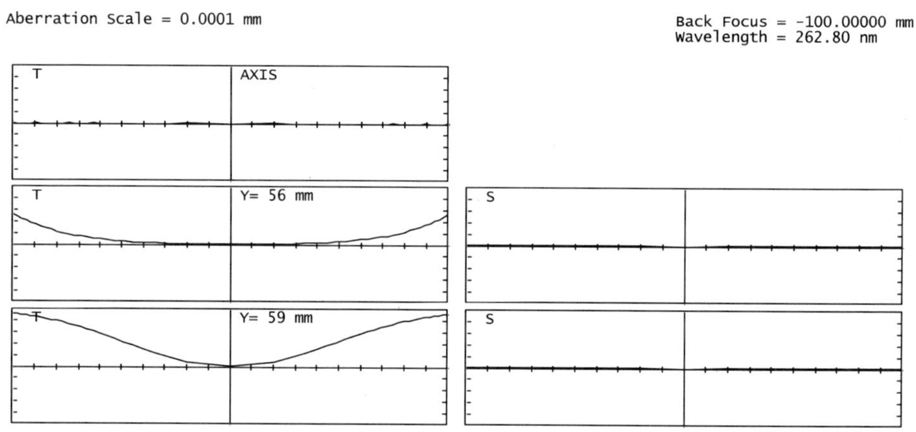

Figure 8.12 Wavefront aberrations of Wynne-Offner 1× relay.

allows looser tolerances on feature sizes and placements on the reticle by a factor equal to the reduction ratio. This is typically 4× or 5×, appreciably lower than the magnification of a microscope objective, since the field size on the wafer may be as large as 26 × 34 mm, and the reticle field size should fit within a standard glass blank size of 152.4 × 152.4 mm.

Relaying an image at other than 1× magnification makes it difficult to conceive of a monocentric optical design, since the object and image cannot coincide. Odd aberrations such as coma and distortion therefore have to be corrected in the same way that they are for other lenses, such as photographic or microscopic objectives. This approach leads to a considerable increase in complexity and expense, but it turns out that over the operational life of a lens the savings in reticle cost are much more significant than the price of the lens.

In 1980, Glatzel[10] described the evolution of early reduction microlithographic lenses from photographic lenses, where a considerable reduction in residual aber-

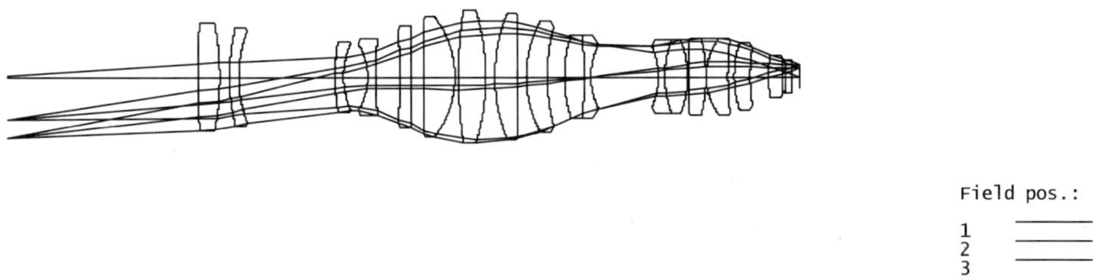

Figure 8.13 Braat's 5× reduction lens.

rations was achieved by the principles that we have already discussed in Chapter 2. The design was "relaxed" in order to reduce intrinsic and induced fifth- and higher-order aberrations, simply by splitting power and bending lens elements for minimum angles of incidence, along with an increase in lens diameters and overall track length to reduce element optical powers.

Although Glatzel did not publish his lens prescriptions, in 1987, Braat described a similar 5× NA 0.38 reduction design, shown here in Fig. 8.13.[1] It may be thought of as a basic double-Gauss form, with an inverse telephoto front end and a Petzval lens back end. It therefore has something in common with both the photographic and microscopic objective designs that we have seen in previous chapters. This is reasonable, since the lens must have the diffraction-limited resolution of a microscope objective over the large image field size of a photographic lens. In addition, for microlithography the distortion and field curvature have to be corrected to much smaller levels than is typical in microscopy. This is because the projection lens has to accurately reproduce the dimensional structure of the reticle over the entire image area and depth of focus, without a loss of image quality or pattern fidelity from the optical axis to the edge of the field. Critical features are just as likely to be at the extreme edge of the field as the center. The reticle and wafer must be flatter than the Rayleigh depth of focus, and no refocusing is allowed across the field. This is perhaps the most stringent image quality requirement of any type of optical system that we have encountered.

It is interesting to observe, in passing, that the forms of Glatzel's and Braat's designs are quite similar, in terms of power distribution, to Shafer's monochromatic lens with zero third- and fifth-order aberrations that we saw in Fig. 2.6.

```
EFL =      77.427
OBJECT DISTANCE = 144.235, CURVATURE = 0, TILT ANGLE = 0
WAVELENGTHS   [nm]
      405.00       409.00       401.00
SURFACES
```

#	SURF SPACE	RADIUS	SEPN	INDEX1	V	CLR RAD	GLASS
0			0.00000	1.000000			
1	S	563.31681	17.00000	1.622470		43.679	S-F8
2	S	-257.96333	7.67380	1.000000		43.018	
3	S	187.80366	7.00000	1.498902		39.845	S-FK5
4	S	113.10068	72.83400	1.000000		37.914	
5	S	180.83281	6.50000	1.637678		28.501	S-SK16

6 S	70.79746	18.50000	1.000000	27.095	
7 S	-54.66632	6.50000	1.682203	27.216	S-SF2
8 S	574.57725	18.22150	1.000000	30.656	
9 S	-405.48705	8.00000	1.650529	38.002	S-F2
10 S	2906.04748	9.90510	1.000000	41.126	
11 S	Plane	24.00000	1.498902	46.660	S-FK5
12 S	-83.47524	2.50000	1.000000	49.414	
13 S	373.42694	21.00000	1.637678	53.610	S-SK16
14 S	-203.04362	7.00000	1.000000	54.138	
15 S	144.52078	17.00000	1.637678	52.000	S-SK16
16 S	1.27910e+004	6.50000	1.000000	50.504	
17 S	99.66910	16.00000	1.650529	44.740	S-F2
18 S	159.78959	13.02880	1.000000	40.483	
19 S	117.27014	15.00000	1.498902	34.041	S-FK5
20 S	-379.42313	5.00000	1.682203	34.040	S-SF2
21 S	58.95879	6.80700	1.000000	25.951	
22#S	Plane	43.80000	1.000000	25.763	
23 S	-67.28343	5.00000	1.682203	25.143	S-SF2
24 S	93.71369	17.00000	1.498902	29.370	S-FK5
25 S	-90.73587	1.00000	1.000000	29.365	
26 S	2148.92017	13.00000	1.629822	30.467	S-BAF4
27 S	-131.38602	0.10000	1.000000	31.401	
28 S	54.94324	16.50000	1.637678	31.622	S-SK16
29 S	271.14967	4.82620	1.000000	29.589	
30 S	94.27736	12.00000	1.637678	27.097	S-SK16
31 S	142.42316	13.99680	1.000000	23.803	
32 S	79.34619	10.40000	1.642338	17.655	S-SSK2
33 S	182.43415	3.26280	1.000000	14.480	
34 S	-75.06042	5.00000	1.682203	13.669	S-SF2
35 S	-741.14700	5.99887	1.000000	12.362	
36 S	Plane			10.005	

LAGRANGE INVARIANT = -3.8000

#	H	U	HBAR	UBAR	D(U/N)	A	ABAR
1	10.96185	0.07600	-32.60177	0.12062	-0.05173	0.09546	0.06275
2	11.63125	0.03938	-30.96042	0.09655	0.01155	-0.00927	0.35138
3	11.90613	0.03582	-29.18502	0.23136	-0.03395	0.09922	0.07596
4	11.92570	0.00280	-27.74249	0.20608	0.05493	0.16224	-0.05878
5	16.06251	0.05680	-14.15806	0.18651	-0.05674	0.14562	0.10822
6	16.06313	0.00010	-13.21963	0.14437	0.14478	0.37173	-0.06936
7	18.74263	0.14484	-11.04832	0.11737	-0.01100	-0.19802	0.31947
8	20.20605	0.22514	-11.12756	-0.01219	0.26889	0.43789	-0.05309
9	27.54431	0.40273	-11.74200	-0.03372	-0.23867	0.33480	-0.00476
10	29.71048	0.27077	-11.99674	-0.03184	0.28952	0.46379	-0.05937
11	34.20309	0.45357	-12.54394	-0.05524	-0.25169	0.45357	-0.05524
12	41.46547	0.30260	-13.42848	-0.03686	0.00386	-0.29100	0.18588
13	41.97982	0.20574	-13.36595	0.02501	-0.15576	0.31816	-0.01078
14	43.69882	0.08186	-12.75252	0.02921	-0.05317	-0.21840	0.15070
15	43.67653	-0.00318	-12.13730	0.08789	-0.06986	0.29903	0.00391
16	41.64296	-0.11962	-10.66905	0.08637	-0.12078	-0.19057	0.14008
17	40.38310	-0.19383	-9.75313	0.14091	0.02593	0.21135	0.04306
18	35.94911	-0.27712	-7.77008	0.12394	-0.14315	-0.08607	0.12431
19	31.89655	-0.31105	-5.51694	0.17293	0.11220	-0.03905	0.12589
20	27.42583	-0.29805	-3.55145	0.13103	0.04566	-0.55509	0.21044
21	26.13736	-0.25769	-2.97277	0.11574	0.02213	0.31225	0.10987
22	25.24522	-0.13106	-1.88166	0.16029	0.00000	-0.13106	0.16029

23	19.50466	-0.13106	5.13915	0.16029	0.15463	-0.42095	0.08391
24	19.70291	0.03965	5.77046	0.12626	0.02327	0.42038	0.31598
25	20.89649	0.07021	8.30743	0.14923	-0.05650	-0.23996	0.08645
26	20.88683	-0.00966	8.48544	0.17801	0.00372	0.00006	0.18196
27	20.76095	-0.00968	9.88546	0.10769	-0.10936	-0.27332	0.05289
28	20.74942	-0.11530	9.89827	0.12813	-0.01748	0.26235	0.30829
29	17.16141	-0.21746	10.03181	0.00809	-0.18298	-0.25247	0.07384
30	15.63747	-0.31576	10.20963	0.03685	0.15859	-0.14990	0.14514
31	12.54872	-0.25740	9.97361	-0.01967	-0.20818	-0.27724	0.08247
32	7.43503	-0.36535	10.14780	0.01245	0.20758	-0.27164	0.14034
33	4.74034	-0.25910	9.70640	-0.04244	-0.25108	-0.38286	0.01768
34	3.40636	-0.40885	9.59047	-0.03553	0.27531	-0.45423	-0.16330
35	2.28317	-0.22464	9.74395	0.03070	-0.24645	-0.38307	0.02952
		-0.37999		0.04267			

#	S1	S2	S3	S4	S5	C1	C2
1	0.005167	0.003397	0.002233	0.009835	0.007932	0.001434	0.000943
2	-0.000012	0.000438	-0.016588	0.021476	-0.185304	0.000148	-0.005601
3	0.003980	0.003047	0.002332	0.025592	0.021378	0.000681	0.000521
4	-0.017244	0.006247	-0.002263	-0.042496	0.016216	-0.001115	0.000404
5	0.019327	0.014363	0.010673	0.031093	0.031038	0.001900	0.001412
6	-0.321354	0.059958	-0.011187	-0.079419	0.016905	-0.004850	0.000905
7	0.008084	-0.013043	0.021042	-0.107123	0.138879	-0.006409	0.010340
8	-1.041807	0.126301	-0.015312	-0.010192	0.003092	-0.015279	0.001852
9	0.736885	-0.010482	0.000149	-0.014036	0.000198	0.014184	-0.000202
10	-1.850219	0.236855	-0.030321	-0.001958	0.004132	-0.021194	0.002713
11	1.770944	-0.215698	0.026272	0.000000	-0.003200	0.008941	-0.001089
12	-0.013558	0.008661	-0.005532	0.057577	-0.033245	0.006954	-0.004442
13	0.661883	-0.022424	0.000760	0.015057	-0.000536	0.010848	-0.000368
14	0.110826	-0.076469	0.052762	0.027692	-0.055512	0.007752	-0.005348
15	0.272836	0.003563	0.000047	0.038905	0.000509	0.010608	0.000139
16	0.182663	-0.134265	0.098690	-0.000440	-0.072218	0.006445	-0.004738
17	-0.046764	-0.009527	-0.001941	0.057102	0.011237	0.013128	0.002674
18	0.038121	-0.055057	0.079518	-0.035617	-0.063405	0.004759	-0.006874
19	-0.005459	0.017596	-0.056719	0.040985	0.050720	-0.000718	0.002314
20	-0.385813	0.146263	-0.055448	-0.002767	0.022069	-0.017515	0.006640
21	-0.056386	-0.019840	-0.006981	-0.099324	-0.037405	-0.014094	-0.004959
22	0.000000	0.000000	0.000000	0.000000	0.000000	0.000000	0.000000
23	-0.534447	0.106536	-0.021237	-0.087035	0.021583	-0.014178	0.002826
24	-0.081024	-0.060903	-0.045779	-0.011202	-0.042830	-0.009529	-0.007163
25	0.067983	-0.024493	0.008824	0.052970	-0.022263	0.002890	-0.001041
26	-0.000000	-0.000001	-0.002571	0.002597	0.076349	0.000002	0.004588
27	0.169607	-0.032824	0.006352	0.042471	-0.009449	0.006850	-0.001326
28	0.024965	0.029337	0.034474	0.102335	0.160765	0.004421	0.005195
29	0.200162	-0.058544	0.017123	-0.020736	0.001057	0.003519	-0.001029
30	-0.055722	0.053954	-0.052242	0.059639	-0.007163	-0.001904	0.001843
31	0.200787	-0.059730	0.017769	-0.039478	0.006458	0.002826	-0.000841
32	-0.113886	0.058836	-0.030396	0.071177	-0.021069	-0.001939	0.001002
33	0.174465	-0.008054	0.000372	-0.030957	0.001412	0.001743	-0.000080
34	-0.193489	-0.069562	-0.025008	-0.078017	-0.037039	-0.002672	-0.000961
35	0.082570	-0.006363	0.000490	0.007901	-0.000647	0.001510	-0.000116
Sum	0.014071	-0.001929	0.000357	0.003608	0.000645	0.000145	0.000135

The plots in Figs. 8.14 and 8.15 show a much more complex pattern of residual aberrations than those of the 1× designs. This is a consequence of the loss of mono-

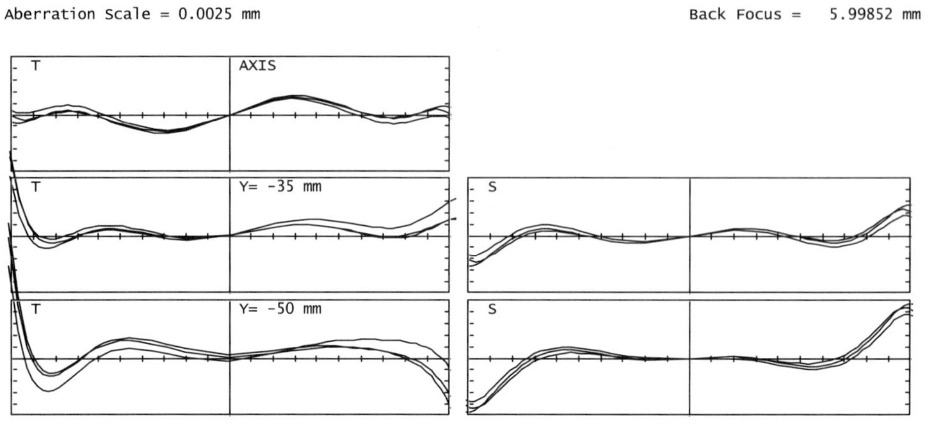

Figure 8.14 Transverse ray aberrations of Braat's 5× reduction lens.

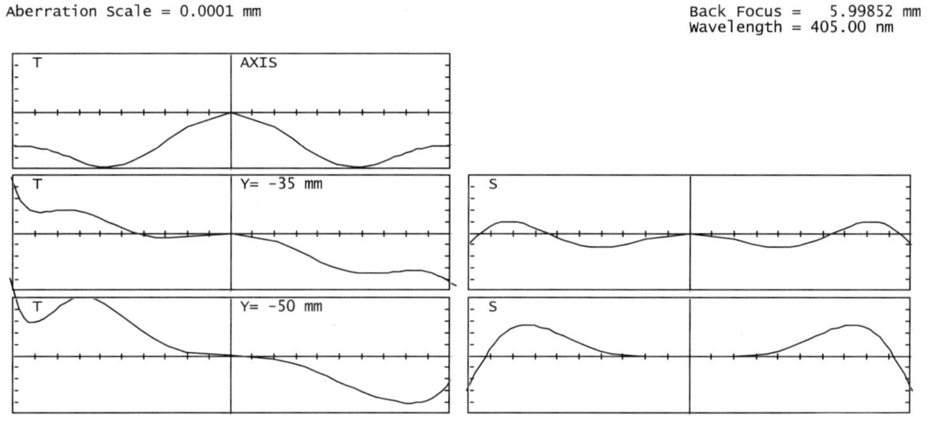

Figure 8.15 Wavefront aberrations of Braat's 5× reduction lens.

centricity, so that sagittal astigmatism and oblique spherical aberration are no longer automatically zero by symmetry, and there is also some residual coma and distortion (the latter visible in Fig. 8.16). Some spherical aberration on-axis deliberately offsets the oblique spherical aberration, so that the on-axis field has a similar magnitude of aberration residuals to the edge of the field. It can also be seen that there is no vignetting, again because in microlithography the diffraction-limited resolution should be exactly the same over the entire field.

Residual chromatic aberrations, including lateral color, are also apparent, but sufficiently small over the relatively narrow spectral range of 401–409 nm, the mercury H-line. Generally, this level of residual aberration is acceptable for imaging features of the order of 1 μm, which was the requirement at the time this lens was designed.

A more recent 5× reduction design by Williamson is shown in Fig. 8.17.[11] It is also a double-Gauss derivative, has a numerical aperture of 0.42, and operates over

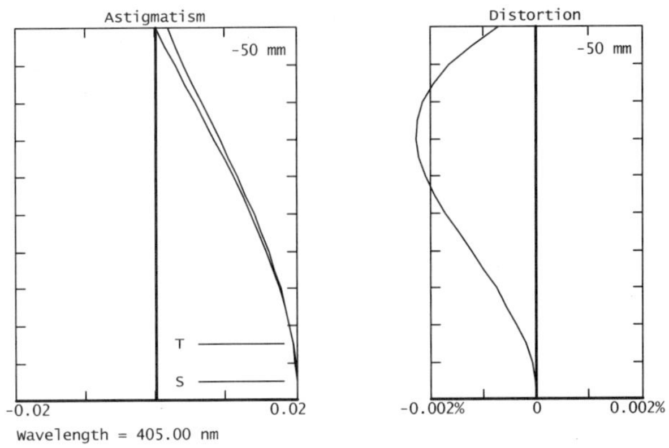

Figure 8.16 Astigmatism and distortion of Braat's 5× reduction lens.

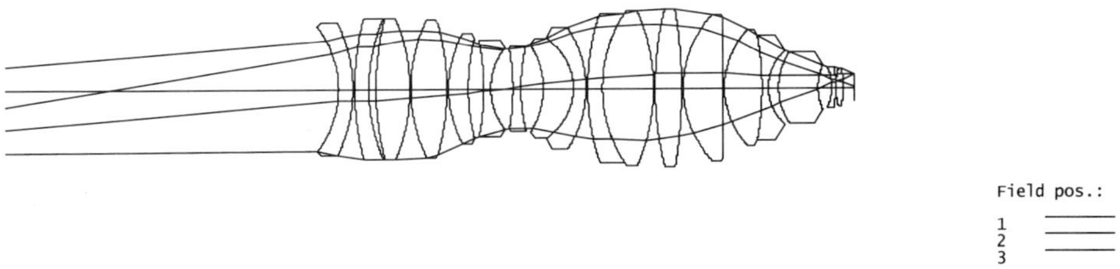

Figure 8.17 Williamson's 5× reduction lens.

a narrow waveband from 0.363 μm to 0.369 μm, within the mercury I-line. Surprisingly, the spectral bandwidth is limited by secondary spectrum. At this wavelength, the higher refractive index and more dispersive glasses have poor transmission, so only a limited number of low-index glass types may be used. The design prescription includes a number of "crown" elements of fused silica. This has excellent transmission and damage resistance, but is not an ideal choice in terms of its partial dispersion, which does not match available "flint" glasses.

From Eq. (8.1), with $k_1 = 0.5$, the minimum resolution of this lens should be of the order of 0.5 μm. With wavefront aberrations less than λ/4 over the entire image of 24-mm diameter, the lens should therefore resolve 24,000/0.5 = 48,000 lines across the image. In 1987, this was the largest number of resolved lines of any flat-field objective that we know of.

The design residual aberrations are shown in Figs. 8.18 and 8.19, and are seen to be quite uniform across the field. Even though the numerical aperture is larger, oblique spherical aberration is much smaller than earlier designs, as are residual astigmatism and distortion, shown in Fig. 8.20.

```
EFL =    135.539
OBJECT DISTANCE = 549.818, CURVATURE = 0, TILT ANGLE = 0
WAVELENGTHS  [nm]
       366.50        370.50        362.50
```

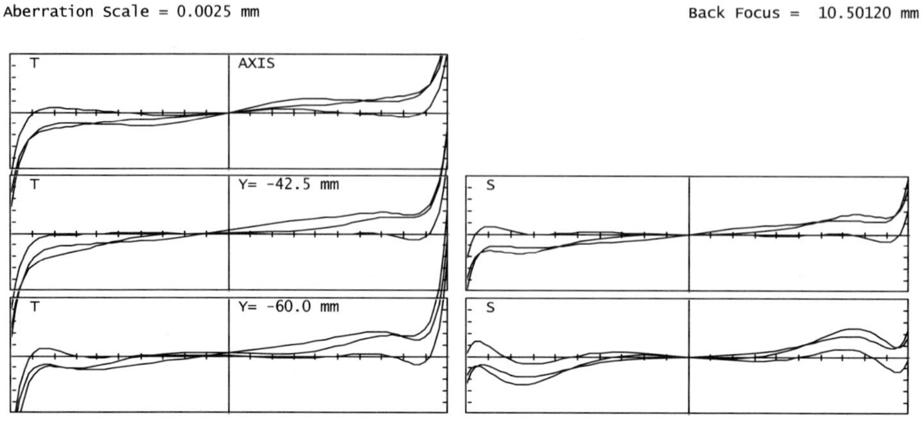

Figure 8.18 Transverse ray aberrations of Williamson's 5× reduction lens.

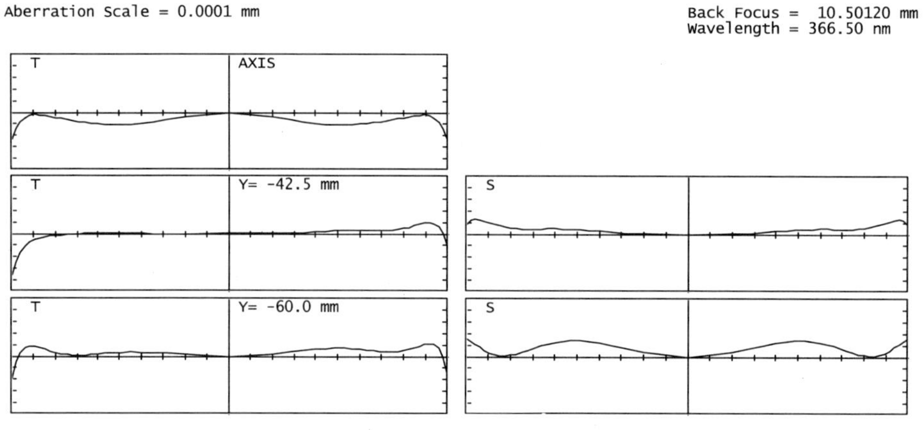

Figure 8.19 Wavefront aberrations of Williamson's 5× reduction lens.

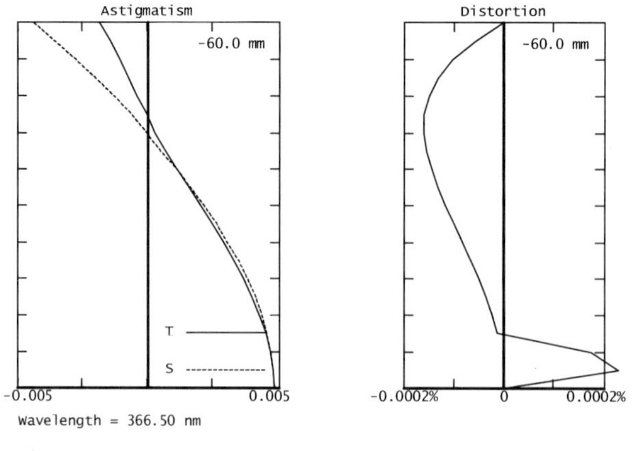

Figure 8.20 Astigmatism and distortion of Williamson's 5× reduction lens.

SURFACES

#	SURF	SPACE	RADIUS	SEPN	INDEX1	V	CLR RAD	GLASS
0				0.00000	1.000000			
1	S		-100.06000	12.57920	1.611162		58.059	S-LF7
2	S		-147.82000	1.50350	1.000000		62.453	
3	S		265.70000	12.48770	1.611162		65.559	S-LF7
4	S		176.60500	6.27660	1.000000		65.277	
5	S		277.98600	29.54470	1.474326		65.657	G-SILICA
6	S		-181.45800	1.00000	1.000000		66.312	
7	S		160.22900	31.23600	1.474326		64.139	G-SILICA
8	S		-238.89000	1.50000	1.000000		62.305	
9	S		112.37600	17.50600	1.474326		51.807	G-SILICA
10	S		195.06600	13.67470	1.000000		46.708	
11	S		-232.05400	6.00000	1.474326		44.417	G-SILICA
12	S		66.33000	11.93630	1.000000		37.901	
13	#S		Plane	6.99200	1.000000		37.896	
14	S		-239.17000	7.00000	1.474326		37.400	G-SILICA
15	S		187.69800	19.54230	1.000000		39.413	
16	S		-80.15920	25.98640	1.503784		40.934	S-FK5
17	S		-57.49600	0.02000	1.577289		46.751	S-PSK50
18	S		-57.49600	11.00000	1.611162		46.757	S-LF7
19	S		-107.19000	1.50000	1.000000		56.116	
20	S		238.89000	12.75000	1.611162		67.385	S-LF7
21	S		137.69000	0.02000	1.577289		69.948	S-PSK50
22	S		137.69000	43.99790	1.503784		69.951	S-FK5
23	S		-191.96300	1.50000	1.000000		71.837	
24	S		327.02660	23.00000	1.474326		73.263	G-SILICA
25	S		-486.00200	1.50000	1.000000		72.956	
26	S		105.47710	35.02790	1.474326		68.931	G-SILICA
27	S		Plane	1.00000	1.000000		65.762	
28	S		82.96000	32.66700	1.536002		54.371	S-BK7
29	S		-283.33030	0.02000	1.577289		49.208	S-PSK50
30	S		-283.33030	5.50000	1.611162		49.194	S-LF7
31	S		47.54700	9.05900	1.000000		34.710	
32	S		62.65800	34.68950	1.474326		33.421	G-SILICA
33	S		53.32500	11.90310	1.000000		20.686	
34	S		-48.61100	3.50000	1.611162		18.970	S-LF7
35	S		-133.09990	0.25000	1.000000		18.550	
36	S		78.23150	6.00000	1.611162		17.815	S-LF7
37	S		-158.02860	10.49989	1.000000		17.167	
38	S		Plane				12.004	

LAGRANGE INVARIANT = -5.0400

#	H	U	HBAR	UBAR	D(U/N)	A	ABAR
1	46.18472	0.08400	-11.64313	0.08795	0.05703	-0.37757	0.20431
2	49.04301	0.22722	-11.51169	0.01045	0.02229	-0.16845	0.14231
3	49.28857	0.16333	-11.41482	0.06443	-0.14408	0.34883	0.02147
4	49.67574	0.03100	-10.71193	0.05629	0.20262	0.50314	-0.00704
5	51.06827	0.22186	-10.37540	0.05362	-0.15988	0.40557	0.01629
6	53.76806	0.09138	-8.94619	0.04837	-0.06781	-0.30214	0.14401
7	53.76223	-0.00582	-8.85148	0.09471	-0.07007	0.32971	0.03946
8	50.26692	-0.11190	-6.28985	0.08201	-0.18889	-0.47520	0.15973
9	49.86975	-0.26478	-6.08975	0.13340	0.04613	0.17899	0.07921
10	44.22634	-0.32237	-4.20060	0.10791	-0.14908	-0.14101	0.12735
11	39.19764	-0.36774	-2.16462	0.14889	0.23542	-0.53665	0.15821
12	38.02715	-0.19508	-1.57671	0.09799	0.11664	0.55762	0.10942

13	37.83994	-0.01568	0.01305	0.13319	0.00000	-0.01568	0.13319
14	37.73028	-0.01568	0.94430	0.13319	0.04289	-0.17344	0.12924
15	38.01109	0.04012	1.58556	0.09161	0.12799	0.35771	0.14751
16	41.04407	0.15520	4.30324	0.13907	0.02750	-0.35683	0.08538
17	48.18365	0.27474	7.17376	0.11046	0.00813	-0.84707	-0.02152
18	48.18967	0.30099	7.17599	0.11113	0.00300	-0.84723	-0.02158
19	51.62482	0.31229	8.40157	0.11142	0.01497	-0.27282	0.05323
20	51.93801	0.20880	8.59898	0.13161	-0.17955	0.42621	0.16760
21	52.53882	0.04712	9.46637	0.06803	0.00646	0.69070	0.22038
22	52.53995	0.05633	9.46779	0.07097	0.01598	0.69071	0.22039
23	55.96005	0.07773	12.89075	0.07780	-0.08166	-0.32148	0.01601
24	55.91510	-0.02997	13.01549	0.08316	-0.02113	0.14101	0.12296
25	54.18242	-0.07533	14.01833	0.04360	-0.11285	-0.27543	0.02176
26	53.93649	-0.16395	14.09423	0.05060	-0.02307	0.34741	0.18423
27	44.27868	-0.27572	13.79062	-0.00867	-0.21949	-0.40650	-0.01278
28	43.87219	-0.40650	13.77784	-0.01278	0.11406	0.12234	0.15330
29	29.19854	-0.44919	11.61285	-0.06627	0.01682	-0.84825	-0.16475
30	29.18985	-0.43473	11.61158	-0.06347	0.01281	-0.84820	-0.16475
31	26.86100	-0.42343	11.27459	-0.06127	-0.07413	0.22799	0.28333
32	23.80863	-0.33694	11.69316	0.04620	0.09901	0.04303	0.23282
33	11.63997	-0.35079	10.69757	-0.02870	-0.17571	-0.19535	0.25345
34	6.71639	-0.41364	11.32655	0.05284	0.28682	-0.55180	-0.18016
35	6.00126	-0.20432	11.75069	0.12118	-0.22994	-0.40184	0.05300
36	5.91207	-0.35675	11.78601	0.14129	0.20153	-0.28118	0.29194
37	4.41152	-0.25009	11.96928	0.03055	-0.26478	-0.44792	-0.07282
		-0.420001		0.002923			

#	S1	S2	S3	S4	S5	C1	C2
1	-0.375495	0.203189	-0.109951	-0.096298	0.111606	-0.032053	0.017345
2	-0.031026	0.026210	-0.022143	0.065185	-0.036362	0.015185	-0.012828
3	0.864144	0.053184	0.003273	0.036265	0.002433	0.031603	0.001945
4	-2.548027	0.035643	-0.000499	-0.054560	0.000770	-0.045942	0.000643
5	1.343000	0.053953	0.002167	0.029398	0.001268	0.016261	0.000653
6	0.332808	-0.158626	0.075606	0.045037	-0.057502	0.012754	-0.006079
7	0.409542	0.049017	0.005867	0.051004	0.006807	0.013917	0.001666
8	2.144067	-0.720669	0.242233	0.034210	-0.092919	0.018754	-0.006304
9	-0.073700	-0.032613	-0.014432	0.072723	0.025795	0.007008	0.003101
10	0.131104	-0.118404	0.106935	-0.041895	-0.058739	0.004896	-0.004422
11	-2.657573	0.783500	-0.230990	-0.035217	0.078483	-0.016515	0.004869
12	-1.379123	-0.270613	-0.053100	-0.123207	-0.034595	-0.016648	-0.003267
13	0.000000	0.000000	0.000000	0.000000	0.000000	0.000000	0.000000
14	-0.048682	0.036276	-0.027031	-0.034169	0.045604	-0.005138	0.003828
15	-0.622527	-0.256719	-0.105866	-0.043540	-0.061612	-0.010675	-0.004402
16	-0.143720	0.034390	-0.008229	-0.106162	0.027371	-0.011592	0.002774
17	-0.281029	-0.007138	-0.000181	-0.013691	-0.000352	-0.005455	-0.000139
18	-0.103675	-0.002640	-0.000067	-0.005889	-0.000152	-0.037274	-0.000949
19	-0.057521	0.011222	-0.002189	0.089893	-0.017111	0.025889	-0.005051
20	1.694010	0.666153	0.261959	0.040335	0.118874	0.040690	0.016001
21	-0.162038	-0.051701	-0.016496	-0.002459	-0.006048	-0.033130	-0.010571
22	-0.400543	-0.127807	-0.040781	-0.005717	-0.014837	-0.004851	-0.001548
23	0.472269	-0.023518	0.001171	0.044331	-0.002266	0.014239	-0.000709
24	0.023495	0.020487	0.017864	0.024990	0.037367	0.006190	0.005398
25	0.463869	-0.036643	0.002895	0.016815	-0.001557	0.011717	-0.000926
26	0.150149	0.079621	0.042222	0.077479	0.063476	0.014711	0.007801
27	1.605891	0.050485	0.001587	0.000000	0.000050	0.014131	0.000444
28	-0.074892	-0.093845	-0.117595	0.106848	-0.013467	0.004963	0.006218
29	-0.353368	-0.068634	-0.013331	-0.001528	-0.002886	-0.000014	-0.000003

30	-0.269050	-0.052258	-0.010150	-0.001195	-0.002204	-0.022604	-0.004390
31	0.103510	0.128634	0.159855	-0.202654	-0.053187	-0.011257	-0.013989
32	-0.004366	-0.023619	-0.127784	0.130427	0.014302	0.000804	0.004352
33	0.078053	-0.101266	0.131382	-0.153255	0.028378	0.001785	-0.002316
34	-0.586568	-0.191512	-0.062528	-0.198218	-0.085133	-0.006812	-0.002224
35	0.222822	-0.029391	0.003877	0.072394	-0.010060	0.004433	-0.000585
36	-0.094200	0.097805	-0.101549	0.123168	-0.022446	-0.003056	0.003173
37	0.234347	0.038098	0.006194	0.060974	0.010919	0.003632	0.000590
Sum	0.005957	0.000252	0.000194	0.001820	0.000069	0.000548	0.000100

8.5 Catadioptric reduction systems

As illumination wavelengths have reduced further into the ultraviolet to improve resolution to 0.25 μm, or less, the number of glasses with acceptable transmission becomes very limited. At the KrF (248 nm) or ArF (193 nm) excimer laser wavelengths, for example, only fused silica and a few crystalline materials may be used. Many reduction lenses at the KrF wavelength have been designed, using only fused silica. These, of course, cannot be achromatized for primary axial color, C_I, although lateral color, C_{II}, can be corrected if the stop position and power distribution are chosen appropriately. Such lenses can therefore give diffraction-limited resolution only over a very narrow portion of the natural excimer laser bandwidth, typically 0.5 pm or less.[12, 13]

Since dispersion increases with reduced wavelength, ArF excimer laser reduction lenses have to be partially achromatized to cover a similar spectral bandwidth. This is achieved by the use a minimal number of elements of calcium fluoride as the "crown," with fused silica as the "flint."

As we have already seen, catadioptric systems have the advantage that some degree of achromatization is possible with only one glass type, and several such nonmonocentric designs have been developed for microlithography. For example, Williamson has described a reduction variant of the Dyson 1× system, which is also afocal or telecentric in both object and image spaces, but with a 4× reduction ratio.[14] In order to avoid obscuration and allow physical separation of object and image, the catadioptric design shown in Fig. 8.21 uses a beamsplitter, which allows a full 30-mm field diameter to be covered on the wafer. As in the Dyson 1× system, the field curvature of the concave mirror corrects that of the positive lens elements.

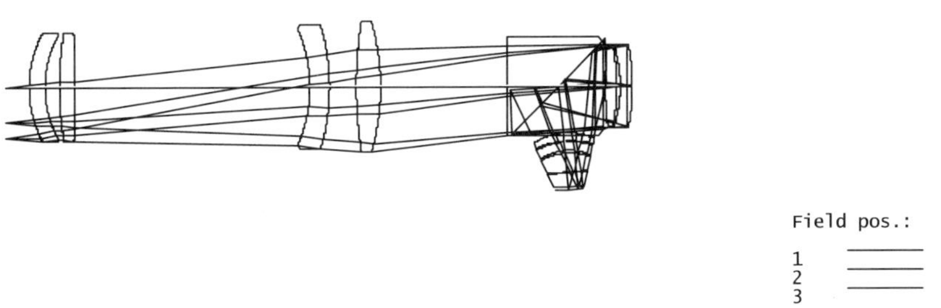

Figure 8.21 Catadioptric 4× reduction lens.

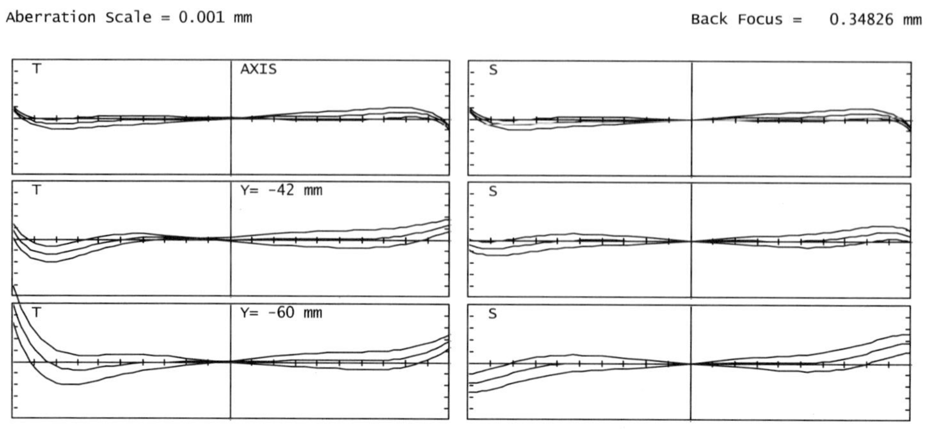

Figure 8.22 Transverse ray aberrations of catadioptric 4× reduction lens.

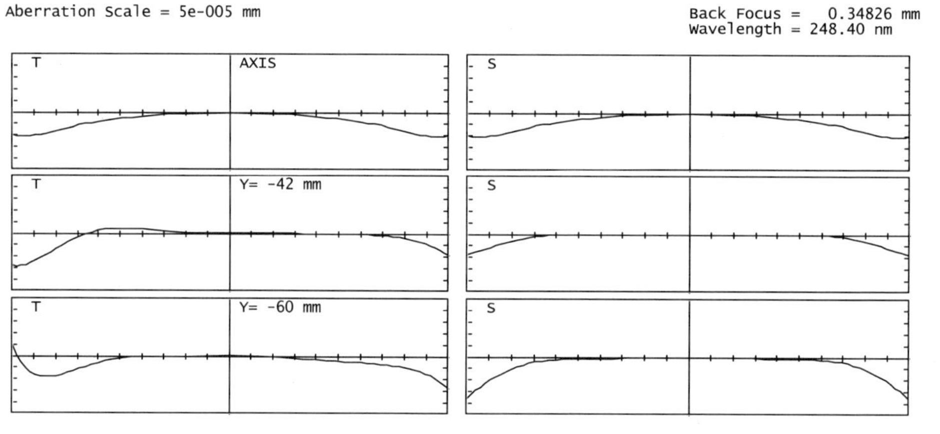

Figure 8.23 Wavefront aberrations of catadioptric 4× reduction lens.

This allows a relatively simple, relaxed design with low refracting power, small higher-order aberrations, and sufficient achromatization to cover the natural spectral width of an excimer laser, of the order of 1 nm. This may be seen from the aberration plots of Figs. 8.22 and 8.23 and the astigmatism and distortion plots in Fig. 8.24. Transmission of the system can be high if a polarizing beamsplitter and quarter-waveplate are used in conjunction with a linearly polarized light source.

```
EFL = 11155.375
OBJECT DISTANCE = 25.0672, CURVATURE = 0, TILT ANGLE = 0
WAVELENGTHS   [nm]
      248.40       248.80       248.00
SURFACES
  # SURF SPACE    RADIUS       SEPN      INDEX1      V     CLR RAD   GLASS
  0                          0.00000   1.000000
  1 S           143.94740   18.00000   1.508277             64.602  G-SILICA
  2 S           136.45060   16.00000   1.000000             62.896
  3 S           420.34260   17.00000   1.508277             63.572  G-SILICA
```

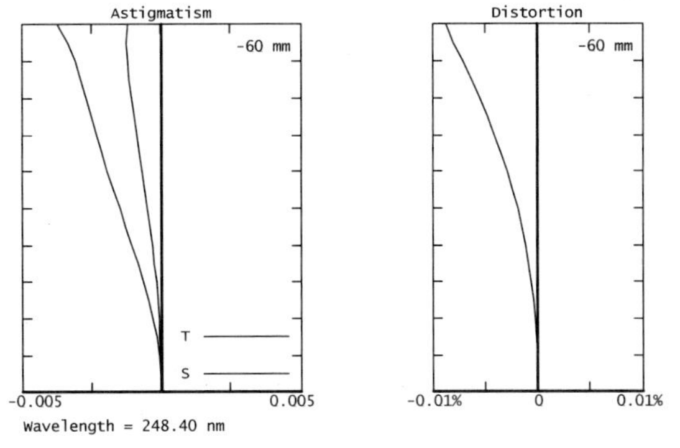

Figure 8.24 Astigmatism and distortion of catadioptric 4× reduction system.

4 S	-1665.77060	263.66480	1.000000	63.889	
5 S	-184.35870	21.70770	1.508277	69.974	G-SILICA
6 S	-279.07840	30.04850	1.000000	73.845	
7 S	540.12080	27.39580	1.508277	77.970	G-SILICA
8 S	-303.68180	140.26570	1.000000	78.086	
9 S	Plane	108.00000	1.508277	59.595	G-SILICA
10 S	Plane	1.80560	1.000000	50.833	
11 S	453.70760	12.50000	1.508277	50.268	G-SILICA
12 S	-283.63070	4.26830	1.000000	49.634	
13 S	-164.55900	6.94900	1.508277	49.327	G-SILICA
14#S	2325.05500	6.67970	1.000000	49.056	
15 SM	-276.29130	-6.67970	1.000000	50.271	
16 S	2325.05500	-6.94900	1.508277	49.980	G-SILICA
17 S	-164.55900	-4.26830	1.000000	48.256	
18 S	-283.63070	-12.50000	1.508277	48.230	G-SILICA
19 S	453.70760	-1.80560	1.000000	47.804	
20 S	Plane	-50.00000	1.508277	47.082	G-SILICA
21 SM	Plane	54.59404	1.508277	77.000	G-SILICA
22 S	Plane	1.29383	1.508277	54.000	G-SILICA
23 S	Plane	0.01005	1.000000	35.561	
24 S	63.81850	10.20072	1.508277	33.783	G-SILICA
25 S	89.37740	0.98600	1.000000	31.857	
26 S	76.80240	10.00009	1.508277	31.357	G-SILICA
27 S	50.81010	3.35201	1.000000	27.563	
28 S	63.43130	22.62636	1.508277	27.299	G-SILICA
29 S	919.65310	1.00001	1.000000	22.586	
30 S	216.33430	16.48244	1.508277	21.800	G-SILICA
31 S	Plane	4.01816	1.000000	17.007	
32 S	Plane			15.150	

LAGRANGE INVARIANT = -6.7500

#	H	U	HBAR	UBAR	D(U/N)	A	ABAR
1	2.82006	0.11250	-60.01864	-0.00074	-0.06742	0.13209	-0.41769
2	4.04382	0.06799	-57.49837	0.14002	0.07253	0.14724	-0.42439
3	5.92551	0.11761	-57.54636	-0.00300	-0.06906	0.13170	-0.13990
4	7.17030	0.07322	-56.79586	0.04415	0.05971	0.10395	0.11801
5	35.71269	0.10825	-34.67021	0.08392	-0.01739	-0.08546	0.27197
6	38.68778	0.13705	-34.83817	-0.00774	0.04538	-0.00238	0.17661
7	42.78194	0.13625	-33.28227	0.05178	-0.09406	0.21546	-0.00984
8	44.52550	0.06364	-31.77287	0.05510	-0.02073	-0.12515	0.24090
9	47.53685	0.02147	-12.65767	0.13628	-0.01203	0.02147	0.13628

10	49.07413	0.01423	-2.89946	0.09035	0.01203	0.02147	0.13628
11	49.11290	0.02147	-2.65339	0.13628	-0.03622	0.12972	0.13043
12	48.83484	-0.02224	-1.49933	0.09232	-0.10632	-0.29324	0.14722
13	48.31810	-0.12106	-0.89350	0.14194	0.13345	-0.41469	0.14737
14	48.44791	0.01868	-0.25227	0.09228	0.02638	0.05960	0.13901
15	48.70687	0.03877	0.67703	0.13912	-0.35258	-0.13752	0.13667
16	46.61072	0.31381	1.57359	-0.13422	0.18034	-0.33386	0.13355
17	45.21187	0.20130	2.19357	-0.08922	-0.03051	0.11077	0.15467
18	44.51199	0.16397	2.79686	-0.14134	0.05683	-0.00704	0.15120
19	42.49197	0.16160	3.92670	-0.09039	-0.18420	-0.38500	0.12328
20	41.96592	0.29134	4.16492	-0.13193	0.16327	-0.29134	0.13193
21	32.30781	0.19316	8.53847	-0.08747	0.00000	-0.29134	0.13193
22	21.76230	-0.19316	13.31386	0.08747	0.00000	-0.29134	0.13193
23	21.51238	-0.19316	13.42703	0.08747	-0.16327	-0.29134	0.13193
24	21.50946	-0.29134	13.42836	0.13193	0.08797	0.04570	0.34235
25	18.38047	-0.30674	13.59731	0.01656	-0.15475	-0.15248	0.25444
26	18.02735	-0.35813	13.69819	0.10231	0.14826	-0.12340	0.28066
27	14.86192	-0.31654	13.77545	0.00773	-0.11889	-0.03626	0.42057
28	13.75992	-0.32876	14.27642	0.14946	0.13578	-0.11183	0.37452
29	7.17401	-0.29107	14.80234	0.02324	-0.24207	-0.42725	0.05933
30	6.73895	-0.43505	14.84558	0.04324	0.23685	-0.40390	0.11186
31	1.81167	-0.29894	14.93693	0.00554	-0.25269	-0.45089	0.00836
		-0.45089		0.00836			

#	S1	S2	S3	S4	S5	C1	C2
1	0.003318	-0.010491	0.033173	0.106665	-0.442191	0.000110	-0.000347
2	-0.006359	0.018327	-0.052824	-0.112526	0.476577	-0.000175	0.000506
3	0.007098	-0.007540	0.008009	0.036528	-0.047310	0.000230	-0.000244
4	-0.004626	-0.005252	-0.005962	0.009217	0.003696	-0.000220	-0.000249
5	0.004535	-0.014432	0.045928	-0.083284	0.118885	-0.000900	0.002863
6	-0.000010	0.000737	-0.054768	0.055017	-0.018525	0.000027	-0.002014
7	0.186800	-0.008531	0.000390	0.028427	-0.001316	0.002717	-0.000124
8	0.014455	-0.027824	0.053560	0.050560	-0.200423	0.001642	-0.003162
9	0.000264	0.001673	0.010622	0.000000	0.067426	0.000301	0.001909
10	-0.000272	-0.001727	-0.010966	0.000000	-0.069606	-0.000311	-0.001971
11	0.029930	0.030094	0.030260	0.033842	0.064454	0.001878	0.001888
12	0.446462	-0.224149	0.112535	0.054134	-0.083678	0.004221	-0.002119
13	-1.108844	0.394051	-0.140034	-0.093305	0.082922	-0.005906	0.002099
14	-0.004541	-0.010591	-0.024700	-0.006604	-0.073009	-0.000851	-0.001985
15	0.324772	-0.322770	0.320779	-0.329815	0.008980	0.000000	0.000000
16	-0.936926	0.374779	-0.149915	-0.006604	0.062609	-0.004587	0.001835
17	0.016925	0.023632	0.032997	-0.093305	-0.084207	-0.001476	-0.002061
18	-0.000125	0.002691	-0.057832	0.054134	0.079459	-0.000092	0.001984
19	1.160138	-0.371475	0.118946	0.033842	-0.048922	0.004822	-0.001544
20	-0.581595	0.263368	-0.119263	0.000000	0.054006	-0.003604	0.001632
21	0.000000	0.000000	0.000000	0.000000	0.000000	0.000000	0.000000
22	0.000000	0.000000	0.000000	0.000000	0.000000	0.000000	0.000000
23	0.298135	-0.135006	0.061136	0.000000	-0.027685	0.001847	-0.000837
24	-0.003952	-0.029603	-0.221764	0.240592	0.141046	0.000290	0.002170
25	0.066129	-0.110352	0.184149	-0.171791	-0.020623	0.000826	-0.001378
26	-0.040699	0.092566	-0.210532	0.199918	0.024141	-0.000656	0.001491
27	0.002323	-0.026946	0.312538	-0.302188	-0.120051	0.000159	-0.001842
28	-0.023366	0.078251	-0.262059	0.242060	0.066974	-0.000454	0.001519
29	0.317009	-0.044024	0.006114	-0.016696	0.001470	0.000903	-0.000125
30	-0.260389	0.072116	-0.019973	0.070974	-0.014125	-0.000802	0.000222
31	0.093067	-0.001725	0.000032	0.000000	-0.000001	0.000241	-0.000004
Sum	-0.000346	-0.000153	0.000577	-0.000205	0.000973	0.000181	0.000110

8.6 Catoptric reduction systems

Even further into the ultraviolet, below about 157 nm, which corresponds to the F_2 excimer laser, even the crystalline materials stop transmitting. In this spectral region the only possible designs are all-reflective, or catoptric. Several reduction variants of Offner's 1× design have been proposed; one by Williamson is shown in Fig. 8.25.[15] Note that the full mirrors are drawn, but in practice only the illuminated off-axis segments are fabricated, to allow beams to pass by them unobstructed.

This design is again approximately afocal, at 4× reduction ratio, and uses four mirrors in a nonmonocentric configuration to form a reduced intermediate image that is reimaged and further reduced to the wafer by two more mirrors. All six mirrors are aspheric, in order to correct the aberrations arising from the loss of monocentricity. Residual aberrations are reasonably small at numerical aperture 0.5, over the unobscured annular field, and are shown in Figs. 8.26, 8.27 and 8.28.

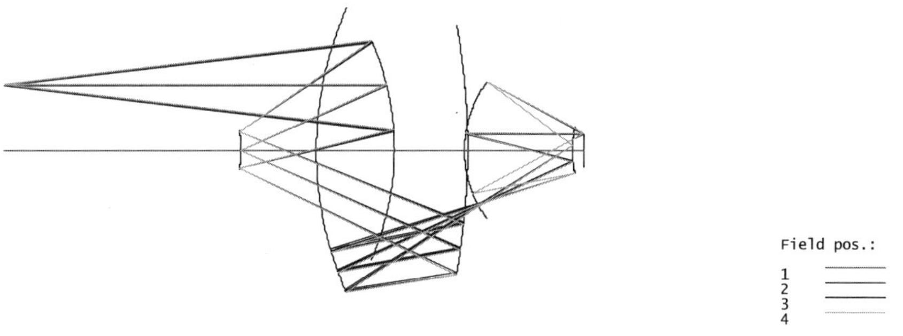

Figure 8.25 Catoptric 4× reduction system.

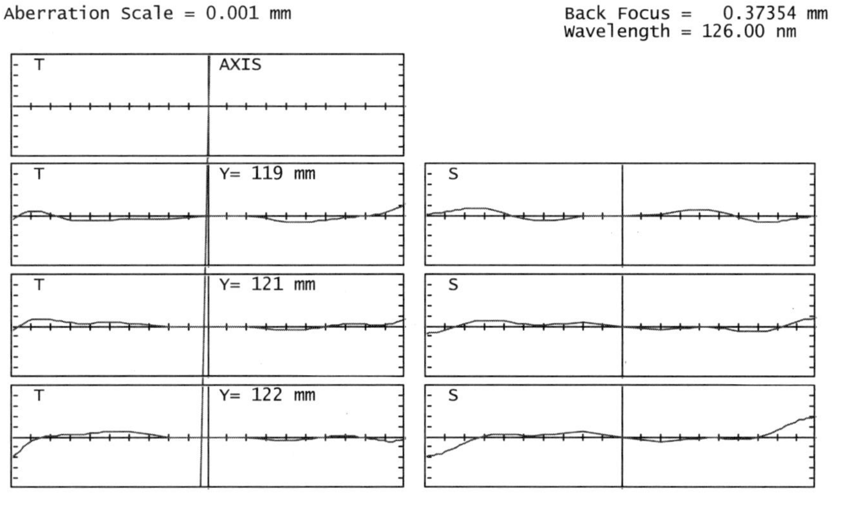

Figure 8.26 Transverse ray aberrations of catoptric 4× reduction system.

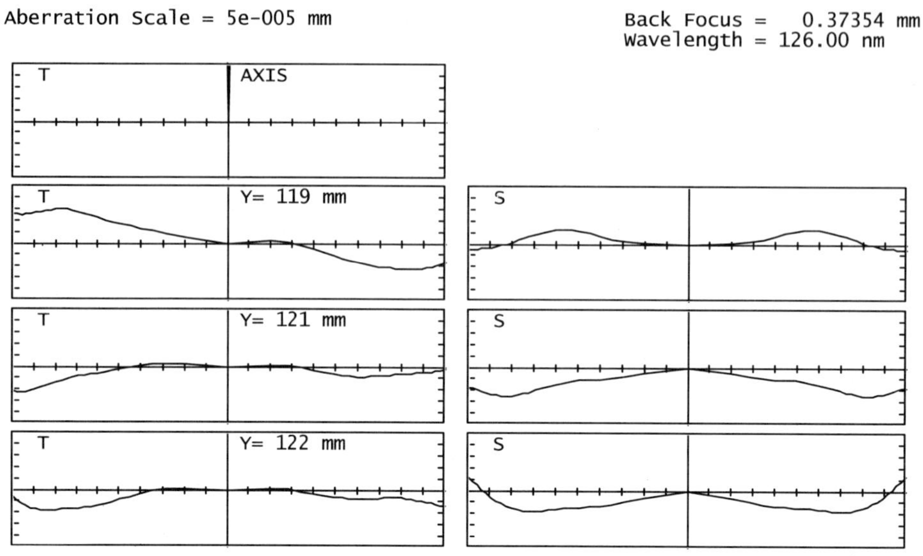

Figure 8.27 Wavefront aberrations of catoptric 4× reduction system.

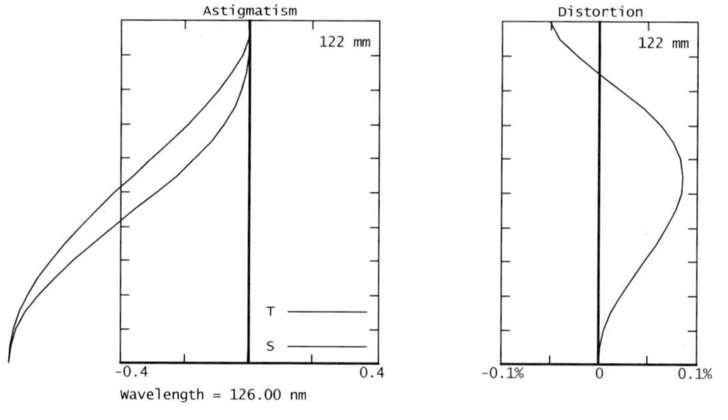

Figure 8.28 Astigmatism and distortion of catoptric 4× reduction system.

There is substantial defocus on-axis, but this part of the field is not illuminated As in the Offner 1× system, the reticle and wafer have to be scanned in synchronism (in the plane of the page in Fig. 8.25), past the narrow off-axis annular field to expose the whole reticle pattern onto the wafer.

Variants of this design have been proposed that operate at a wavelength of 13.4 nm, using complex multilayer reflecting coatings and an extreme ultraviolet (EUV) source. At an NA of 0.25 and k_1 factor of 0.5, such a design can be made diffraction-limited and therefore should be able to print features as small as 0.025 μm (25 nm) over a 26-mm field width (perpendicular to the page in Fig. 8.25). This corresponds to 26,000/0.025 = 1,040,000 lines across the image.

```
EFL = -3016.551
OBJECT DISTANCE = 672.354, CURVATURE = 0, TILT ANGLE = 0
WAVELENGTHS  [nm]
      126.00
SURFACES
  # SURF SPACE    RADIUS          SEPN       INDEX1      V      CLR RAD  GLASS
  0                            0.00000    1.000000
  1 AM        -543.09112 -264.53789    1.000000             201.798
  2#AM        -290.05060  390.23706    1.000000              35.244
  3 AM       -1351.41604 -260.00000    1.000000             231.045
  4 AM         687.36348  260.00000    1.000000             262.218
  5 S        -1351.41604   -5.69917    1.000000             121.246
  6 S          226.10416  186.28625    1.000000             103.862
  7 AM         224.99210 -186.28625    1.000000              43.664
  8 AM         226.10416  186.28625    1.000000             127.159
  9 S          224.99210   19.62646    1.000000              40.004
 10 S            Plane      0.37354    1.000000              30.716
 11 S            Plane                                       30.552
ASPHERIC SURFACE   1 CC =    0.000000  (SPHERE)
 A4 = 2.24506e-012   A6 = -1.06529e-015    A8 =  1.40153e-020
                                          A10 = -1.48525e-026
ASPHERIC SURFACE   2 CC =    0.000000  (SPHERE)
 A4 = 2.70239e-008   A6 = 6.29102e-012    A8 =  1.39941e-015
                                          A10 = -1.11823e-019
                                          A12 =  3.43046e-023
ASPHERIC SURFACE   3 CC =    0.000000  (SPHERE)
 A4 = 3.47651e-010   A6 = 1.86467e-016    A8 = -3.27086e-020
                                          A10 =  1.15234e-025
ASPHERIC SURFACE   4 CC =    0.000000  (SPHERE)
 A4 = -6.10242e-011  A6 = 7.8245e-016     A8 = -8.98169e-021
                                          A10 = -4.15911e-026
ASPHERIC SURFACE   5 CC =    0.000000  (SPHERE)
 A4 = 1.67615e-007   A6 = 1.75232e-011    A8 = 8.40253e-016
                                          A10 = 8.38291e-020
                                          A12 = 6.03283e-023
ASPHERIC SURFACE   6 CC =    0.000000  (SPHERE)
 A4 = 5.24492e-010   A6 = 8.43072e-015    A8 =  4.12933e-019
                                          A10 = -9.7652e-024
                                          A12 =  5.1581e-028

LAGRANGE INVARIANT =     15.2500

  #      H          U         HBAR       UBAR      D(U/N)      A         ABAR
  1  84.04424    0.12500 122.30153   0.00045   -0.30950  -0.02975  -0.22475
  2  35.23613    0.18450   3.27482   0.44994    0.24297  -0.06302  -0.43865
  3  58.05023    0.05846-163.49725  -0.42736   -0.08591   0.01551  -0.30638
  4  50.91373    0.02745-211.70037   0.18540   -0.14814  -0.10152   0.12259
  5   5.26029   -0.17559 -99.74922   0.43058    0.00000  -0.17948   0.50439
  6   6.26101   -0.17559-102.20317   0.43058    0.00000  -0.14790  -0.02144
  7 -26.44903   -0.17559 -21.99179   0.43058   -0.23511  -0.29315   0.33284
  8-102.95696    0.41070  21.80260  -0.23509    0.91070   0.04465   0.13866
  9  -9.81329    0.50000  29.67085   0.04224    0.00000   0.45639   0.17411
 10  -0.00000    0.50000  30.49982   0.04224    0.00000   0.50000   0.04224
                 0.50000             0.04224

  #     S1        S2        S3        S4        S5        C1        C2
  1  0.023025  0.173931  1.313893 -0.856440  3.455646  0.000000  0.000000
  2 -0.034001 -0.236665 -1.647297  1.603600 -0.304153  0.000000  0.000000
```

```
  3  0.001199 -0.023694  0.468130 -0.344176 -2.449002  0.000000  0.000000
  4  0.077734 -0.093870  0.113355 -0.676680  0.680259  0.000000  0.000000
  5  0.000000  0.000000  0.000000  0.000000  0.000000  0.000000  0.000000
  6  0.000000  0.000000  0.000000  0.000000  0.000000  0.000000  0.000000
  7 -0.534378  0.606732 -0.688881  2.067295 -1.565047  0.000000  0.000000
  8  0.186937  0.580537  1.802870 -2.057127 -0.789601  0.000000  0.000000
  9  0.000000  0.000000  0.000000  0.000000  0.000000  0.000000  0.000000
 10  0.000000  0.000000  0.000000  0.000000  0.000000  0.000000  0.000000
     ASPHERIC SURFACE  1
    -0.001792 -0.002608 -0.003795  0.000000 -0.005523
     ASPHERIC SURFACE  2
     0.666532  0.061947  0.005757  0.000000  0.000535
     ASPHERIC SURFACE  3
    -0.063165  0.177904 -0.501063  0.000000  1.411233
     ASPHERIC SURFACE  4
    -0.006561  0.027280 -0.113432  0.000000  0.471652
     ASPHERIC SURFACE  5
    -1.312417 -1.091246 -0.907347  0.000000 -0.754439
     ASPHERIC SURFACE  6
     0.942935 -0.199680  0.042285  0.000000 -0.008954

Sum -0.053953 -0.019431 -0.115525 -0.263529  0.142607  0.000000  0.000000
```

References

1. J. Braat, "Quality of microlithographic projection lenses," in *SPIE Proc.* Vol. 811, Bellingham, WA (1987).
2. D.G. Flagello et al., "Optical lithography into the millennium: sensitivity to aberrations, vibration, and polarization," in *SPIE Proc.* Vol. 4000, Bellingham, WA (2000).
3. J. Dyson, *JOSA* **49**(7), 713 (1959).
4. A. Offner, "New concepts in projection mask aligners," *Opt. Eng.* **14**(2), 130–132 (1975).
5. A. Offner, U.S. Patent 3,748,015 (1971).
6. A. Offner, U.S. Patent 4,293,186 (1979).
7. C.G. Wynne, "A unit-power telescope for projection copying," in *Optical Instruments and Techniques,* 429, Oriel Press, London (1970).
8. C.G. Wynne, "Monocentric telescopes for microlithography," *Optical Engineering* **26** (4) (1987).
9. C.G. Wynne, U.S. Patent 4,796,984 (1987).
10. E. Glatzel, "New lenses for microlithography," in *SPIE Proc.* Vol. 237, Bellingham, WA (1980).
11. D.M. Williamson, "The monochromatic quartet explained," in *Lens Design,* Warren Smith, Ed., SPIE Press, Bellingham, WA (1992).
12. R. Mercado and T. Matsuyama, "Microlithographic lenses," in *SPIE Proc.* Vol. 3482, Bellingham, WA (1998).
13. W. Ulrich, "Trends in optical design of projection lenses for UV and EUV lithography," in *SPIE Proc.* Vol. 4146, Bellingham, WA (2000).

14. D.M. Williamson, "Catadioptric microlithographic reduction lenses," *Proc. International Optical Design Conference*, Vol. 22, OSA (1994).
15. D.M. Williamson, "Evolution of ring field systems in microlithography," in *SPIE Proc.* Vol. 3482, Bellingham, WA (1998).

CHAPTER 9
ZOOM LENSES

In this chapter we will consider some general aspects of zoom lens design. In particular we will see why several different types of zoom lenses exist, and discuss the techniques for designing them.

9.1 General principles

In all zoom lenses there are at least two moving groups of lenses to satisfy the following two conditions:

a. The image size must change—either the focal length or the magnification varies;

b. The image must stay in focus.

Very often, one moving group (the "variator") produces the change in magnification, and the other (the "compensator") moves to maintain focus.

9.1.1 Control of chromatic aberration

In many zoom systems, each group is made achromatic. In this way, color correction is maintained over the complete zoom range. If each group of lenses was a thin achromat, the color correction would be exact. However, since the groups are necessarily thick, the color correction is not exact, but the principle often applies.

9.1.2 Field curvature

The Petzval sum for a complete system is given by[1]

$$P = -\sum c \times \delta\left(\frac{1}{n}\right), \tag{9.1}$$

and this is clearly not affected by movements of the lenses. Therefore, if the Petzval sum is corrected at any zoom position, it is corrected at all zoom positions.

9.1.3 Minimization of movements

If the change of magnification is produced by one lens group, the object-image distance for that group will vary. For a thin lens of focal length f', at a magnification m (remember, m is usually negative) the object-image distance is

$$f'\left(2 - m - \frac{1}{m}\right). \tag{9.2}$$

Therefore, for a given zoom range, the movement of the compensator will be minimized if the magnification varies symmetrically around a magnification of 1. For example, if the zoom range is 4:1, the magnification of the variator should ideally be from 0.5 to 2.

9.2 Two-component zooms

Since a zoom lens must have a variable focal length, and since this is achieved by moving lenses along the optical axis, the simplest possible type of zoom lenses are two-component systems. In order to correct Petzval sum, it is necessary to have at least one negative component, so there are two classes of two-component zooms, some with a plus-minus configuration, and some with a minus-plus configuration.

9.2.1 Minus-plus plastic disposable zoom

In a recent patent from Kodak (U.S. Patent 5,473,473), Estelle and Lewis show designs for a plastic zoom lens for disposable cameras. The examples included in the patent cover both the plus-minus and the minus-plus configurations.[2] The aspheric coefficients in Fig. 9.1 have been reoptimized to correct a possible error

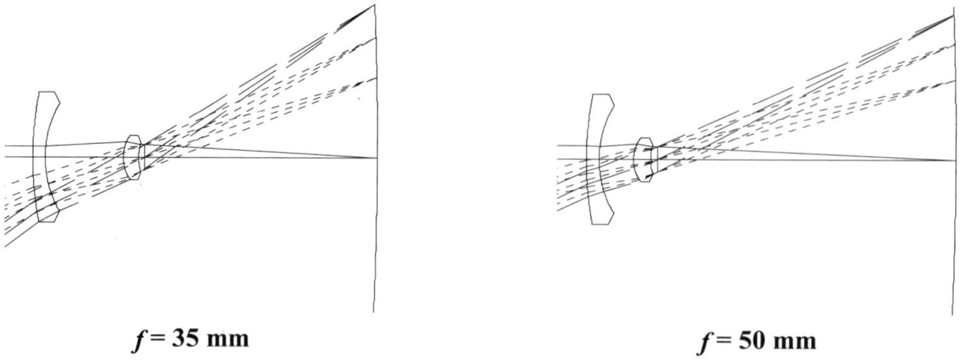

Figure 9.1 Minus-plus plastic disposable zoom.

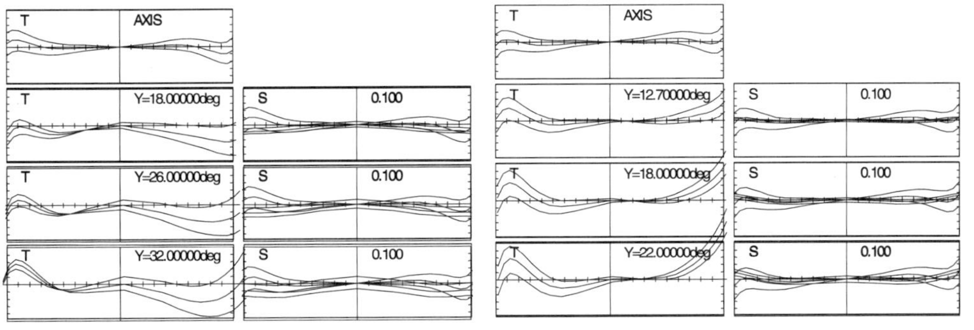

Figure 9.2 Transverse ray aberrations for focal lengths 35 and 50 mm.

in the patent specification. Compared with the single disposable lens described in Chapter 4, the use of two components, each aspherized, allows the correction of coma and astigmatism, as well as spherical aberration. Field curvature is not correctable, and the film is therefore curved onto a cylinder, upon which the aberrations of Fig. 9.2 are calculated. Again, axial color is not correctable. The design has a long back focus (long enough for a 35-mm SLR!) and an impressively large front lens.

```
WAVELENGTHS   [nm]
    587.60       656.30       486.10
SURFACES
# SURF SPACE  RADIUS        SEPN       INDEX1       V      CLR RAD  GLASS
0                           0.000    1.000000
1 S             44.073      2.000    1.493800    57.67      9.099  G-PMM
2 A    Z        13.276     13.108    1.000000               7.501
3 A              4.869      2.665    1.493800    57.67      3.059  G-PMM
4 A              6.970      1.000    1.000000               2.042
5#S    Z       Plane       38.404    1.000000               1.786
6 S           -459.600                                     21.564

ASPHERIC SURFACE  2 CC =  -0.769178  A4 = 4.90061e-05   A6 =  1.66533e-007
ASPHERIC SURFACE  3 CC =   0.595318  A4 = -.000139181   A6 = -1.16452e-005
ASPHERIC SURFACE  4 CC =   6.125970  A4 = 0.000147821   A6 = -0.000196982

ZOOM POSITION 1  EFL =   35.019  SEPARATION 2 = 13.108  SEPARATION 5 = 38.404
ZOOM POSITION 2  EFL =   41.745  SEPARATION 2 =  8.939  SEPARATION 5 = 42.346
ZOOM POSITION 3  EFL =   49.857  SEPARATION 2 =  5.407  SEPARATION 5 = 47.101
```

9.2.2 Plus-minus plastic disposable zoom

The design in Fig. 9.3 shows the basic principles of the plus-minus zoom lens. From simple paraxial optics the back focus will always be small in this layout, and the lens is therefore compact. In this layout the aperture stop and the shutter will be in front of the lens, which may not be aesthetically desirable. Transverse ray aberrations for this design, again on a curved image, are shown in Fig. 9.4.

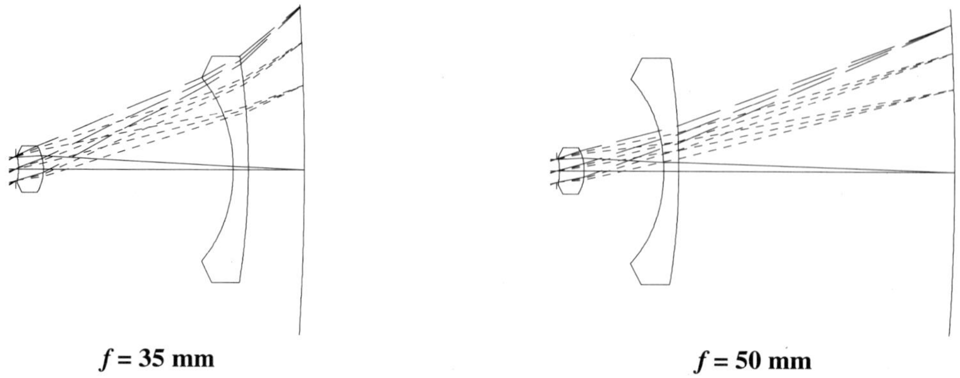

Figure 9.3 Plus-minus plastic disposable zoom.

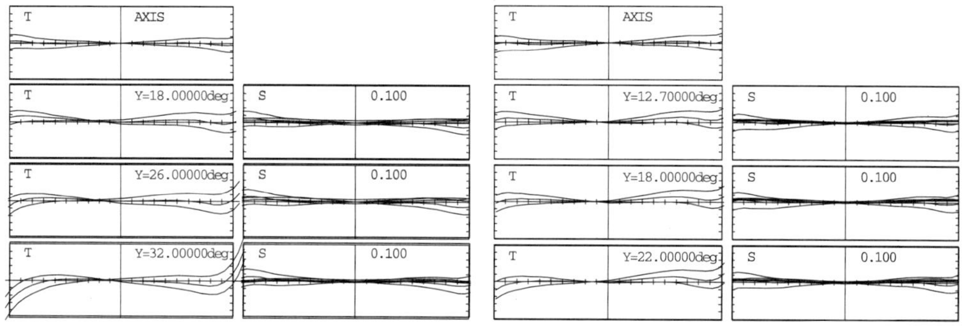

Figure 9.4 Transverse ray aberrations for focal lengths 35 and 50 mm.

WAVELENGTHS [nm]
587.60 656.30 486.10

SURFACES

# SURF SPACE	RADIUS	SEPN	INDEX1	V	CLR RAD	GLASS
0		0.000	1.000000			
1#S	Plane	0.340	1.000000		1.756	
2 A	-10.355	3.300	1.493800	57.67	1.856	G-PMM
3 A Z	-6.757	25.751	1.000000		3.119	
4 A	-22.412	2.000	1.493800	57.67	12.448	G-PMM
5 S Z	-100.678	7.572	1.000000		15.322	
6 S	-370.000				22.285	

ASPHERIC 2 CC = -34.2514 A4 = -0.00481453 A6 = 0.000515517 A8 = -5.59759e-5
ASPHERIC 3 CC = 0.0000 A4 = -0.000241489 A6 = -1.82039e-5
ASPHERIC S CC = 0.0000 A4 = -2.09391e-5 A6 = -2.34944e-8

ZOOM POSITION 1 EFL = 34.997 SEPARATION 3 = 25.751 SEPARATION 5 = 7.572
ZOOM POSITION 2 EFL = 42.650 SEPARATION 3 = 16.628 SEPARATION 5 = 22.484
ZOOM POSITION 3 EFL = 49.963 SEPARATION 3 = 10.522 SEPARATION 5 = 36.782

The layout of these two simple zoom lenses shows how more complex lenses can work. We have already seen in Chapter 3 that if the aspheric single lenses are replaced by doublets with spherical surfaces, the doublets can usually be designed to have the same third-order aberrations as the single lenses, but they can also be corrected for chromatic aberration. It is also necessary in a normal camera with a flat film to correct the Petzval sum, and something more is needed. If the positive component is replaced by a Cooke triplet, field curvature should be correctable.

9.2.3 A typical minus-plus zoom

These principles are illustrated in the next design, by Ogino (Minolta), a typical system for 35-mm single-lens reflex cameras. The focal length varies from 36 to 68 mm. The design in Fig. 9.5 retains one aspheric surface on the first glass element, and has been reoptimized from U.S. Patent 4,560,253.

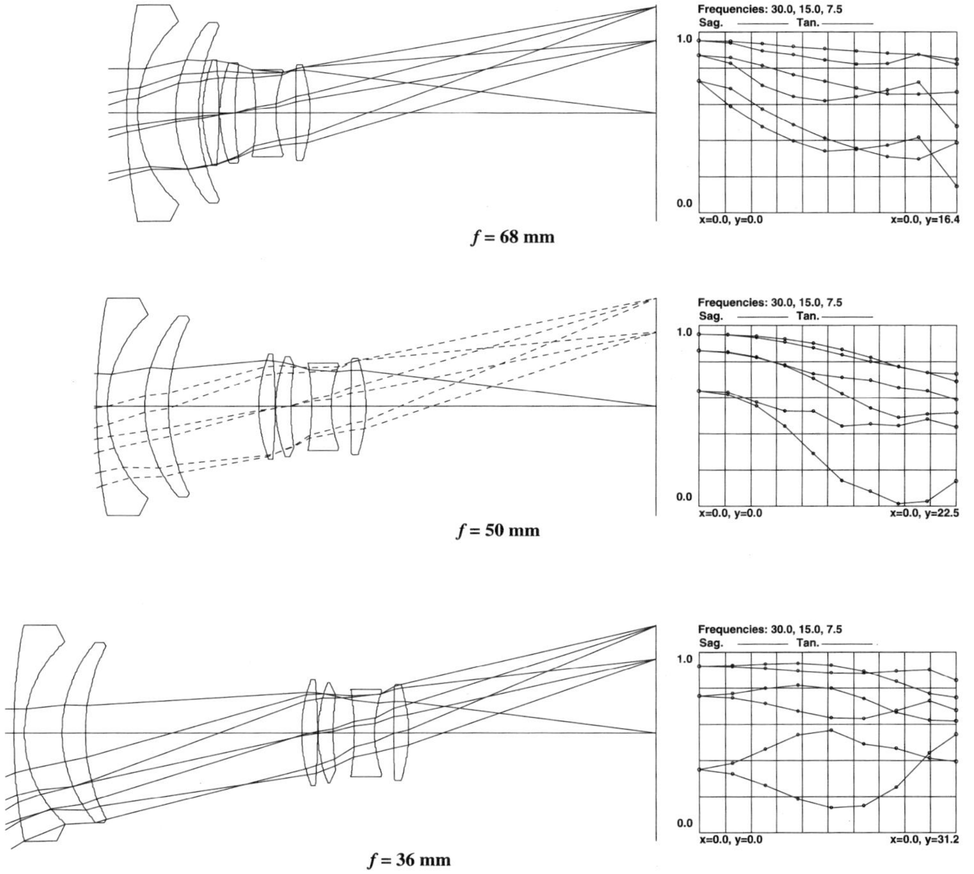

Figure 9.5 Ogino minus-plus zoom.

WAVELENGTHS [nm]

587.60 656.30 486.10

SURFACES

# SURF	SPACE	RADIUS	SEPN	INDEX1	V	CLR RAD	GLASS
0			0.000	1.000000			
1	A	113.015	2.000	1.716995	47.94	20.543	S-LAF3

2	S		23.687	7.000	1.000000		17.345	
3	S		26.826	4.300	1.805176	25.38	17.150	S-SFL6
4	S	Z	36.783	0.800	1.000000		16.434	
5	S		28.471	3.000	1.607288	59.44	9.979	S-SK7
6	S		-119.721	0.150	1.000000		9.923	
7	S		26.351	2.800	1.691000	54.69	9.532	S-LAK9
8#	S		54.138	3.900	1.000000		9.078	
9	S		-51.938	3.700	1.805176	25.38	8.300	S-SFL6
10	S		23.517	3.600	1.000000		7.908	
11	S		276.169	2.800	1.749499	34.94	8.799	S-LAFN7
12	S	Z	-29.165	64.084	1.000000		9.129	
13	S		Plane				20.539	

ASPHERIC SURFACE 1 CC = -2.745020 A6 = 1.01126e-009

ZOOM POSITION 1 EFL = 68.200 SEPARATION 4 = 0.800 SEPARATION 12 = 64.084
ZOOM POSITION 2 EFL = 50.001 SEPARATION 4 = 16.860 SEPARATION 12 = 53.787
ZOOM POSITION 3 EFL = 36.000 SEPARATION 4 = 40.270 SEPARATION 12 = 45.865

9.2.4 A typical plus-minus zoom

The example shown in Fig. 9.6 is a design for a compact camera, from example 1 in U.S. Patent 4,906,077 by Takayuki Ito, of Asahi. The front component is a weak aspheric plastic lens.

WAVELENGTHS [nm]
587.60 656.30 486.10

SURFACES

#	SURF	SPACE	RADIUS	SEPN	INDEX1	V	CLR RAD	GLASS
0				0.000	1.000000			
1	S		20.319	1.879	1.491757	57.43	7.507	G-PMM
2	A		29.887	2.100	1.000000		6.833	
3	S		-20.317	6.552	1.739992	28.28	6.789	O-PBH3
4	S		-30.192	3.695	1.000000		6.730	
5	S		33.138	4.178	1.588994	48.57	5.863	O-BAM6
6	S		-11.078	1.300	1.805175	25.42	5.474	O-TIH6
7	S		-19.923	0.000	1.000000		5.325	
8#	S	Z	Plane	11.877	1.000000		5.157	
9	S		-54.991	3.010	1.805175	25.42	8.679	O-TIH6
10	S		-18.275	2.203	1.000000		8.996	
11	S		-14.431	1.350	1.833995	37.15	8.982	O-LAH60
12	S		-100.764	2.289	1.000000		10.300	
13	S		-20.812	1.400	1.833995	37.15	10.338	O-LAH60
14	S	Z	-46.298	8.946	1.000000		11.820	
15	S		Plane				21.973	

ASPHERIC 2 CC = 0.000000 A4 = 2.12334e-5 A6 = -2.26565e-8 A8 = 2.15426e-9

ZOOM POSITION 1 EFL = 36.002 SEPARATION 8 = 11.877 SEPARATION 14 = 8.946
ZOOM POSITION 2 EFL = 49.995 SEPARATION 8 = 6.697 SEPARATION 14 = 23.253
ZOOM POSITION 3 EFL = 67.979 SEPARATION 8 = 3.172 SEPARATION 14 = 42.025

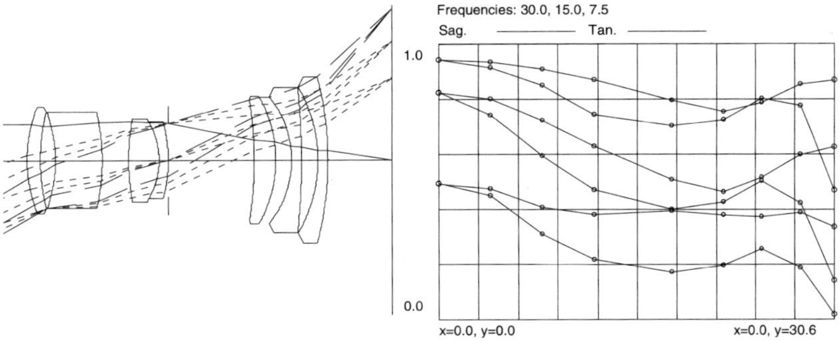

f = 36 mm

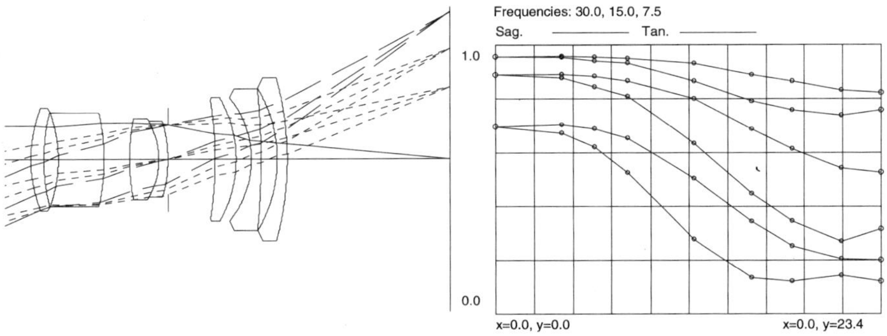

f = 50 mm

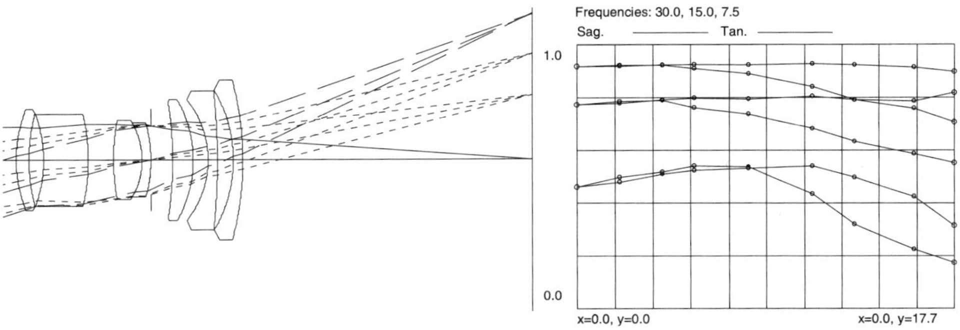

f = 68 mm

Figure 9.6 Ito plus-minus zoom.

9.3 Three-component zooms

Three-component zoom systems are able to give larger zoom ranges than two-component systems, and can solve the problem of the variable stop diameter if the system is one in which the first two lenses move and the third lens is stationary. In this case, the stop is near the third lens, has a constant distance from the image surface, and can therefore have a zoom-independent aperture control. Several combi-

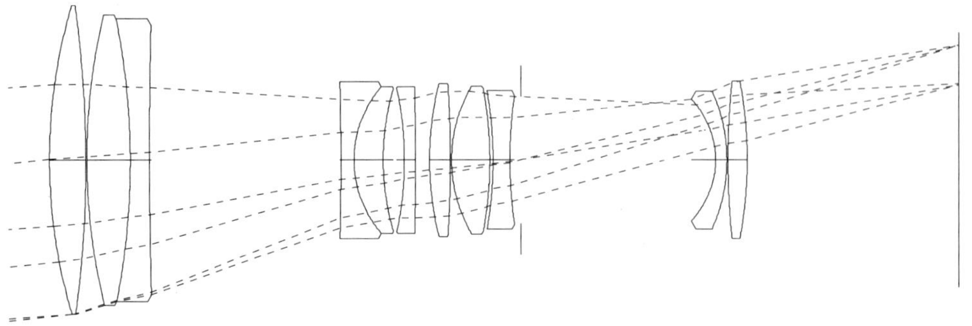

***f* = 193 mm**

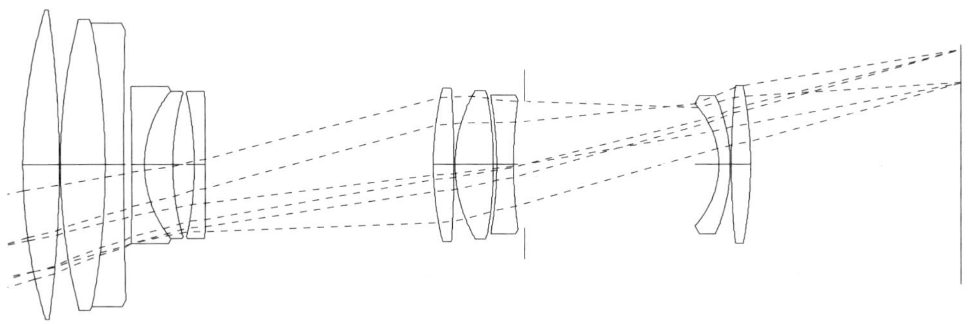

***f* = 70.6 mm**

Figure 9.7 Betensky three-component zoom.

nations of positive and negative components are possible, but the plus-minus-plus system seems to be most common.

We show in Fig. 9.7 one example for 35-mm cameras, by Betensky (Vivitar).[3,4] The focal length varies from 71 to 193 mm, and the third component is fixed.

WAVELENGTHS [nm]

587.60 656.30 486.10

SURFACES

#	SURF	SPACE	RADIUS	SEPN	INDEX1	V	CLR RAD	GLASS
0				0.00000	1.000000			
1	S		100.61000	6.50000	1.500000	66.40	28.000	User-USER
2	S		-208.00000	0.20000	1.000000		27.908	
3	S		108.10000	8.10000	1.534000	62.00	26.338	User-USER-2
4	S		-130.00000	3.50000	1.797000	26.80	25.703	User-USER-3
5	S	Z	965.06000	34.46000	1.000000		24.418	
6	S		-608.86000	2.30000	1.794000	45.18	14.234	User-USER-4
7	S		21.02000	5.10000	1.847000	24.84	13.337	User-USER-5
8	S		47.40000	3.81000	1.000000		13.030	
9	S		-61.93000	2.00000	1.815000	44.90	13.016	User-USER-6
10	S	Z	-717.67000	2.55000	1.000000		13.288	
11	S		58.55000	3.70000	1.742000	49.10	13.881	User-USER-7
12	S		-206.21000	0.20000	1.000000		13.838	
13	S		27.13000	6.80000	1.487000	68.40	13.423	User-USER-8
14	S		-73.05000	0.59000	1.000000		12.910	

15 S	-74.44000	3.00000	1.805000	25.50	12.569	User-USER-9
16 S	83.46000	2.00000	1.000000		11.741	
17#S	Plane	34.59000	1.000000		11.518	
18 S	-16.48000	2.00000	1.843000	38.70	11.000	User-USER-10
19 S	-31.29000	0.20000	1.000000		12.546	
20 S	144.41000	3.40000	1.796000	25.50	14.040	User-USER-11
21 S	-79.57000	37.88072	1.000000		14.288	
22 S	Plane				20.912	

```
ZOOM POSITION 1  EFL =   192.875   SEPARATION 5 = 34.460   SEPARATION 10 =  2.550
ZOOM POSITION 2  EFL =    70.636   SEPARATION 5 =  1.500   SEPARATION 10 = 41.000
```

Since the variator is the only component that changes the magnification, in almost all cases its magnification changes approximately symmetrically about the unit magnification case. This minimizes the movements and therefore the size of the outer components. In this example, the magnification of the variator changes from about –0.67 to about –1.77.

Before discussing other zoom systems, we should briefly mention the principle of optical compensation. It is already clear that in the two- and three-component zoom systems described above, a nonlinear movement of at least one of the lenses is needed. This is mechanically complex, and the need for these nonlinear movements can be removed by using optical compensation.

Suppose that we have three lenses, and lens 1 and 3 are moving while lens 2 is stationary. It is possible to find solutions for the powers and separations such that the back focus is constant at three zoom positions, and varies approximately cubically with zoom. We will not give examples of optical compensation, but the following U.S. patents may be of interest.

2,566,485	Cuvillier Pan Cinor 3:1 zoom range
2,902,901	2.3:1 zoom range
2,913,957	2.3:1 zoom range

It would appear that optical compensation is no longer popular, presumably because designers have found that better aberration correction and exact control of focus are possible using mechanical compensation.

9.4 Four-component zooms

If we use four lens groups, we can keep lens 1 and lens 4 fixed with zoom, and move only lens 2 and lens 3 to zoom. This means that lens 1 can be used to focus, independently of zoom position, and the stop can be near lens 4, where it is also independent of zoom. This is a very favorable arrangement, and we give just two examples.

In British Patent 685,945, H.H. Hopkins gives a very clever example of how the correction of aberrations at three zoom positions can be considered, as shown in Fig. 9.8.[5] It is noteworthy that this lens was designed before computers.

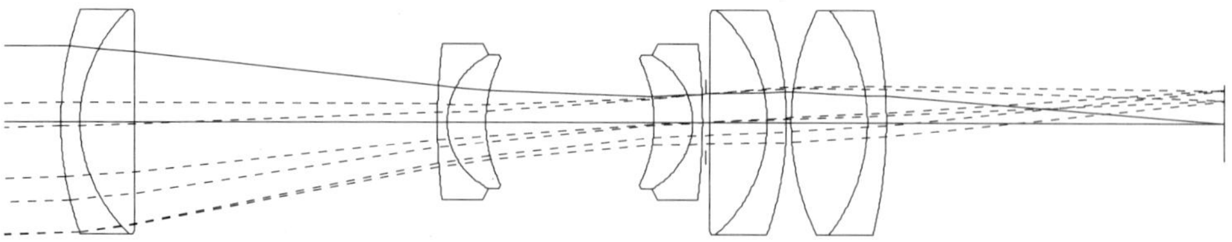

telescope section at √5 magnification

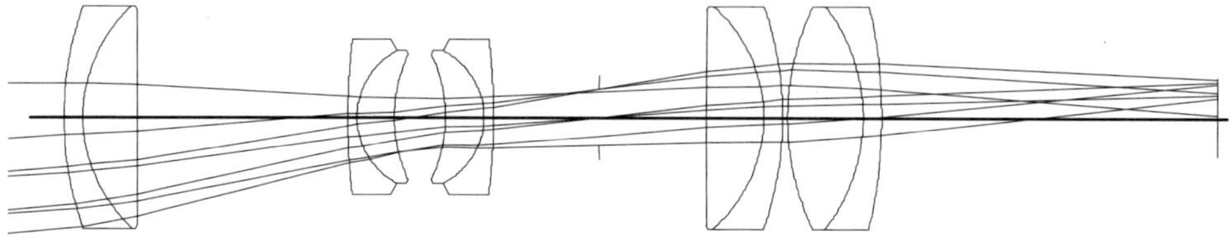

telescope section at 1:1

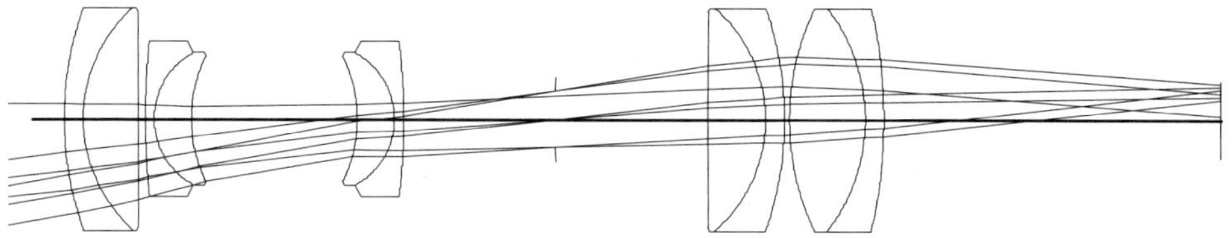

telescope section at 1/√5 magnification

Figure 9.8 H.H. Hopkins four-component zoom.

WAVELENGTHS [nm]

587.60	656.30	486.10

SURFACES

#	SURF	SPACE	RADIUS	SEPN	INDEX1	V	CLR RAD	GLASS
0				0.00000	1.000000			
1	S		182.70000	10.00000	1.625358	35.55	59.197	S-F7
2	S		80.29000	30.00000	1.572498	57.53	55.896	S-BAK1
3	S	Z	Plane	160.66000	1.000000		55.015	
4	S		357.00000	5.00000	1.572498	57.53	41.603	S-BAK1
5	S		40.64000	20.00000	1.626058	38.95	35.667	S-BASF1
6	S	Z	77.84000	88.42000	1.000000		33.781	
7	S		-77.84000	20.00000	1.626058	38.95	20.198	S-BASF1
8	S		-40.64000	5.00000	1.572498	57.53	20.205	S-BAK1
9	S	Z	-357.00000	1.93500	1.000000		20.017	
10#	S	Z	Plane	1.93500	1.000000		15.140	
11	S		Plane	30.00000	1.572498	57.53	28.923	S-BAK1
12	S		-80.29000	10.00000	1.625358	35.55	31.476	S-F7
13	S		-182.70000	2.50000	1.000000		33.063	
14	S		142.10000	40.00000	1.516798	64.14	33.808	S-BK7
15	S		-99.40000	10.00000	1.698947	30.06	32.813	S-SF15
16	S		-261.20000	177.18046	1.000000		32.926	
17	S		Plane				20.876	

```
EFL = 505.909  D(3) = 160.660  D(6) = 88.420  D(9) =  1.935  D(10) =  1.935
EFL = 233.319  D(3) = 112.750  D(6) = 27.460  D(9) = 56.375  D(10) = 56.375
EFL = 103.358  D(3) =   3.870  D(6) = 88.420  D(9) = 80.330  D(10) = 80.330
```

The first four doublets form an afocal zoom system and, moreover, lens 1 and lens 4 are identical, as are lens 2 and lens 3. Doublet 5 focuses the collimated light to a real image. As we see, all lenses are achromatic doublets, which automatically corrects C_1 and C_2 at all zoom positions, thin-lens-wise. The Petzval sum is of course independent of zoom in all zoom lenses, so we never need to consider variation of Petzval sum with zoom.

Considering the afocal system alone, there can be a unit magnification position (zoom position 2) in which the afocal system is symmetric; if the stop is in the center of the afocal system, the odd aberrations are automatically zero, so the only aberrations to consider are S_1, S_3, and S_4, and we have already seen that S_4 is constant with zoom.

Considering one end of the zoom range (zoom position 1), the system is not symmetric, so we need to correct all of the Seidel aberrations if possible. The other end of the zoom range (zoom position 3) is simply the mirror image of position 1, so it does not have to be designed separately.

Basically, therefore, we need only consider S_1 and S_3 in zoom position 2, and S_1, S_2, S_3, and S_5 in zoom position 1. We also need to design lens 5, the final doublet, but that is not difficult. The design process has been greatly simplified compared with the horrendous problem of thinking about seven Seidel aberrations in three zoom positions. These simple principles, though very elegant, do not achieve the performance that is required in modern zoom lenses, and more complex designs are now universal. This design does however, illustrate three useful general principles:

1. The plus-minus-minus-plus arrangement is very suitable for large zoom ranges.

2. Each component is separately achromatic.

3. If the zoom section is corrected for constant aberration (but not zero) the final fixed lens can correct the aberrations of the zoom section.

A more modern example is described in U.S. Patent 4,763,998 by Tsui et al., and is shown in Fig. 9.9. This has a zoom range of 11.5 to 1 and is intended for small video cameras. It follows the same principles as the Hopkins zoom lens, but the rear fixed component is clearly more complex. Focusing is done by moving the front group; this is a common method of focusing, but it forces the designer to increase the diameter of the front lens, or causes vignetting, or limits the range of focus.

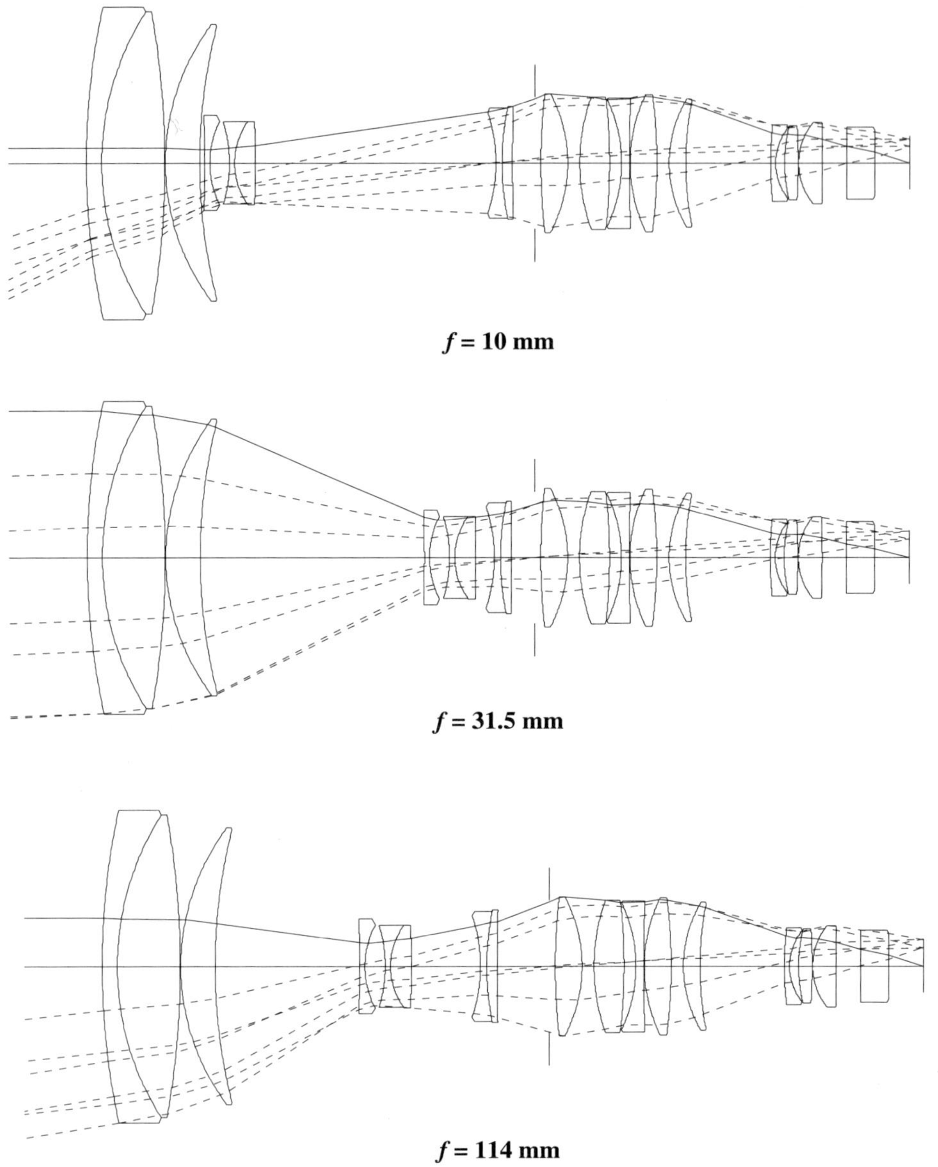

Figure 9.9 Tsui four-compenent zoom.

```
WAVELENGTHS  [nm]
     587.60       656.30       486.10
SURFACES
  # SURF SPACE    RADIUS        SEPN      INDEX1       V     CLR RAD  GLASS
  0                           0.00000   1.000000
  1 S           145.23300     3.00000   1.805176    25.42   30.541  S-SF6
  2 S            55.60227    12.20000   1.622988    58.04   29.579  S-SK15
  3 S          -171.78947     0.10000   1.000000            29.515
  4 S            44.80228     6.80000   1.622988    58.04   27.065  S-SK15
  5 S      Z    108.09614     1.00000   1.000000            26.650
```

6 S		-275.94960	1.00000	1.834000	37.20	9.263	User-GLASS
7 S		18.27300	3.80000	1.000000		8.131	
8 S		-27.65162	1.00000	1.785900	44.20	7.934	User-LAF10
9 S		14.09063	4.00000	1.846656	23.82	8.052	S-SF57
10 S	Z	-310.16384	46.31000	1.000000		8.132	
11 S		-31.60796	1.20000	1.713000	53.80	9.960	User-USER
12 S		56.29710	2.00000	1.846656	23.82	10.820	S-SF57
13 S	Z	560.37400	4.24000	1.000000		11.022	
14#S		Plane	1.20000	1.000000		12.894	
15 S		119.54100	5.20000	1.696798	55.39	13.480	S-LAKN14
16 S		-29.16168	2.30000	1.000000		13.609	
17 S		38.77655	5.80000	1.516798	64.14	12.959	S-BK7
18 S		-117.98387	2.30000	1.000000		12.465	
19 S		-28.94210	1.50000	1.846656	23.82	12.411	S-SF57
20 S		485.06233	0.10000	1.000000		12.776	
21 S		31.11849	5.20000	1.516798	64.14	13.392	S-BK7
22 S		-104.68921	2.40000	1.000000		13.346	
23 S		24.98840	3.20000	1.516798	64.14	12.548	S-BK7
24 S		66.52836	16.20000	1.000000		12.289	
25 S		64.87830	1.00000	1.883000	40.80	7.476	User-GLASS-2
26 S		11.30668	1.70000	1.000000		6.937	
27 S		30.39576	2.60000	1.517419	52.18	6.994	S-KF6
28 S		-117.55579	0.10000	1.000000		7.251	
29 S		13.02200	4.55000	1.517419	52.18	7.990	S-KF6
30 S		-444.75735	4.55000	1.000000		7.805	
31 S		Plane	5.45000	1.516798	64.14	7.011	S-BK7
32 S		Plane	6.75049	1.000000		6.398	
33 S		Plane				5.236	

EFL = 10.000	D(5) = 1.000	D(10) = 46.310	D(13) = 4.240
EFL = 31.571	D(5) = 28.270	D(10) = 13.340	D(13) = 9.940
EFL = 114.545	D(5) = 43.480	D(10) = 3.550	D(13) = 4.520

Other recent video zoom lenses also have four groups, but only groups 2 and 4 move to zoom.[6] The example in Fig. 9.10 is from U.S. Patent 5,296,969 by Mihara (Olympus), and uses two aspherics. Groups 1 and 3 are fixed, and group 4 moves to focus. This reduces the size of the front component.

WAVELENGTHS [nm]
587.60 656.30 486.10
SURFACES

# SURF	SPACE	RADIUS	SEPN	INDEX1	V	CLR RAD	GLASS
0			0.000	1.000000			
1 S		32.104	1.000	1.846655	23.82	12.721	S-SF57
2 S		20.132	4.500	1.568728	63.06	12.098	S-PSK2
3 S		419.848	0.150	1.000000		12.015	
4 S		22.600	3.100	1.603108	60.57	11.461	S-SK14
5 S	Z	108.487	0.900	1.000000		11.277	
6 S		35.034	0.700	1.772497	49.58	5.280	O-LAH66
7 S		6.839	2.850	1.000000		4.301	
8 S		-9.211	0.700	1.487489	70.38	4.186	S-FK5
9 S		12.045	2.050	1.846655	23.82	4.089	S-SF57
10 SU	Z	105.236	17.746	1.000000		3.969	
11 S		Plane	1.400	1.000000		4.285	
12 A		6.363	3.864	1.677899	55.32	5.215	O-LAL12

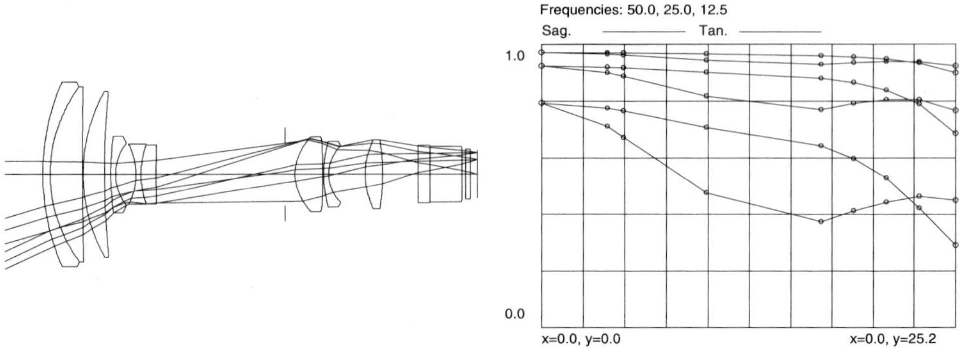

f = 6.7 mm

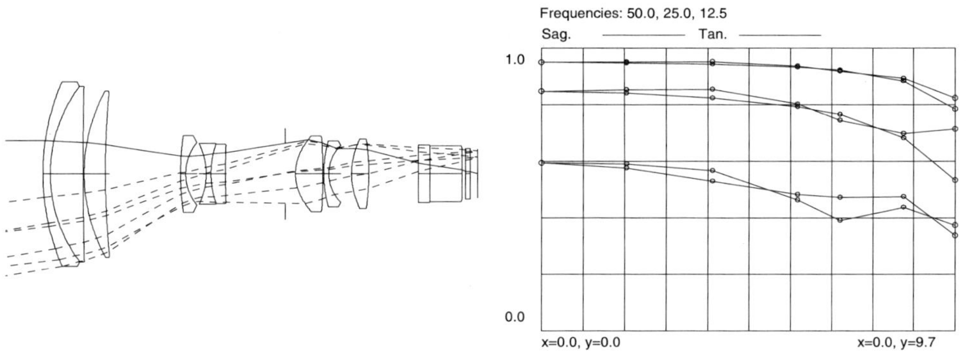

f = 18.384 mm

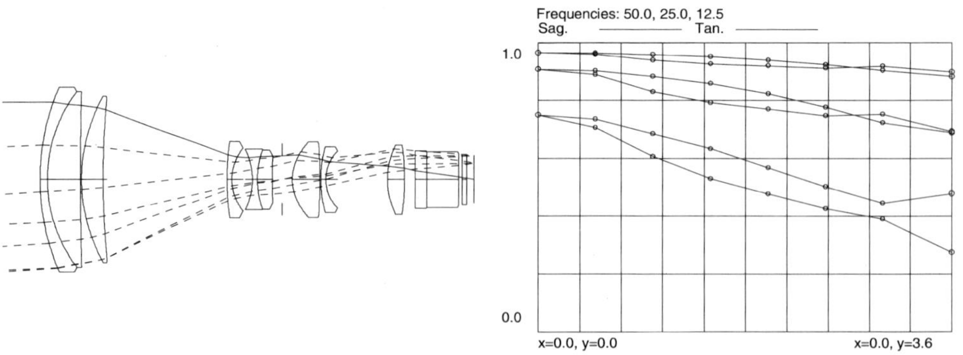

f = 50.433 mm

Figure 9.10 Mihara four-compent zoom.

13 S		-95.104	0.150	1.000000		4.899	
14 S		16.789	0.700	1.846655	23.82	4.524	S-SF57
15 S	Z	5.515	4.933	1.000000		3.922	
16 A		10.078	2.350	1.589128	61.24	4.758	S-SK5
17 S	Z	-32.992	4.977	1.000000		4.720	
18 S		Plane	1.600	1.516330	64.12	3.956	O-BSL7
19 S		Plane	4.400	1.547710	62.81	3.834	O-BAL21
20 S		Plane	0.500	1.000000		3.505	
21 S		Plane	0.600	1.487489	70.38	3.447	S-FK5
22 S		Plane	0.996	1.000000		3.401	
23 S		Plane				3.301	

```
ASPHERIC SURFACE 12 CC =  -0.764779    A4 = 2.91406e-005
ASPHERIC SURFACE 16 CC =  -1.361050    A4 = -8.48372e-006

EFL =  6.700  D(5) =    .900   D(10) = 17.746   D(15)= 4.933   D(17) = 4.977
EFL = 18.384  D(5) = 10.436   D(10) =  8.210   D(15)= 3.100   D(17) = 6.810
EFL = 50.433  D(5) = 17.346   D(10) =  1.300   D(15)= 8.410   D(17) = 1.500
```

Figure 9.11 shows our last example of this type of video zoom design, from U.S. Patent 5,424,869 by Nanjo (Sony). Because of their relatively high expansion coefficients, the three plastic lenses are single elements, rather than being parts of cemented doublets. In other respects this design is similar to U.S. Patent 5,296,969. Again, only groups 2 and 4 move to zoom. Groups 1 and 3 are fixed, and group 4 moves to focus.

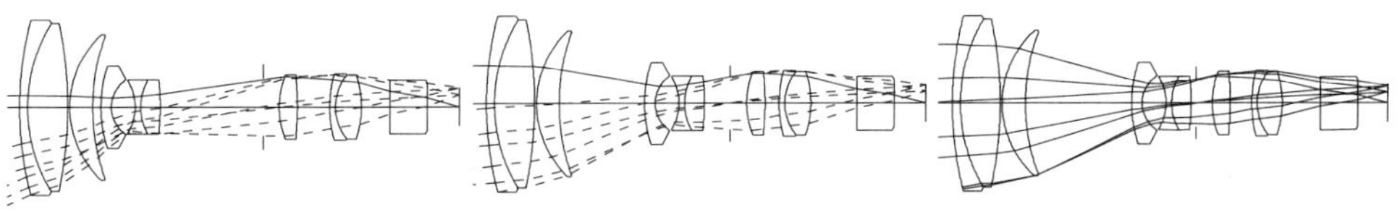

f = 5.3 mm *f* = 19.142 mm *f* = 30.743 mm

Figure 9.11 Nanjo four-component zoom.

```
WAVELENGTHS  [nm]
    587.60      656.30      486.10
SURFACES
```

# SURF	SPACE	RADIUS	SEPN	INDEX1	V	CLR RAD	GLASS
0			0.000	1.000000			
1 S		43.803	1.000	1.805175	25.42	10.953	S-SF6
2 S		20.488	4.920	1.516798	64.14	10.503	S-BK7
3 S		-52.953	0.250	1.000000		10.398	
4 S		12.580	2.830	1.491757	57.43	9.139	G-PMM
5 A	Z	27.563	1.000	1.000000		8.932	
6 S		16.531	1.300	1.491757	57.43	5.142	G-PMM
7 A		3.945	2.960	1.000000		3.413	
8 S		-6.312	0.700	1.723723	38.10	3.296	S-BASF51
9 S		6.890	2.420	1.846655	23.82	3.361	S-SF57
10 S	Z	-35.183	12.760	1.000000		3.343	
11 S		Plane	2.000	1.000000		3.563	
12 S		10.650	2.260	1.491757	57.43	3.940	G-PMM
13 A	Z	-28.576	3.967	1.000000		4.002	
14 S		15.856	0.700	1.846655	23.82	4.123	S-SF57

15	S		6.409	3.200	1.589128	61.24	3.975	S-SK5
16	S	Z	-12.449	3.439	1.000000		3.990	
17	S		Plane	4.600	1.516798	64.14	3.336	S-BK7
18	S		Plane	3.941	1.000000		2.908	
19	S		Plane				2.348	

```
ASPHERIC SURFACE  5 CC = 0.000  A4 = 9.26576e-006  A6 =  4.45927e-008
                                                   A8 = -2.56324e-010
ASPHERIC SURFACE  7 CC = 0.000  A4 = -0.000135126  A6 = -0.000140435
                                                   A8 =  1.5808e-005
                                                  A10 = -9.6286e-007
ASPHERIC SURFACE 13 CC = 0.000  A4 = 0.000350662   A6 =  5.80323e-006
                                                   A8 = -2.09823e-007

ZOOM POSITION 1  EFL =   5.300  D(5) = 1.000  D(10) = 12.760  D(13) = 3.967
                                                              D(16) = 3.439
ZOOM POSITION 2  EFL =  19.142  D(5) =10.520  D(10) =  3.240  D(13) = 1.710
                                                              D(16) = 5.698
ZOOM POSITION 3  EFL =  30.743  D(5) =13.160  D(10) =  0.600  D(13) = 2.688
                                                              D(16) = 4.719
```

9.5 Zoom relays

Another example of a zoom system is the variable magnification copying lens. Several of these are described in the patent literature, but the design shown in Fig. 9.12, by Hirano (Fuji), is one of the more simple examples. The design operates over a range of magnification from –0.615 to –1.414, including, of course, a magnification of –1.0. It is required to have a constant object-image distance at all magnifications, and is symmetric at the –1.0 position. This symmetry is, perhaps surprisingly, not seen in all such designs. As the lens changes magnification, the zoom movements are symmetric in the sense that the configuration at –1.5 is the reverse of the configuration at –0.66.

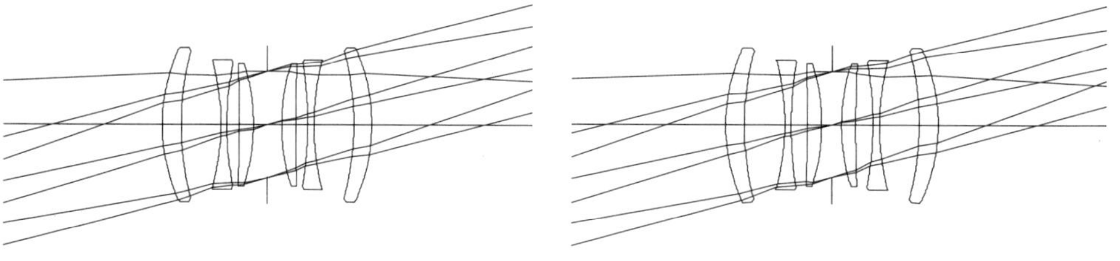

Figure 9.12 Zoom lens for photocopier, at magnifications of –1 and –0.66.

```
WAVELENGTHS   [nm]
      587.60      656.30      486.10
SURFACES
```

#	SURF SPACE	RADIUS	SEPN	INDEX1	V	CLR RAD	GLASS
0			0.00000	1.000000			
1	S	105.52800	10.70500	1.603108	60.57	44.604	S-SK14
2	S	166.71100	22.78000	1.000000		42.347	

3 S		-140.25600	3.46100	1.603417	38.01	37.174	S-F5
4 S	Z	214.24900	7.00000	1.000000		35.703	
5 S		-843.00000	8.20000	1.784429	43.93	35.320	S-LAFN10
6 S	Z	-114.09700	8.25000	1.000000		35.187	
7#S	Z	Plane	8.25000	1.000000		30.289	
8 S		114.09700	8.20000	1.784429	43.93	35.187	S-LAFN10
9 S	Z	843.00000	7.00000	1.000000		35.320	
10 S		-214.24900	3.46100	1.603417	38.01	35.703	S-F5
11 S		140.25600	22.78000	1.000000		37.174	
12 S		-166.71100	10.70500	1.603108	60.57	42.347	S-SK14
13 S		-105.52800	630.50116	1.000000		44.604	
14 S		Plane				214.707	

```
ZOOM POSITION 1  D(0) = 630.501  D(4) = 7.000  D(6) =  8.250  D(7) = 8.250
                                                               D(9) = 7.000
ZOOM POSITION 2  D(0) = 512.519  D(4) = 8.339  D(6) =  6.911  D(7) = 6.911
                                                               D(9) = 8.339
ZOOM POSITION 3  D(0) = 696.886  D(4) = 7.418  D(6) =  7.832  D(7) = 7.832
                                                               D(9) = 7.418
ZOOM POSITION 4  D(0) = 794.480  D(4) = 9.643  D(6) =  5.607  D(7) = 5.607
                                                               D(9) = 9.643
```

9.6 Zoom telescopes

A third class of zoom system consists of variable magnification telescopes. Not many of these are described in the literature; there are some infrared zoom telescopes[7] such as the example by Noyes (U.S. Patent 3,947,084), shown in Fig. 9.13.

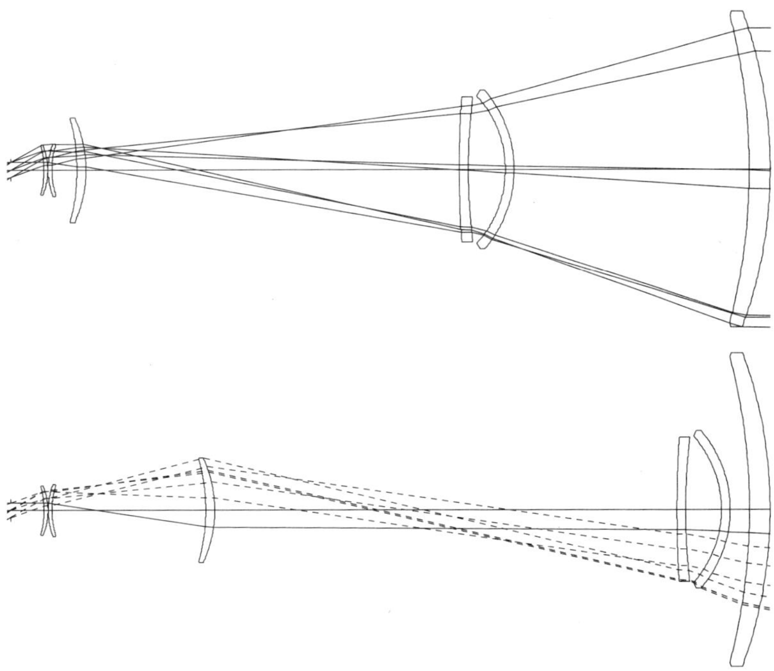

Figure 9.13 Noyes zoom infrared telescope.

```
WAVELENGTHS   [nm]
    10000.00    12000.00     8000.00
SURFACES
  # SURF SPACE     RADIUS         SEPN        INDEX1     V     CLR RAD   GLASS
  0                             0.00000     1.000000
  1#S              Plane       32.96920     1.000000              7.500
  2 S            -97.68840      4.06400     4.003286             24.702  User-GE
  3 S            -66.04000      0.25400     1.000000             25.419
  4 S             80.92440      3.55600     4.003286             26.375  User-GE
  5 S     Z       86.18220     25.00000     1.000000             25.669
  6 S           -195.24980      8.89000     4.003286             52.089  User-GE
  7 S     Z     -130.45440    374.30000     1.000000             53.341
  8 S           1165.88540      8.89000     2.406509             73.016  User-ZNSE
  9 S            620.47120     36.47440     1.000000             73.662
 10 S           -112.95380      8.89000     4.003286             74.888  User-GE
 11 S     Z     -123.92660    237.10000     1.000000             80.421
 12 S           -634.64440     21.59000     4.003286            156.505  User-GE
 13 S           -461.05276      0.00000     1.000000            160.021
 14 S              Plane                                        160.744

ZOOM POSITION 1   D(5) =  25.000  D(7) =  374.300  D(11) = 237.100
ZOOM POSITION 2   D(5) = 154.100  D(7) =  463.200  D(11) =  19.100
```

9.7 Zoom modules

It is probably clear that complex zoom lenses are not designed very quickly, and it is common practice to reuse zoom lens designs in various simple ways. For example, if we have a four-component zoom lens, we can design a new fourth component without any change in the first three components; in this way we can change the focal length and therefore also the *f*/number without any redesign of the front components. Another option is to scale the design; this changes the focal lengths without changing the *f*/number or the field angle.

If we want to change the field angle, we can redesign the front component to give a different focal length without making large changes to the rear three components, thus giving the same range of image diameters. This is not as simple as redesigning the back component, as the front component operates at a varying aperture whereas the rear component (in a simple four-component zoom) operates under conditions that do not change with zoom.

References

1. M.J. Kidger, *Fundamental Optical Design,* SPIE Press, Bellingham, WA (2002).
2. J.B. Caldwell, "Tracking zoom lens developments via the patent literature," in *SPIE Proc.* Vol. 4487, Bellingham, WA (2001).
3. E. Betensky, "The modern zoom lens for 35-millimeter photography," in *SPIE Proc.* Vol. 0531, Bellingham, WA (1985).
4. E. Betensky, "Zoom lens principles and types," in *Lens Design*, Warren Smith, Ed., SPIE Press, Bellingham, WA (1992).

5. H.H. Hopkins, "An analytical technique for stable aberration correction in zoom systems," in *SPIE Proc.* Vol. 0399, Bellingham, WA (1983).
6. K. Tanaka, "Recent development of zoom lenses," in *SPIE Proc.* Vol. 3129, Bellingham, WA (1997).
7. I.A. Neil, "Zoom lenses for the thermal infrared,"in *SPIE Proc.* Vol. 0399, Bellingham, WA (1983).

CHAPTER 10
DECENTERED AND ASYMMETRIC SYSTEMS

In the past, almost all optical design was concerned with systems having rotational symmetry, such as camera lenses, microscopes, etc. Other systems that did not have rotational symmetry (such as prismatic binoculars) could often be simulated assuming rotational symmetry, and techniques were developed which implicitly assumed that decentered and asymmetric systems (in the sense of having no axis of symmetry) were rarely required. Within the last couple of decades, this situation has entirely changed, and we are now faced with many instruments and devices that do not have an axis of symmetry.

The variety is enormous, and this chapter will not attempt to explain the design of specific decentered systems, but will instead discuss the general techniques common to many of these devices. Some of the comments below apply to systems with cylindrical or toroidal surfaces; although these systems do not have an axis of symmetry (they may have a plane of symmetry), they are not strictly decentered systems. However, we will use the term decentered to include all systems that are not rotationally symmetric.

10.1 General properties of decentered systems

Since decentered systems do not have an axis of symmetry, several of the assumptions that are familiar need to be revised. First, paraxial optics, which we can assume to be valid within a narrow cylinder close to the optical axis, is not valid if there is no optical axis. Obviously, in some simple cases, some of the results of paraxial optics will apply, but in complex situations none of the results are valid without modification. For example, we are used to the concepts of focal length and magnification. In the case of many anamorphic systems we deliberately set out to design systems with different focal lengths or magnifications in the two principal azimuths. By the term *principal azimuths*, we usually mean the plane of symmetry, and another plane orthogonal to it. This assumes that there is a plane of symmetry, but this is usually the case.

Further, in the symmetrical case we are familiar with the use of paraxial ray tracing to determine the image surface position, but we cannot do this in the decentered case. While we could use various pseudoparaxial techniques, we normally determine the back focus by real ray tracing, using two rays close together.

Another problem concerns astigmatism calculations. We are familiar with the concept of a tangential section (usually the y-z plane) in a centered system, and we can still use this concept in decentered systems with plane symmetry; but in the case of generalized decentered systems the concept becomes too confused to be useful. Still another problem concerns distortion calculations. Normally we define distortion as the error in image size compared with the image size predicted by paraxial optics. When paraxial optics is not valid, a completely new definition is required.

10.2 Coordinate systems

Let us first consider the case of centered systems. We adopt a coordinate system in which the *z*-axis defines the optical axis, the *y*-axis is orthogonal to the *z*-axis, and is usually shown in the plane of any diagrams, and the *x*-axis is orthogonal to the *y*- and *z*-axes. The system is assumed to be a right-handed system, so that in a normal diagram showing a section in the *y*-*z* plane, a positive value of *x* implies a point beneath the plane of the diagram.

Distances *D* along the axis between the vertices of successive surfaces are defined by the separations, with a positive separation implying that a surface is to the right of the preceding surface. Therefore, in the case of reflecting surfaces, where a secondary mirror is to the left of the primary mirror, the separation will normally be negative, as shown in Fig. 10.1.

Now consider the case of systems with plane symmetry, e.g., systems with only *y*-decentrations. Any surface may now have a *y*-decentration, but all of these decentrations are referred to a common axis, i.e. there are no local axes that differ from the common axis. The separations are now interpreted as the difference between the *z*-coordinates of the vertices of successive surfaces.

If a surface is tilted, we first apply the decentration (if there is one) and then rotate the surface about an *x*-axis passing through the new vertex. The sign convention is that a positive rotation of α represents a surface whose normal has a positive gradient, in the usual sense. An example is shown in Fig. 10.2.

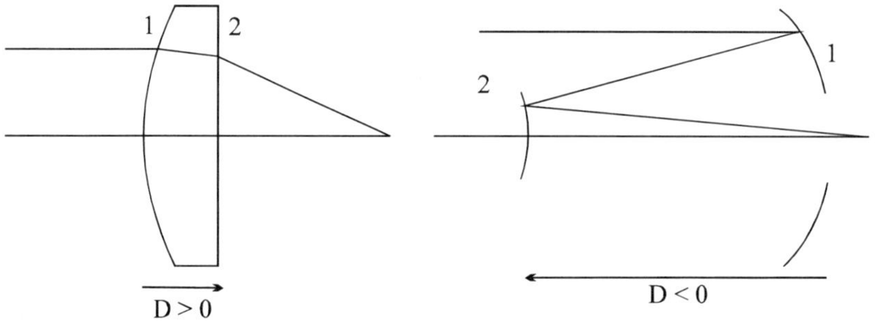

Figure 10.1 Refracting and reflecting separations.

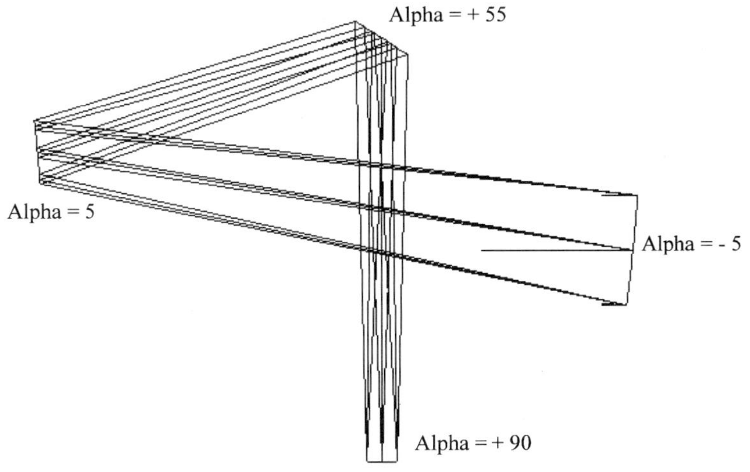

Figure 10.2 Tilted system with plane symmetry.

In the diagram above:

$\alpha_1 = -5$ deg,

$\alpha_2 = +5$ deg,

$\alpha_3 = +55$ deg,

$\alpha_4 = +90$ deg.

The case of three-dimensional decentrations is more complex. We now have to be clear about the order in which the various transformations take place. First, we apply any y- and x-decentrations. (The order of these does not matter.) Then we apply the α-rotation, i.e. rotation about the x-axis, in the same way as in the 2D case. This gives us new y- and z-axes, which we shall refer to as y' and z'. Next we apply the β-rotation, which is a rotation about the y' axis. This gives us new x- and z-axes, which will be referred to as x'' and z''. The sign convention here is consistent with the convention for α.

Consider for a moment a plane mirror with $\alpha = +45$ deg, as shown in Fig. 10.3. Remember that this means that the normal to the surface has a positive gradient. A ray incident along the z-axis will be reflected in the direction of negative y.

Our convention for β is similar, and is shown in Fig. 10.4. A positive β will reflect a ray in the direction of negative x. (Note that this convention for β is not the same as some other optical design programs. Code V, for example, uses a convention for β that is the opposite of ours, but in other respects I believe that the Code V conventions and ours are identical.)

Finally we apply the γ-rotation, which is a rotation about the z'' axis. This gives us new y- and x-axes, which we can refer to as y''' and x'''. The sign convention here is that a small positive γ will transform a point with positive y'' and zero x'' to a point with positive x'''.

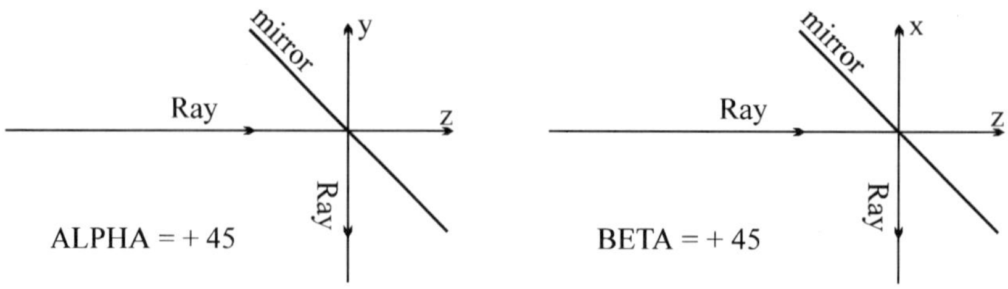

Figure 10.3 Tilted plane mirror.

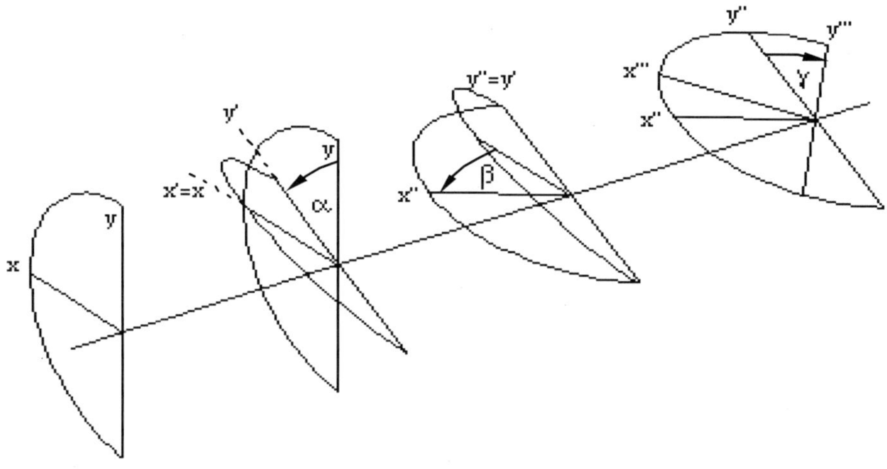

Figure 10.4 α, β, γ rotations.

Optical design programs typically carry out these coordinate transformations in turn and then use standard ray-tracing equations to determine the ray direction after refraction, reflection, or diffraction. The programs then transform back to the global *x*, *y*, *z* coordinate system before tracing through the next surface.

10.3 Interpretation of results

The first thing that the program will do before any analysis or optimization is to determine the "paraxial" back focus. In the case of decentered systems this cannot be done by paraxial ray tracing, so the program traces two real rays. The first ray is incident along the *z*-axis with *x* and *y* initially zero, and the second ray has a small *y*-coordinate and an *M* direction cosine that is chosen so that the two rays come from an object point whose *z*-coordinate is implicitly defined by the paraxial ray data. The point of intersection of these two rays determines the location of the vertex of the image surface.

There are several points to note:

1. If the system has astigmatism, the focal position that is selected will be appropriate for the *y-z* plane, but not for the *x-z* plane.

2. It is assumed that an "axial" ray and a second ray close to it will be transmitted through the system, and will give a sensible image position.

3. The tilt of the image surface is not determined by the program, unless you use the optimization module to optimize for the tilt.

The final transverse ray aberrations are computed in the local plane of the image surface (i.e., in the x''', y''', z''' coordinate system). This needs to be considered when interpreting results if the image surface has decentrations.

10.4 New-axis surface

In some cases, the calculations of global positions of surfaces can become quite difficult, and we have therefore introduced, in SIGMA-2100, the option to define a new axis. A "new-axis" surface is defined by the letter Z in the surface label. The new-axis surface defines the direction of a new axis, with reference to the previous axis, and the position is defined with reference to the vertex of the new-axis surface. In practice, there are some systems that can be set up much more easily with the new-axis surface, while other systems are easier with the global convention. Two examples of the KUTTER system will be shown in Section 10.8. The first example uses the global coordinate system, while the second uses the new-axis system.

10.5 Toroids

Toroidal surfaces are produced by rotating a circle about an axis that does not pass through the center of the circle, as shown in Fig. 10.5. In our programs we assume

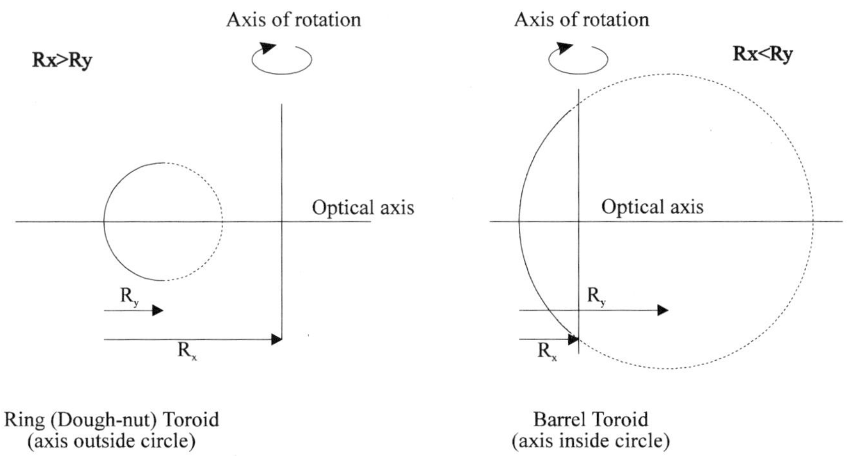

Figure 10.5 Toroidal surfaces.

that the axis of rotation is parallel to the y-axis (although the toroidal surface, like any surface, can be given a γ-rotation). The surface therefore has two principal curvatures, in the y-z plane (c_y) and in the x-z plane (c_x).

If c_x is greater than c_y, then the surface is barrel shaped; but if c_y is greater than c_x, then the surface is doughnut shaped.

Note that it is not enough to specify the two principal curves. You also need to know the direction of the axis of symmetry, which denotes whether the surface is barrel shaped or doughnut shaped.

10.6 Offset surfaces (or off-axis surfaces)

Sometimes we come across an aspheric surface for which the axis of symmetry does not correspond to the mechanical center. For example, as illustrated in Fig. 10.6, sometimes a decentered section of a paraboloidal mirror is used as a collimator, with the advantage that there is no central obstruction.

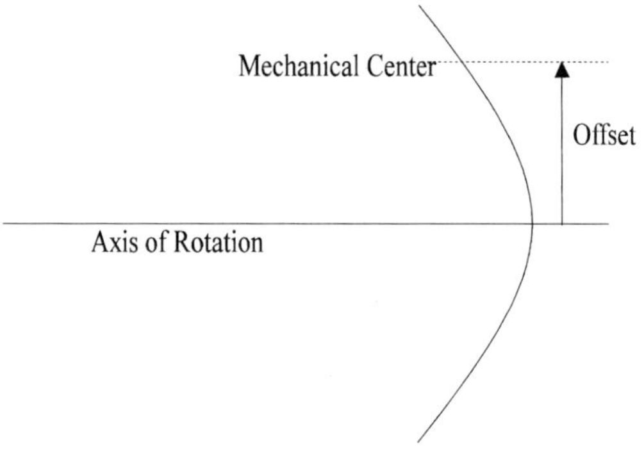

Figure 10.6 Offset surface.

To deal with these cases, we can define an "offset" parameter. This is the distance from the vertex of the surface (i.e., the axis of symmetry) to the physical center of the surface, measured in a direction perpendicular to the axis of rotation. This concept only applies in the case of aspheric surfaces, since they have a unique axis of symmetry and therefore a unique vertex. Spherical surfaces, on the other hand, do not have a unique axis of symmetry (any axis through the center of curvature is as good as another), and so there is strictly no need to use the concept of offsets in the case of spherical surfaces. Having said that, however, there may be cases where it might be more convenient to define a spherical surface as offset rather than decentered.

10.7 Convention for mirrors

Most books on geometrical optics will tell you that we can consider reflection as a special case of refraction by considering that $n' = -n$. This convention can be used

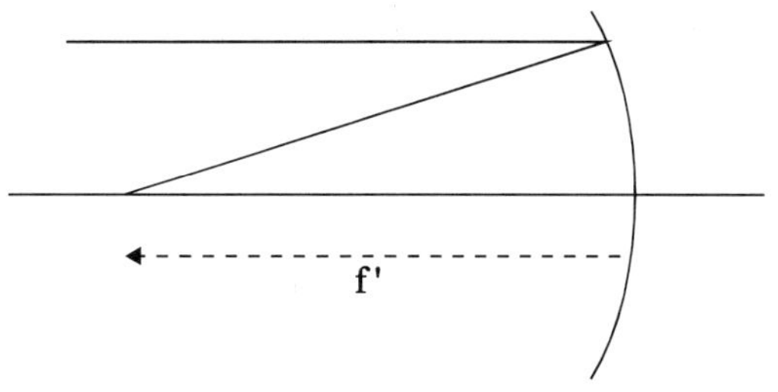

Figure 10.7 Convention for mirror.

in the case of centered reflecting systems, but in the case of decentered reflecting systems it causes great confusion. We suggest that it is preferable to denote the refractive index as positive, and to define the surface as a mirror by including an "M" in the surface label.

Some confusing points remain, however. Suppose that we have a concave (centered) mirror with a negative curvature. The focus is to the left of the mirror in Fig. 10.7, so f' is clearly negative.

However, since we define EFL as f'/n', EFL would be negative if we consider n' to be positive, and it will be positive if we consider n' to be negative. In a simple case like this there is little confusion, but in the case of more complex decentered reflecting systems, the situation will be more difficult to predict. The message is: don't worry too much about whether the EFL is positive or negative.

10.8 Kutter system

The first example, KUTTER, is included to illustrate how to set up systems with decentrations and tilts. In the version of the design shown in Fig. 10.8, we use a simple system of global coordinates.

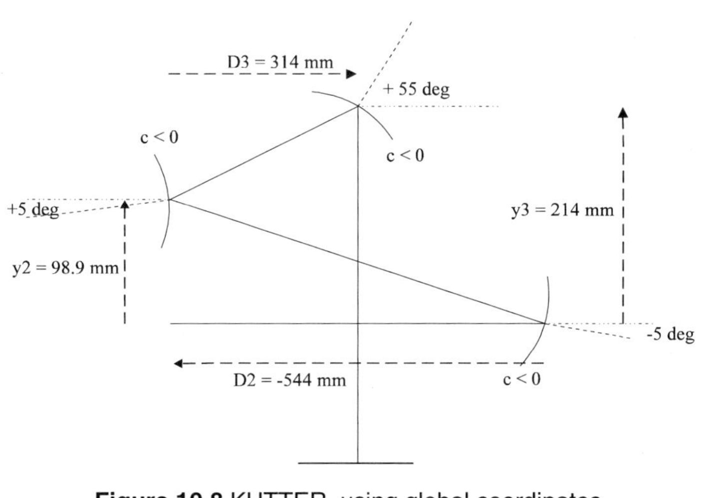

Figure 10.8 KUTTER, using global coordinates.

```
EFL = -1561.414
SURFACES
  # TYPE           CURVE          SEPN      INDEX1     DISPN       CLRAD  GLASS
  1 DSM     -0.000475000     0.00000  1.000000  0.000000    55.289
  2 DSM     -0.000367000 -544.02500  1.000000  0.000000    32.144
  3 DSM     -0.000069000  314.60900  1.000000  0.000000    29.753
  4 DS       0.000000000     0.02500  1.000000  0.000000    13.779

DECENTERED SURFACES
 No        Dx           Dy           Alpha        Beta        Gamma
  1      0.0000       0.0000       -5.1560       0.0000       0.0000
  2      0.0000      98.9940        4.8960       0.0000       0.0000
  3      0.0000     214.1610       55.0520       0.0000       0.0000
  4      0.0000    -212.7495       90.0000       0.0000       0.0000

Defocus =      0.000000
  L         M          X          Y         OPD      COLOR     S      T     DIST(%)
 0.00000  0.00000  0.00000  -0.00000   - AXIAL RAY -
          0.03543           -0.01665  0.00015  0.00000
          0.02478           -0.00583  0.00004  0.00000
         -0.02458            0.00488  0.00003  0.00000
         -0.03508            0.01461  0.00013  0.00000
 0.03533  0.00008 -0.01308   0.00663  0.00009  0.00000
 0.02473  0.00004 -0.00346   0.00325  0.00001  0.00000

CHIEF RAY 0.01087          -13.62096                    -1.800 -2.878
          0.04627            0.06589 -0.00142  0.00000
          0.03562            0.05593 -0.00077  0.00000
         -0.01369           -0.07536 -0.00092  0.00000
         -0.02417           -0.10582 -0.00187  0.00000
 0.03525  0.01094  0.04759   0.00037 -0.00098  0.00000
 0.02468  0.01090  0.03898   0.00018 -0.00051  0.00000

CHIEF RAY -0.01090          13.63205                     1.706  3.205
          0.02457           -0.11081  0.00194  0.00000
          0.01390           -0.07574  0.00094  0.00000
         -0.03551            0.09348  0.00108  0.00000
         -0.04601            0.14704  0.00233  0.00000
 0.03541 -0.01083 -0.07643   0.01294  0.00121  0.00000
 0.02479 -0.01087 -0.04778   0.00634  0.00056  0.00000
```

With the new-axis surface, the same system can be set up in a very different way, as shown in Fig. 10.9. Both methods have advantages, as we will see.

```
EFL = -1561.407                                              File KUTTER3

SURFACES
  # TYPE           CURVE          SEPN      INDEX1     DISPN       CLRAD  GLASS
  1 DSM     -0.000475000     0.00000  1.000000  0.000000    55.289
  2 DSZ      0.000000000     0.00000  1.000000  0.000000    55.344
 ** *** ***  New Axis
  3 DSM      0.000367000  552.95800  1.000000  0.000000    32.134
  4 DSZ      0.000000000     0.00000  1.000000  0.000000    30.898
 ** *** ***  New Axis
  5 DSM     -0.000069000  335.02600  1.000000  0.000000    29.781
  6 DSZ      0.000000000     0.00000  1.000000  0.000000    24.052
```

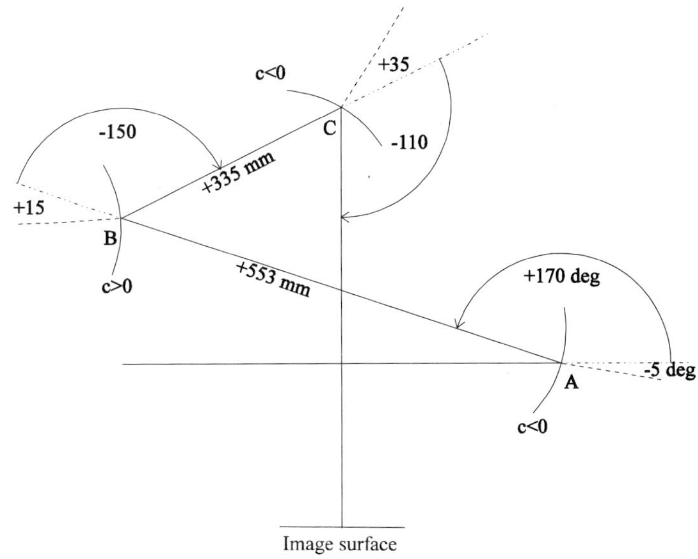

Figure 10.9 KUTTER, using new-axis surfaces.

```
** *** ***  New Axis
 7 DS       0.000000000  426.90581  1.000000  0.000000   13.779

DECENTERED SURFACES
 No       Dx          Dy          Alpha        Beta        Gamma
  1       0.0000      0.0000      -5.1560      0.0000      0.0000
  2       0.0000      0.0000     169.6880      0.0000      0.0000
 **       New Axis    *.****      *.****      *.****      *.****
  3       0.0000      0.0000      15.2080      0.0000      0.0000
  4       0.0000      0.0000    -149.5840      0.0000      0.0000
 **       New Axis    *.****      *.****      *.****      *.****
  5       0.0000      0.0000      34.9480      0.0000      0.0000
  6       0.0000      0.0000    -110.1040      0.0000      0.0000
 **       New Axis    *.****      *.****      *.****      *.****
  7       0.0000     -0.0000

Defocus =      0.000000
  L        M          X          Y         OPD       COLOR    S      T    DIST(%)
  0.00000 -0.00000   0.00000  -0.00000  - AXIAL RAY -
           0.03543             0.01666  0.00015  0.00000
           0.02477             0.00583  0.00004  0.00000
          -0.02459            -0.00488  0.00003  0.00000
          -0.03508            -0.01460  0.00013  0.00000
 -0.03533  0.00007  -0.01308  -0.00662  0.00009  0.00000
 -0.02473  0.00004  -0.00347  -0.00325  0.00001  0.00000

CHIEF RAY  0.01086            13.62090                     1.800  2.878
           0.04627            -0.06589 -0.00142  0.00000
           0.03562            -0.05593 -0.00077  0.00000
          -0.01370             0.07537 -0.00092  0.00000
          -0.02418             0.10582 -0.00187  0.00000
 -0.03525  0.01094   0.04759  -0.00037 -0.00098  0.00000
 -0.02468  0.01090   0.03898  -0.00018 -0.00051  0.00000
```

CHIEF RAY	-0.01091		-13.63199			-1.706 -3.205
	0.02456		0.11081	0.00194	0.00000	
	0.01389		0.07574	0.00094	0.00000	
	-0.03551		-0.09347	0.00108	0.00000	
	-0.04601		-0.14704	0.00233	0.00000	
-0.03541	-0.01083	-0.07644	-0.01294	0.00121	0.00000	
-0.02479	-0.01087	-0.04778	-0.00634	0.00056	0.00000	

After surface 1, which has a tilt of – 5.156 deg, we define a new axis at an angle of +169.688 deg. Note that this is 180 – 2 × 5.156, and it is anticlockwise and positive. The new-axis surface is surface 2. We now have a distance of +553 mm to the second mirror (surface 3); the distance is positive because we are going along the new axis in the direction indicated. Relative to the new axis, surface 3 has a tilt of +15.208 deg, and the new axis has an angle of –180 + 2 × 15.208 = 149.584 deg. Note that this rotation is clockwise and negative. This new-axis surface is surface 4.

Relative to the new axis, the third mirror (surface 5) has a tilt of +34.948 deg, and the rotation to the next new axis is –180 + 2 × 34.948 = –110.104 deg. Again it is clockwise and negative. This new axis is optional, but it has been added for clarity. This new axis surface is surface 6.

10.9 Single parabolic mirror

We sometimes think that the analysis and design of centered systems can be difficult, but decentered systems are even more difficult! This section explains how an ultrasimple decentered system can be analyzed, and compares results from different versions of the same design.

We consider a single parabolic mirror, shown in Fig. 10.10, with rays traced on axis and at $y = +5$, $y = -5$, $x = +5$ and $x = -5$ deg. The vertex curvature of the mirror is –0.005, so its focal length is –100 mm.

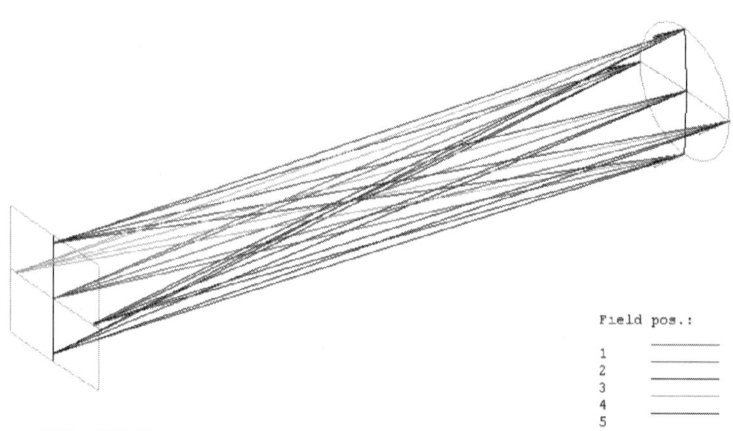

Figure 10.10 Parabolic mirror.

The ray trace below shows that the image height (radius) is 8.74886 mm, and $S = 0$ and $T = 0.760$ mm, at an angle of 5 deg off-axis. Note that $S = 0$ because $S3 = -S4$, in this case.

```
EFL =   -100.000

  # TYPE          CURVE          SEPN      INDEX1     DISPN      CLRAD  GLASS
  1 EM      -0.005000000     0.00000  1.000000  0.000000    10.000
  2 SD       0.000000000 -100.00000  1.000000  0.000000    36.762
CONIC    SURFACE    1 CC =   -1.000000   (PARABOLA)

  L        M          X          Y        OPD      COLOR     S      T    DIST(%)
         0.09975              -0.00000 -0.00000  0.00000
        -0.09975               0.00000 -0.00000  0.00000
         0.00000               0.00000  0.00000  0.00000
  0.09975  0.00000  -0.00000   0.00000 -0.00000  0.00000
 -0.09975  0.00000   0.00000   0.00000 -0.00000  0.00000

CHIEF RAY -0.08716             8.74886                      0.000  0.759

CHIEF RAY  0.08716            -8.74886                      0.000  0.760

CHIEF RAY
 -0.08716  0.00000   8.74886   0.00000

CHIEF RAY
  0.08716  0.00000  -8.74886   0.00000
```

Of course, there is lots of coma, as well as astigmatism, off axis, which shows in the spot diagrams in Fig. 10.11. So far, this system is centered.

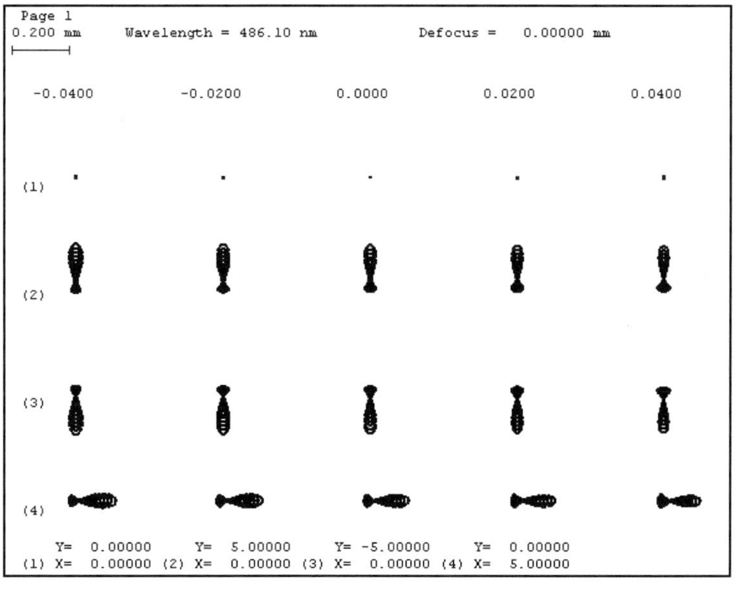

Figure 10.11 Spot diagrams of parabola

10.9.1 Alpha rotations

Now take the same system and set $\alpha 1 = 5$ and $\alpha 2 = 10$. When we ray trace, SIGMA-2100 will trace a ray along the axis and another ray close to it, from the same object point, displaced in the *y*-direction. The intersection of these two rays will be the chosen focus. Note that if there is astigmatism, this process will set the image point at what was the tangential focus on the previous page, because the surface is tilted in the alpha direction. In this case, EFL = –99.619 mm.

```
EFL =    -99.619

SURFACES
  # TYPE          CURVE          SEPN      INDEX1     DISPN      CLRAD  GLASS
  1 EMD      -0.005000000     0.00000  1.000000  0.000000    10.000
  2 SD        0.000000000   -98.09956  1.000000  0.000000    36.762
CONIC   SURFACE    1 CC =   -1.000000  (PARABOLA)

DECENTERED SURFACES
 No       Dx          Dy          Alpha        Beta        Gamma
  1      0.0000      0.0000      5.0000      0.0000      0.0000
  2      0.0000    -17.2976     10.0000      0.0000      0.0000

  L        M         X          Y        OPD      COLOR     S       T   DIST(%)
         0.10035             -0.06572  0.00218  0.00000
        -0.09991             -0.06646 -0.00222  0.00000
         0.00000              0.00000  0.00000  0.00000
 0.09937 0.00065   0.07690   -0.02145 -0.00380  0.00000
-0.09937 0.00065  -0.07690   -0.02145 -0.00380  0.00000

CHIEF RAY -0.08716            8.71500                     -0.007 -0.007

CHIEF RAY  0.08716           -8.71500                     -1.543  1.508

CHIEF RAY
-0.08716  0.00000   8.71500  -0.00000

CHIEF RAY
 0.08716  0.00000  -8.71500  -0.00000
```

How does this EFL relate to the results on the original system? With a single mirror, the EFL is simply the distance from the vertex of the mirror to the focal point. From the first ray trace, it is clear that this distance should be

– SQR(100^2 + 8.74886^2) + T/COS(5)

= – SQR(100^2 + 8.74886^2) + 0.759/.99619 = –99.619 mm.

This agrees with the ray trace results, so the calculation of EFL is consistent with the ray trace of the centered system. Note that the focal length is smaller simply because the *T*-focus is closer to the mirror than the paraxial focus.

We can also calculate EFL from the image distance (98.09956) and *Y* at the image surface (17.2976). This gives 99.613 mm. Note that this is not exactly the same as the printed EFL because the image position is actually found by tracing two close rays, and this is affected by aberrations. Since there is significant coma, the results are slightly different, but the difference is insignificant.

Note also that the image size is the same for all chief rays. Since all of the chief rays are simply reflected at the vertex of the mirror, it is clear that they are reflected back at exactly 5 deg, and the image height is simply 99.613×tan(5), for all four chief rays.

10.9.2 Beta rotations

Now set β1 = 5 and β2 = 10. When we ray trace, SIGMA-2100 will trace one ray along the axis and another ray close to it, from the same object point, always displaced in the *y*-direction. The intersection of these two rays will be the focus. Note that if there is astigmatism, this process will now set the image point at the sagittal focus of the centered system, because the surface is tilted in the β-direction. In this case, EFL is –100.382 mm.

```
EFL =   -100.382

SURFACES
  # TYPE           CURVE          SEPN       INDEX1     DISPN       CLRAD  GLASS
  1 EMD       -0.005000000     0.00000   1.000000  0.000000     10.060
  2 SD         0.000000000   -98.85696   1.000000  0.000000     36.762
CONIC    SURFACE    1 CC =   -1.000000   (PARABOLA)

DECENTERED SURFACES
 No        Dx            Dy           Alpha          Beta          Gamma
  1       0.0000       0.0000         0.0000        5.0000        0.0000
  2     -17.4311       0.0000         0.0000       10.0000        0.0000

  L         M          X          Y         OPD       COLOR     S     T   DIST(%)
  0.00065  0.09937  -0.02196   0.00010   0.00000   0.00000
  0.00065 -0.09937  -0.02196  -0.00010   0.00000   0.00000
  0.00000  0.00000   0.00000   0.00000   0.00000   0.00000
  0.10035  0.00000  -0.14329   0.00000   0.00606   0.00000
 -0.09991  0.00000   0.01077   0.00000   0.00163   0.00000

CHIEF RAY
  0.00000 -0.08716  -0.00000   8.78228

CHIEF RAY
  0.00000  0.08716  -0.00000  -8.78228

CHIEF RAY
 -0.08716  0.00000   8.78228   0.00000

CHIEF RAY
  0.08716  0.00000  -8.78228   0.00000
```

Why is the focal length now larger than 100 mm? Since we are effectively finding the sagittal focus, and since in the original system the sagittal astigmatism is zero, EFL should now be

– SQR(100^2 + 8.74886^2) + S/COS(5)

= – SQR(100^2 + 8.74886^2) + 0 = –100.382 mm,

which is correct. Note that the focal length is larger simply because the *S*-focus is farther from the mirror than the paraxial focus. I originally thought that the difference in focal lengths was because the system is anamorphic, but we can see from the ray trace results on pages 204 and 205 that at least in this case, this was incorrect.

In this example at least, the variation in focal lengths is simply due to the different positions of the *S*- and *T*-focus at different angles. So, if we increase α, the focal length will decrease; but if we apply the same tilt to β, the focal length will decrease, and at large values of α or β, these changes will be very large. There will, I am sure, be other cases where the optical system is anamorphic, and a rotation of the optical system about the *z*-axis will produce a change in focal length that is due to the anamorphic effect. This example, however, does not illustrate this effect.

10.9.3 Alpha and beta rotations

It is interesting to try equal α and β tilts. So set $\alpha 1 = 5$, $\beta 1 = 5$, $\alpha 2 = 10$, $\beta 2 = 10$ and ray trace. Note that *Dx*2 will not be exactly equal to *Dy*2, because the α and β tilts are not equivalent in this case. This is because the α–tilt is about the *x*-axis, but the β-tilt is now about the new *y*-axis, which is rotated 5 deg relative to the global *y*-axis. However, this effect is small at these angles. Since we are almost exactly halfway between the *S*- and *T*-foci, we should get the "best" focus, ignoring coma.

But note that the focal length is now –100.000 mm. This is not a coincidence. Why is it –100 mm?

```
EFL =   -100.000

SURFACES
  # TYPE          CURVE           SEPN      INDEX1     DISPN      CLRAD  GLASS
  1 EMD     -0.005000000     0.00000  1.000000  0.000000    10.000
  2 SD       0.000000000   -96.96103  1.000000  0.000000    36.762
CONIC   SURFACE   1 CC =   -1.000000   (PARABOLA)

DECENTERED SURFACES
 No       Dx           Dy          Alpha         Beta         Gamma
  1       0.0000       0.0000       5.0000       5.0000       0.0000
  2     -17.2966     -17.2308      10.0000      10.0000       0.0000

  L          M          X          Y          OPD        COLOR      S     T     DIST(%)
 -0.00076   0.10129   -0.09828   -0.06611   0.00220   0.00000
  0.00073  -0.09821    0.05537   -0.06588  -0.00223   0.00000
 -0.00067   0.00132    0.00000    0.00000   0.00000   0.00000
  0.09927   0.00424   -0.06396   -0.10017   0.00224   0.00000
 -0.10019  -0.00030   -0.06722    0.05343  -0.00221   0.00000

CHIEF RAY
  0.00065  -0.08583   -0.13238    8.74576
```

```
CHIEF RAY
 -0.00199  0.08846    0.13241   -8.74780

CHIEF RAY
 -0.08781 -0.00001    8.74727    0.13239

CHIEF RAY
  0.08648  0.00263   -8.74628   -0.13238
```

To understand this, remember that the S field is flat. Also, the radius of the Petzval surface will be –100 mm. Remembering that the distances from the Petzval surface to the *T* and *S* image surfaces are in a ratio of 3:1, the distance from the Petzval surface to the "best" image surface will be two times the distance from the Petzval surface to the sagittal surface. Therefore, the image surface that is midway between the *S* and *T* surfaces will have the same curvature as the Petzval surface, but of the opposite sign; i.e. $r = 100$ mm, and this is the image surface that the ray trace has focused on.

This argument assumes that there are no high-order aberrations, and that the effect of an α-tilt and a β-tilt are the same. At large angles these assumptions will not be true, but as we can see, when $\alpha = \beta = 5$, EFL –100 mm. When $\alpha = \beta = 10$, EFL = –99.996 mm, and when $\alpha = \beta = 20$, EFL = –99.793 mm.

10.10 Scanning systems

The "new-axis" surface is very useful in the design of scanning systems. A typical situation is shown in Fig. 10.12. The basic principle is that each configuration represents a different position of the scanning surface, but the amount of computation needed to set up each configuration is reduced to the minimum.

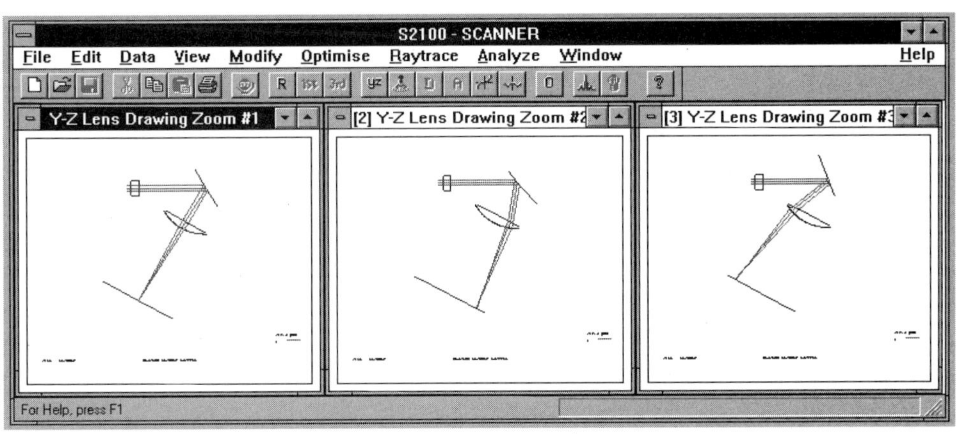

Figure 10.12 Scanning system.

```
ZOOM POSITION 1
EFL =    -43.000

WAVELENGTHS   [nm]          587.60       656.30       486.10

SURFACES
 # TYPE           CURVE          SEPN     INDEX1     DISPN      CLRAD  GLASS
 1 S          0.000000000     0.00000  1.000000  0.000000     4.000
 2 S          0.000000000     4.00000  1.516798  0.008057     4.000  S-BK7
 3 SDZU       0.000000000    50.00000  1.000000  0.000000    10.000
** *** ***    New Axis
 4 SMR        0.000000000   -20.00000  1.000000  0.000000    11.547
 5 SDZU       0.000000000    20.00000  1.000000  0.000000    10.000
** *** ***    New Axis
 6 SDZ        0.000000000   -17.32050  1.000000  0.000000    10.000
** *** ***    New Axis
 7 S          0.000000000    20.00000  1.000000  0.000000    12.000
 8 S         -0.045000000     4.00000  1.516798  0.008057    12.000  S-BK7
 9 SD         0.000000000    42.99980  1.000000  0.000000    20.000

RECTANGULAR APERTURE   4 CLEAR APERTURE X    20.00 CLEAR APERTURE Y    11.55
DECENTERED SURFACES
No       Dx          Dy           Alpha        Beta        Gamma
 3       0.0000     10.0000      30.0000      0.0000      0.0000
**       New Axis    *.****       *.****      *.****      *.****
 5       0.0000      0.0000     -30.0000      0.0000      0.0000
**       New Axis    *.****       *.****      *.****      *.****
 6       0.0000    -10.0000    -120.0000      0.0000      0.0000
**       New Axis    *.****       *.****      *.****      *.****
 9       0.0000     -0.0000
```

For configuration 2, $\alpha(3) = +40$ and $\alpha(5) = -40$. For configuration 3, $\alpha(3) = +20$ and $\alpha(5) = -20$.

The scanning surface is one face of a prism that is rotating at high speed with a known center of rotation. To a first approximation, the optical system can be simulated by simply setting up a centered system and assuming that the aperture stop is placed at the point where the rays are reflected. But this ignores the fact that the point of reflection changes for different scan angles, and more complex data are needed to correctly model the scanner.

A simple way to model scanning systems correctly is shown in Fig. 10.13. Surfaces 1 and 2 are a fixed lens and surface 4 is the reflecting surface, which rotates about point A. Surfaces 7, 8, and 9 represent a fixed imaging lens and the image surface. Surfaces 3, 5, and 6 are dummy surfaces that are used in the construction of this data file. We assume that the scanner is a hexagonal reflecting prism, with a diameter (measured from the centers of opposing sides) of 40 mm. Each face of the hexagon therefore has a clear radius of $20 \times \tan(30) = 11.547$ mm.

We define surface 3 to be a new-axis surface, with its center at the center of rotation; in configuration 1, which will be the reference scan position, it has $\alpha = +30$ deg. The reflecting surface (surface 4) is then –20 mm along this axis; the surface itself has a rectangular aperture, with CLRAD(Y) = 11.547 mm. CLRAD(X) is not specified. We then return to the center of rotation, and surface 5 is a second

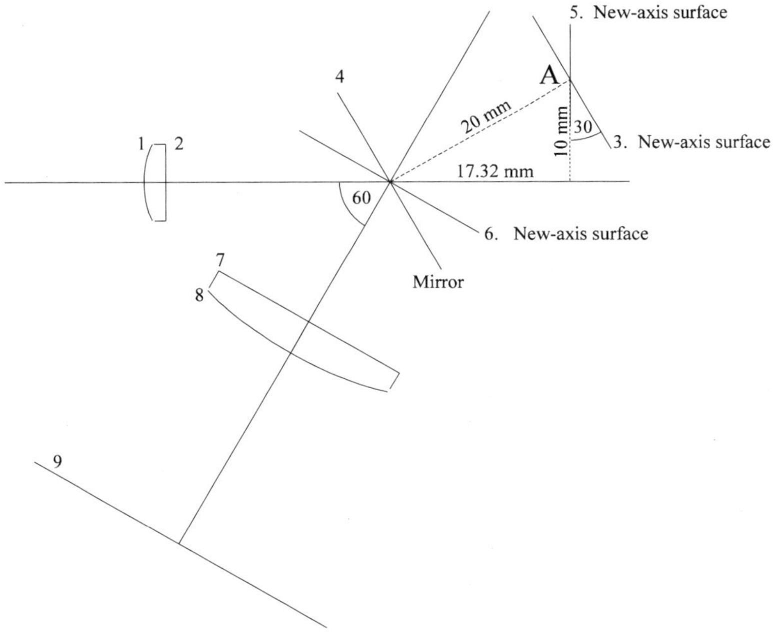

Figure 10.13 Modeling a scanning mirror.

new-axis surface, parallel to the direction of the original axis but displaced from it. Surface 5 therefore has $\alpha = -30$ deg. Surface 6 is a new-axis surface pointing along the axis of surfaces 7, 8, and 9. It has $\alpha = -120$ deg. Surfaces 7, 8, and 9 are centered on this new axis.

To represent different scan angles, we simply change the values of $\alpha 3$ and $\alpha 5$ so that, for example, at scan position 2, $\alpha 3$ is + 40 deg and $\alpha 5$ is – 40 deg. No other changes are needed. Finally, to ensure that all calculations are referred to the same image position, the back focus (Sepn (9)) is picked up from configuration 1 using space pickups, in all other configurations. The paraxial ray labels for all configurations after the first must be "PZN" so that the back focus is not recomputed at each configuration. If the system is to be optimized, note that several separations are negative. Therefore, the "edge" thickness limits for optimization should also be made negative, so that incorrect edge thickness violations do not happen.

10.11 The "active" side of a surface

SIGMA-2100 uses a method of ray tracing that enables the user to define which side of a surface is the "active" side; i.e., which side of the surface is the object. The active side is determined by the direction of the surface normal. This is a vector perpendicular to the surface at the point of incidence, pointing in the positive z-direction, in the local coordinate system for the surface. For surfaces following an even number of reflections and before any reflections, the surface normal should point in the direction of the incident radiation. For surfaces following an odd num-

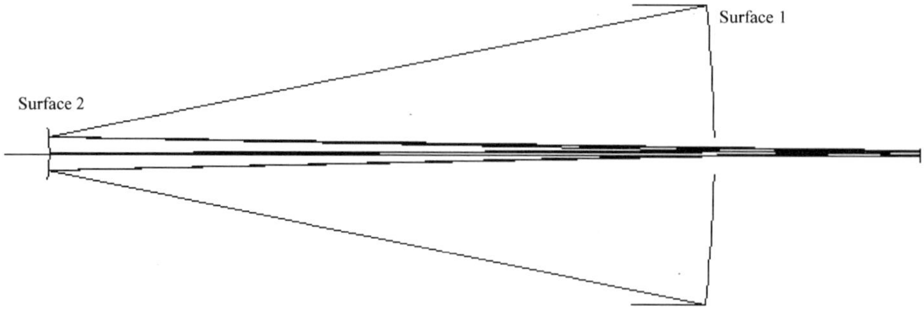

Figure 10.14 Ritchey-Crétien telescope.

ber of reflections, the surface normal should point in the direction opposite to the incident radiation.

With this convention for a simple two-mirror telescope, such as the Ritchey-Chrétien telescope shown in Fig. 10.14, the incident light travels from left to right in the object space. Surface 1 is a centered surface, and therefore its surface normal also points from left to right, which is correct. Surface 2 is also centered and its normal also points from left to right; the incident light is traveling from right to left, which is the opposite direction to the surface normal. Since this surface follows an odd number of reflections (one reflection in this case), this is also correct.

Most texts on geometrical optics say that in ray tracing, the direction of propagation is ignored. In simple ray tracing techniques this is true, but the algorithms that are used in SIGMA-2100 are more complex and do take account of the direction of propagation. The assumption is that in the object space, light is traveling from left to right; every surface must be consistent with this assumption, as far as the active side is concerned. Of course, sometimes we deliberately reverse an optical system, usually so that we can ray-trace from the long conjugate to the short conjugate. This is permitted—indeed there is no way that any program could "know" whether we have done this.

10.12 X-ray telescopes

X-ray telescopes conventionally work at almost grazing incidence, and several types exist. With SIGMA-2100 we have included examples of what are known as "Wolter type-1" and "Wolter type-2" telescopes, although in our examples the angles of incidence have been reduced to make the concept more clear.

10.12.1 WOLTER2 example

Figure 10.15 illustrates a Wolter type-2 system. The active side of surface 1 is on the left; i.e., the surface is concave. The active side of surface 2 is on the right, as in the Ritchey-Chrétien telescope.

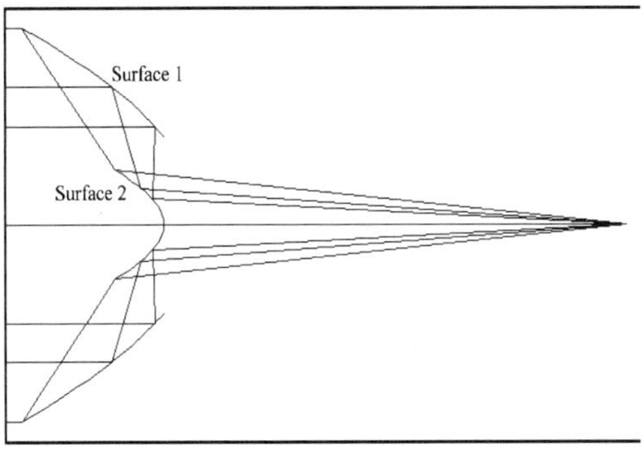

Figure 10.15 Wolter type-2 system.

The fact that some rays are going from right to left after the first reflection, while others are going from left to right is of no significance, although some other optical design programs cannot ray-trace this example correctly!

The data for the Wolter type-2 telescope is a simple centered system, and is shown below.

```
EFL =   9999.618
WAVELENGTHS   [nm]
      587.60        656.30        486.10
SURFACES
  # TYPE          CURVE          SEPN      INDEX1     DISPN      CLRAD   GLASS
  1 S         0.000000000     0.00000   1.000000   0.000000  1000.000
  2 EMI      -0.001934529  1100.00000   1.000000   0.000000  1000.000
  3 EMI      -0.008515643  -201.26300   1.000000   0.000000   245.000
  4 S         0.000000000  2213.02000   1.000000   0.000000     0.000

CONIC    SURFACE    2 CC =   -1.000000  (PARABOLA)
INTERNAL OBSTR   2 RADIUS =  250.000
CONIC    SURFACE    3 CC =   -1.108940
INTERNAL OBSTR   3 RADIUS =     0.000
```

10.12.2 WOLTER1 example

The Wolter type 1 telescope is shown in Fig. 10.16. In this example, surface 1 is a new-axis surface, for reasons explained below, but which are not relevant to the basic point concerning the active side of the mirror surfaces.

```
EFL =    199.997
WAVELENGTHS  [nm]           587.60       656.30       486.10
SURFACES
  # TYPE          CURVE          SEPN      INDEX1     DISPN      CLRAD   GLASS
  1 DSZ       0.000000000     0.00000   1.000000   0.000000   105.000
 ** *** ***   New Axis
  2 EM       -0.116186122   650.00000   1.000000   0.000000   105.000
  3 DEM       0.331895029   -68.60890   1.000000   0.000000    23.588
  4 DS        0.000000000    -1.47202   1.000000   0.000000     0.796
```

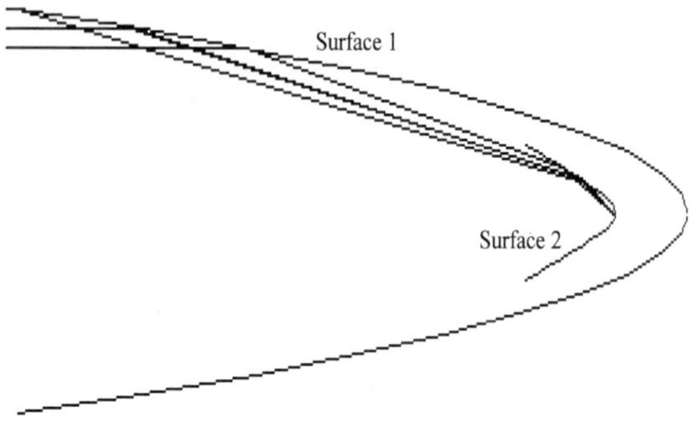

Figure 10.16 Wolter type-1 system.

```
CONIC    SURFACE    2 CC =   -1.000000 (PARABOLA)
CONIC    SURFACE    3 CC =   -1.095900
DECENTERED SURFACES
 No        Dx           Dy          Alpha        Beta        Gamma
  1       0.0000     -95.0000
 **       New Axis      *.****       *.****       *.****       *.****
  3       0.0000      0.0000    180.0000      0.0000       0.0000
  4       0.0000     -0.0000
```

After reflection at the first mirror, which is the concave parabolic mirror, it reflects off the inside of the second mirror; i.e., this surface is concave, not convex. Since light is going from left to right, and surface 3 is an even-numbered reflection, its normal must point from right to left, which means that it must be given an alpha-rotation of +/– 180 deg. Therefore, the sign of the curvature must be reversed, so c is therefore positive. Surface 3 (the second mirror) is a hyperbola, and light is being reflected perfectly from one focus to the other. We can see from Fig. 10.17 that the image is to the left of the mirror, as it must be.

In any decentered optical system, SIGMA-2100 computes the focal length and the back focus by tracing a ray along the initial axis and another ray close to it, displaced in the y-direction. In WOLTER2, a ray along the optical axis and another

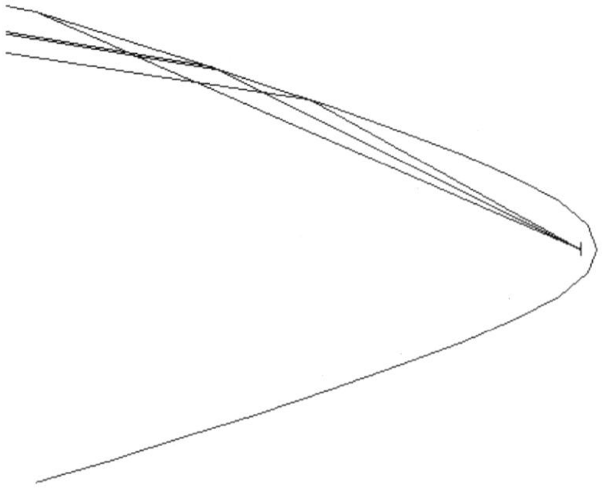

Figure 10.17 Hyperboloid mirror.

ray close to it will successfully find the correct image plane. However, in WOLTER1, a ray along the axis of symmetry will not be transmitted because after reflection at surface 1 the ray approaches surface 2 from the convex (i.e. nonactive) side.

In WOLTER1, rays therefore have to be traced away from the axis of symmetry, so that the rays approach the second surface from the correct side. For this reason, the first surface of WOLTER1 is a new-axis surface, which moves the axis 95 mm below the original axis. The axial ray is therefore a ray that is 95 mm above the axis of the two mirrors, and it is transmitted correctly.

Surface 1 in WOLTER1 also has another but less important function, concerned only with lens drawings. SIGMA plots a ray from a point on the left of the vertex of surface 1, but with a steeply curved surface; this point is to the right of the point of intersection. Therefore, by placing surface 1 a distance 650 mm in front of the first mirror, the rays are drawn from a point on the left of the point of intersection.

What happens if the second mirror is reversed? If the surface is incorrectly oriented, the ray-tracing algorithm will find the wrong intersection point. Remember that a ray intersects a circle or an ellipse or a hyperbola either at two points or none. In any given case, only one of these intersections is correct, so if the orientation of the surface is wrong, the wrong point will be found.

In the case of a parabola, it is possible for there to be only one intersection of a ray with a surface; in this case, if the orientation is wrong, there will be no intersection point at all, and a ray trace error will be generated. Similarly, in the case of a plane there is only one intersection point, and if the orientation of the surface is incorrect a ray-trace error will be generated.

In WOLTER1, if the alpha rotation of the second mirror is removed and the curvature is made negative, the result is shown in Fig. 10.18. What has happened is that the active side is the convex side, and the rays are therefore reflecting off the other arm of the hyperbola.

What happens if the primary mirror has the wrong orientation? A ray parallel to the optical axis intersects a centered parabola at only one point, and if the parabola has the wrong orientation, the ray-trace algorithm will fail to find an intersection point. In this case, SIGMA-2100 will output error messages.

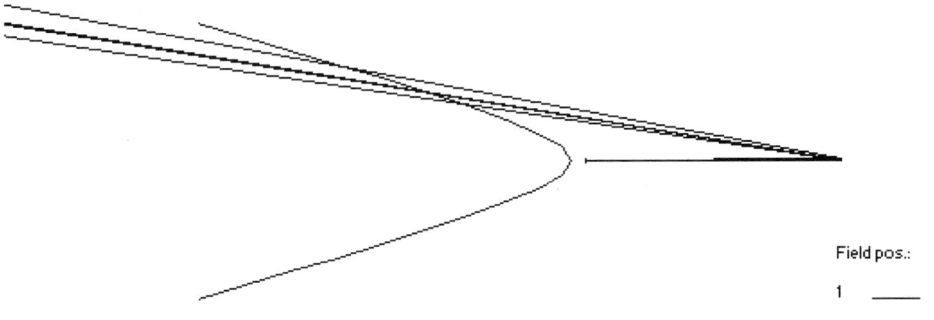

Figure 10.18 Wrong orientation of hyperboloid mirror.

CHAPTER 11
DESIGN FOR MANUFACTURABILITY

We can distinguish between several characteristics that contribute to easy and economical manufacture:

1. Simplicity of design
2. Design for large tolerances
3. Avoidance of unnecessary problems
4. Correct tolerances
5. Reasonable performance specification
6. Optomechanical design.

These topics are, of course, all interconnected, and it is slightly artificial to separate them. In this chapter we will discuss only the first four characteristics. Optomechanical design is really a separate subject, and is not considered here. For those interested, we recommend three references by P. Yoder and R.E. Hopkins.[1,2,3]

11.1 Tolerancing

The fundamentals of tolerance sensitivity calculations and budgeting have been described by W.J. Smith.[4] Optical component specifications have been discussed by several authors, including R.E. Parks.[5] Here we shall only briefly discuss some general aspects of tolerancing.

The performance of most imaging lens systems is sensitive to small changes in the optical parameters. This will have been realized during optimization, when quite modest changes in the parameters generate large changes in performance and where the parameter changes have to be "damped" to keep them approximately within the range of linearity of the differentials of the aberrations. The effect of the manufacturing errors is likely to be disastrous if they are not limited by tolerances.

Tolerances have a significant impact on the cost of manufacture. The aim is to find the optimum compromise between maintaining performance, which implies tightening tolerances, and reducing cost, which implies loosening them. Sometimes, two or more quite different designs will have a similar nominal performance and complexity. A tolerance analysis may show that one design is superior in that it is less sensitive to manufacturing errors.

Tolerancing is not usually a "one-shot" process. On completion of the nominal design, the method for mounting the lens elements and components is not usually known, but analysis of the sensitivity of the system to lens tilts and decenters can give strong indicators to an optimum mechanical design layout. Furthermore, the first analysis may show that performance is likely to be maintained if a few tolerances are tight, and it may be necessary to consult the optical and mechanical workshops to confirm that these tolerances are achievable and that the cost will not be prohibitive. Second and subsequent tolerancing runs will progressively refine the tolerance budget according to feedback from the mechanical designers, and the optical and mechanical production engineers, until the budget represents a true model of a manufacturable lens. Finally, the information gathered during tolerancing will be useful during manufacture when one may have to decide whether elements or components that are outside tolerance can actually be used (possibly in conjunction with other components with "favorable" errors).

Generally, the optical performance of a system is much more sensitive to manufacturing errors than other system requirements (e.g., mechanical constraints). Therefore, the tolerances will be selected primarily with regard to the acceptable variation in optical performance. The optical designer will have designed the nominal system to meet a specification and will have made some allowance for the effect of manufacturing errors. He or she will know which of the aberrations have to be restrained if the specification is to be met. With modern software, tolerancing has become a natural extension of optical design, and it is part of the designer's function.

11.2 Simplicity of design

It is clear that, other things being equal, a simple design will be cheaper than a more complex one, but we must also take care not to underdesign a lens. Ultimately, this becomes a question of confidence and experience, as the penalty for underdesign is worse than the penalty for overdesign, at least in the short-term. In other words, if a lens is not good enough, it cannot be used, in which case the entire design and manufacturing effort is wasted. If it is too good, the end user may be very happy, but it is probably more expensive than it need be.

However, if a lens is just on the limits of the required performance, it is possible that a more complex design with better performance will permit much looser tolerances. An example of this was discussed by M. Kreitzer of Opcon Associates at the Optical Society of America annual meeting in 1990. The paper described the case of a 35-mm *f*/3.5 compact camera lens requiring 30% MTF at 20 c/mm. He assumes the following tolerances:

Glass thicknesses	+/– 0.05 mm
Air spaces	+/– 0.025 mm
Refractive index	+/– 0.001 (normal)
Abbe number (*V*)	+/– 0.8% (normal)

Kreitzer showed that at least four configurations can give the required nominal performance, but some have more depth of focus than others. However, when a tilt of 0.005 mm is assumed, a triplet has almost zero depth of focus, but a quadruplet is very much better. Kreitzer's conclusion was that, for very large quantity production (e.g., in Japan), the quadruplet design is much preferred. Of course, if a lower performance is acceptable, the triplet would be cheaper, and we do find triplets on cheaper cameras.

11.3 Air spaces

When we have a small air space, such as in a doublet, there are four ways of dealing with it. These are listed below in order of preference:

1. Cementing. If the surfaces can be cemented, the refractive index difference between the cement and the glass is probably less than 0.1, whereas the air-glass refractive index difference will be at least 0.5. This results in much larger angles of incidence, as we can see in Fig. 11.1.

   If the Seidel surface aberrations of a cemented doublet are compared with a split doublet, they are always found to be much smaller. The tolerances on the cemented surfaces will therefore be looser than if there is an air space. Although the nominal aberrations might be larger when the lens is cemented, the tolerances still usually make the cemented case preferable. Of course, cementing has other advantages: the lens cannot come out of adjustment, and antireflection coating of the cemented surfaces is not needed (although it has been used on high-index glasses, where the refractive index of the cement is much lower than that of the glass). On the other hand, if glasses are cemented inaccurately, they cannot easily be separated, so no later adjustment is possible.

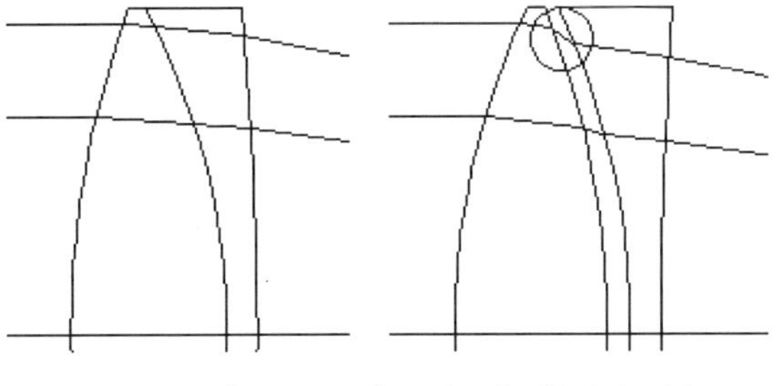

Figure 11.1 Cemented and split *f*/2.5 doublets.

2. Edge contact. If an air space is absolutely necessary, edge contact is preferred to the use of spacers, as long as a reasonable chamfer can be used to avoid a knife edge on the flint component.
3. Thick spacer. If edge contact is not possible, a spacer will probably be used, but it is easier to manufacture if this is thick.
4. Thin spacer. This choice, to be avoided if possible, is the worst case, because a very thin spacer is difficult to make and handle, although sometimes shim material may be used.

11.4 Glass components

We can do several things to make glass components easier to fabricate.

1. Minimum thicknesses. Quite obviously, a lens that is too thin will create manufacturing difficulties. It can distort during polishing, due to inadequate stiffness or uneven heating during the polishing process. Minimum thicknesses, which depend on the diameter of the lens, are suggested as follows:

Diameter (mm)	Minimum center thickness for negative lenses (mm)	Minimum edge thickness for positive lenses (mm)
5.0	1.0	0.5
10.0	1.5	1.0
50.0	5.0	2.0
100.0	8.0	3.0
200.0	12.0	4.0

However, there is a wide variation of opinion concerning acceptable minimums, and the above values are intended as a guide only. Of course, you should remember that any lens must have a clear radius larger than the maximum ray height, to allow for edging and mounting. This has to be taken into account when choosing the edge thickness.
2. Not too thick. Very thick lenses can be expensive, either because of glass cost, or because polishing is more expensive, or both. We should only use thick lenses when there is a real advantage.
3. Monocentric meniscus components. If a meniscus lens has two surfaces that are nearly monocentric, edging to generate a centered component can become very difficult. Remember that lenses are "edged" by rotating the lens about an axis that contains the two centers of curvature of the surfaces, as indicated in Fig. 11.2.

If the surfaces are nearly monocentric, the two centers of curvature become very close to each other, as shown in Fig. 11.3. In this case the axis is not well defined, and a large quantity of glass may need to be removed from the edges to center the element.[6] Such lenses are therefore usually ground and polished so that the surfaces are reasonably well centered before the edging process. This will clearly make the lens more expensive.

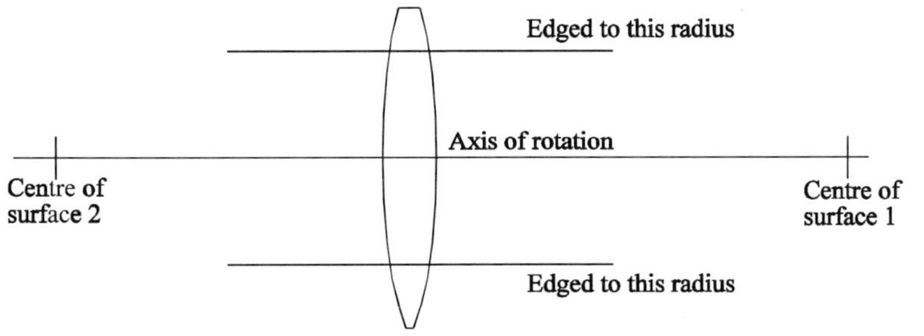

Figure 11.2 Edging a lens.

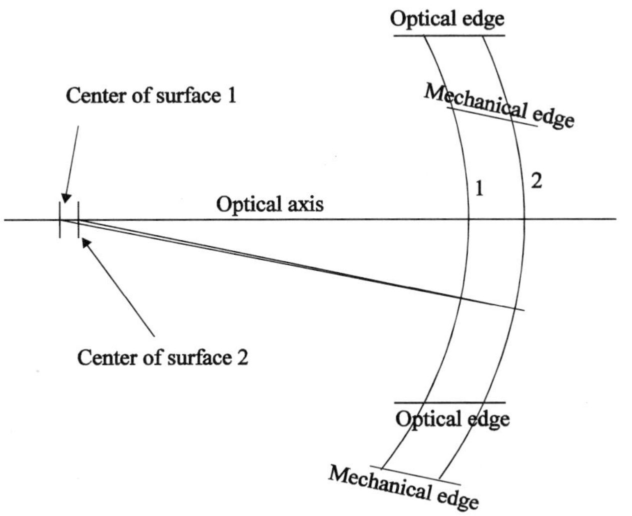

Figure 11.3 Lens with nearly monocentric surfaces.

If the surfaces cannot be made to depart from monocentricity, it may be necessary to split the component into two (obviously more expensive) or perhaps to introduce an air space in the middle of the component (also more expensive). This is often seen in lenses using thick meniscus elements, such as the double Gauss.

4. Shallow curves. When lenses are made in reasonable quantities, it is very common for several lenses to be polished together on a "block." If the curvatures are small, it is possible to polish more lenses at the same time, thus obviously reducing costs.
5. Avoidance of nearly equi-convex or equi-concave lenses. If the two curves on a glass are nearly but not exactly the same, it will be very difficult to distinguish between the two surfaces. In this situation, it is much better to make them identical, so that there is no need to tell them apart. In addition, only one tool and test plate is needed.
6. Plane surfaces. If a surface has a very small curvature, it is almost always better to make it plane if at all possible.

11.5 Glass choice

Two points are worth mentioning here:

1. Staining. Where possible, avoid glasses with poor staining properties.

2. Cost and size. It is obviously true that material cost becomes more important if the glass component is large. Therefore, it is generally desirable for large components to be made of cheap glass, but of course this is not always possible, especially when correcting secondary spectrum. Furthermore, the use of a cheap, low-index glass might force the use of steep curves, therefore increasing higher-order aberrations and making the tolerancing more difficult.

11.6 Mirror surfaces

As we know, in many situations the use of reflecting surfaces in place of refracting surfaces is advantageous. But remember that the required surface figure depends on the refractive index difference at a surface. At a refracting surface, this index difference is typically 0.6, but for a mirror it is 2, and for a rear-silvered mirror it is about 3. Therefore, mirror surfaces must be polished to a better figure than refracting surfaces. On the other hand, a mirror has smaller curvature than a refracting surface of the same focal length.

11.7 Redesign for actual "melt" data

Glass manufacturers provide refractive index data for each batch of glass that they produce, and in some cases it is necessary to redesign for the actual refractive index data. In a prototype lens there is no reason not to do this, but in quantity production, redesign for each melt of glass can become expensive. In addition, redesign for each melt implies that there will be no interchangeability of components between batches, and components therefore need careful identification so that incompatible components are not used together. Therefore, in quantity production, it might be better to specify a higher grade of glass (i.e., a tighter tolerance on refractive index and dispersion) than to require frequent redesigns. However, there will be cases where redesign for the actual refractive index data is essential, even with the best quality glass.

11.8 Use of existing tools and test plates

In almost all cases, it is possible to use existing tools and test plates for most surfaces in a lens. There are several strategies for doing this. In general, we fix a few curvatures to correspond to test plates, and then reoptimize with the remaining curves and some separations. Gradually, all surfaces are made to correspond to test plates; when a large range of test plates is available, it is possible to use existing test plates almost 100% of the time. You should also remember that in many cases, paraxial properties such as focal length do not have to be precisely controlled, and they can often be ignored at the end of this test-plate fitting process, as there may be so few variables that focal length does not change.

11.9 Selective assembly and adjustment after assembly

In some cases, it is desirable to use selective assembly. For example, sometimes the total thickness of a doublet is more critical than the individual thicknesses of the two glasses (a common situation). In this case, we try to use thick crowns with thin flints, or vice versa, by selective assembly. Of course, this process increases costs, but it can be cheaper than specifying a higher precision for the separate components. It has also been suggested that if one component of a cemented doublet is too thin, it is possible to compensate by using extra cement when the two glasses are assembled!

Adjustment after initial assembly is also possible. For example, in a double-Gauss lens, it has been well known for many years that an adjustment of the central air space can be used to control astigmatism without affecting the other aberrations very much. To take another example, some very high quality lenses are made with adjusting screws built into the design, so that the centration of components can be adjusted after initial assembly and testing.

In some circumstances, surfaces might be (accidentally) polished to be slightly toroidal, so that the lens shows axial astigmatism. In these cases, it has been suggested that two astigmatic glasses should be cemented together, with the correct orientation, so that the astigmatism of the doublet is corrected.

11.10 General points

Lenses in which there are large surface aberration contributions usually are more difficult to make than lenses in which the surface contributions are small. The Petzval lens, in which we have two separated doublets, is a good example of a lens that should be easy to make. Each doublet is (or can be) free from spherical aberration and is achromatic; these aberrations, at least, are well distributed across the surfaces of the lens. If it is possible to choose a "relaxed" design type to ease tolerances, this is obviously desirable.

References

1. P. Yoder, *Mounting Optics in Optical Instruments*, SPIE Press, Bellingham, WA (2002).
2. P. Yoder, Ed., *Selected Papers on CD-ROM, Vol. 5, Optomechanical Design*, SPIE Press, Bellingham, WA (1999).
3. R.E. Hopkins, "Lens mounting and centering," in *Applied Optics and Optical Engineering*, Vol. VIII Ed. by R. Shannon and J. Wyant, Academic Press (1980).
4. W.J. Smith, "Fundamentals of establishing an optical tolerance budget," in *SPIE Proc.* Vol. 531 Bellingham, WA (1985).
5. R.E. Parks, "Optical component specifications," in *SPIE Proc.* Vol. 237, Bellingham, WA (1980).
6. W.J. Smith, "The problem of the 'concentric' meniscus element: a possible solution to the lens designer's dilemma," *Optical Engineering* **27** (12), 1039–1041 (1988).

INDEX

A

Abbe sine condition, 121, 133
Abbe V-value, 101, 109, 217
aberrations
  astigmatism, 10, 17, 18, 23, 38-39, 40, 42, 54, 55, 56, 70, 73, 75-77, 80-81, 83, 85, 88, 90, 93, 98, 136, 138, 141, 143, 145, 147, 152, 159, 175, 193, 197, 203-205, 221
  axial color, 5, 53, 163, 175
  balancing, 9
  chromatic, 10, 11, 12, 22, 31, 53, 68, 78, 101, 104, 109, 113, 114, 117, 118, 124, 127, 141, 145, 148, 150-151, 158, 173
  coefficients, 37, 39, 42
  coma, 23, 38-39, 41, 53, 54, 68, 71, 75, 78, 85, 120-123, 146-147, 175, 203
  distortion, 10, 15, 23, 30, 39, 53, 71, 73, 76-77, 81, 83, 87, 90, 93, 159
  field curvature, 3, 17, 23, 31, 38, 57, 70, 73, 75, 81, 89, 98, 155, 163, 173, 175, 177
  fifth-order, 4, 9, 23, 37, 38, 41, 42, 45, 146-147, 149, 151, 155
  higher-order, 9, 37, 52, 119, 164, 220
  lateral color, 5, 53, 71, 98, 111, 115, 135-136, 140-141, 158, 163
  monochromatic, 45, 78, 125
  Petzval sum, 4, 54, 80, 177
  polynomials, 37-38
  third-order, 9, 23, 37, 38, 41, 45, 177
  Seidel, 1, 4, 5, 37, 39, 41, 57, 74, 148, 183, 217
  seventh-order, 9, 32, 42, 45
  spherical, 38, 54, 56, 68, 75, 141, 145, 147, 175
  oblique spherical, 39, 78, 80, 153, 158-159
achromatic doublet, 53, 57, 68
active side of a surface, 209-212
additive damping, 11
afocal system, 54, 121, 183
air spaces, 18, 20, 217-218
angle of incidence, 3, 4, 11, 58
angle of refraction, 4, 196, 198
angular magnification, 8
anti-reflection coatings, 3, 26, 82
aperture stop, 15, 23, 59, 175
aplanatic
  surface, 133-135, 141
  lens, 69
apochromatic doublets, 101-104
apochromatic triplets, 104-105
aspheric
  coefficients, 11, 12, 174
  surface, 95, 127, 128, 167, 177, 198
  element, 62-64, 67, 123
  singlet, 53, 176
astigmatism, 10, 17, 18, 23, 38-39, 40, 42, 54, 55, 56, 70, 73, 75-77, 80-81, 83, 85, 88, 90, 93, 98, 136, 138, 141, 143, 145, 147, 152, 159, 175, 193, 197, 203-205, 221
axial color, 5, 53, 163, 175

B

back focal length, 8, 20, 60-61, 67, 84-86, 89, 136, 175, 181, 193, 196, 209, 212
beam expander, 118, 119, 121, 122, 123
beam waist, 117, 118
beamsplitter, 136, 146, 153, 164
Biogon, 73, 84-85, 89
buried surfaces, 78

C

calcium fluoride, 20, 108, 139, 163
catadioptric, 23, 113, 152, 163-166
catalog components, 67, 68
catoptric, 148, 167
cemented doublet, 24, 65, 75, 187, 217
cemented triplet, 71
cementing, 217
centration, 221
chief ray, 53, 41, 149, 152
chromatic aberrations, 10, 11, 12, 22, 31, 53, 68, 78, 101, 104, 109, 113, 114, 117, 118, 124, 127, 141, 145, 148, 150-151, 158, 173
collimator, 129-130, 198
coma, 23, 38-39, 41, 53, 54, 68, 71, 75, 78, 85, 120-123, 146, 154
compensator, 173-174, 181
concentric, 69, 145
conic, 54
conic constant, 54
confocal parameter, 118
Conrady D-d formula for chromatic aberration, 10, 11, 31
constraints, 12, 19, 23, 25, 31, 34
Contax, 73, 84-85
Cooke triplet, 177
coordinate systems, 194, 196-197, 209
crown glass, 22, 23, 58, 67, 107, 139, 163
curvature, 3, 4, 5, 10, 11, 23, 31, 32, 117, 146
curved field, 142, 145
crystals, 21

D

damped least squares (DLS), 1, 7, 11, 23, 28, 29, 42, 51
damping factor, 7, 26-28, 32-33
defocus, 3, 9, 101, 168
diffraction limit, 126-127, 133, 136, 140, 155, 163, 168
diffractive optics (DOE), 109
dispersion, 12, 20-22, 53, 78, 109-110, 114, 141, 142, 150, 163, 220
    partial, *See* partial dispersion.
disposable camera, 73, 97-99, 174-176
distortion, 10, 15, 23, 30, 39, 53, 71, 73, 76-77, 81-93, 135-145, 154, 194
double Gauss lens, 13-15, 29, 39, 73, 78-81, 106, 125-126, 138, 155, 158, 221

E

endoscopes, 16-19, 54
elliptical coma, 38
entrance pupil, 17, 53, 121
exit pupil, 17, 121, 124
eyepieces, 3, 10, 124, 135-136, 141

F

F-theta lenses, 124-126
f/number, 73, 75, 78, 80, 82, 91, 95, 98, 101, 102, 108, 190
field
    angle, 10, 15, 62, 73, 106, 145, 190
    curvature, 3, 17, 23, 31, 38, 57, 70, 73, 75, 81, 89, 98, 155, 163, 173, 177

flattener, 70, 95, 105, 135-136, 141
lens, 16, 17, 114
flint glass, 15, 21, 23, 67, 136, 159, 163
fluorite, 20, 139
focal length, 8, 20, 30, 31, 54, 60, 61, 74, 91, 115, 173, 190, 193, 206, 221
four-component zooms, 181-188, 190
fused silica, 68, 153, 159, 163

G

Galilean telescope, 118
Gauss doublet
Gaussian beam, 117
geometrical optics, 2, 3, 117, 198
glass boundary conditions, 20
glass choice, 55, 220
glass map, 21, 22
global optimization, 51
global coordinates, 197, 199

H

hologon, 15-16
Hoya glass, 21
hyperboloid mirror, 212-213

I

index of refraction, 4, 20
induced aberrations, 4, 44
intrinsic aberrations, 44
inverse telephoto lens, 115

K

Kutter system, 197, 199

L

Lagrange multipliers, 25
laser diode, 117, 127, 129
lateral color, 5, 53, 71, 98, 111, 115, 135-136, 140-141, 158, 163
Leica, 73, 80, 86
Leitz, 73
Levenberg-Marquardt method, 1, 6
liquids, 21, 108
Lister objective, 133, 135-136, 141

M

magnification, 8, 20, 24, 173, 193
marginal ray, 4, 10, 121, 149, 152
melt data, 220
meniscus singlet, 69
merit function, 2-8, 10, 25-27, 29, 31-34, 45, 47
microscope objective, 108
mirror, 89, 113, 117, 124, 152, 167, 183, 194-199, 204-205, 209-213, 220
modulation transfer function, 2, 3, 73, 76, 77, 81, 83, 85, 88, 90, 93, 95, 97
monochromatic quartet, 13-15
monocentric principle, 145
monocentric meniscus components, 218-219
MTF, *See* modulation transfer function.
multiplicative damping, 7, 11-12

N

new-axis surface, 197, 200-202, 207-209, 211, 213
nonlinearity, 4, 5, 11, 101
normal to a surface, 195, 209-210
numerical aperture, 38, 53, 117-118

O

object space, 53-54
obscuration, 163
offset surfaces, 198
Ohara glass, 21
oil-immersion, 141-143
optical compensation, 181
optical disk objectives, 126-129
optical glass, 20, 35, 108, 115
optical path difference (OPD), 120

optical path length
optimization, 26, 29, 51

P

parabolic mirror, 198, 202, 212
paraxial optics, 52, 175, 193-194
paraxial raytracing, 23, 24, 93, 196
partial dispersion, 102, 108, 159
patents, 51, 52, 74, 78, 93, 106, 107, 126, 127, 174-189
Petzval
  lens, 105, 133, 155
  sum, 4, 54, 80, 134-135, 137, 143, 173-174, 183, 187
  surface, 207
physical constraints, 12, 19, 23
planar, 78-80
plastics, 21, 97, 174-178, 187
point and shoot camera, 73, 95-96
principal azimuth, 193
pupil, 17, 20, 37, 38, 114, 124

Q

quarter-waveplate, 164

R

rangefinder cameras, 84-86
ray
  chief, 41, 53, 149, 152
  paraxial, 4, 12, 24, 41
  marginal, 4, 41, 123
Rayleigh resolution formula, 120, 133, 145
raytracing, paraxial, 23, 24, 193, 196
reflecting surface, 4, 145-146, 208
refracting surface, 4, 114, 145-146, 152, 220
refractive index, 4, 20
relative partial dispersion, 101, 115

S

sagittal
  field curvature, 32, 38
  focus, 205
  oblique spherical aberration (SOBSA), 38, 40, 80-81, 146, 151, 158
scanning systems, 207-209
Schott glass, 20-21
Schupmann lens, 1112-114
Seidel aberrations, 1, 4, 5, 9, 145-146
selective assembly, 221
singlet, 53-55, 58, 68, 75
sign convention, 194-195
sine condition, 121, 133
single lens reflex cameras, 73, 85, 87, 89, 175, 177
skew rays, 1
SLR, *See* single lens reflex cameras.
solves, 23
Sonnar, 82-84
spatial frequency
spherical aberration
sphero-chromatism, 102, 104, 135
starting point, 5, 32, 51, 59
Strehl ratio, 2
Super-Angulon, 73, 86-88

T

tangential
  field curvature, 32, 75
  focus, 204
  oblique spherical aberration (TOBSA), 38, 40, 56-58, 80-81, 146, 151
telecentric, 27, 53, 54, 124, 145, 163
telephoto lens, 73, 91, 92, 93, 95, 108-109, 137, 141
tessar, 42, 73, 75-77
test plates, 219, 221
third-order aberrations, 5, 9
three-component zooms, 179-181
tolerancing, 215-216, 220
toroids, 197-198
transverse ray aberration, 38
transverse secondary spectrum, 115

triplet, 58, 60, 62, 63, 65, 68, 73, 74, 76, 93, 111, 125
two-component zooms, 181

U

unit magnification systems, 17, 18, 59, 111, 145-146, 181, 183

V

variables, 6, 9, 10, 24, 25, 34, 51, 57, 62, 67, 74, 75, 121
variator, 173-174, 181
vignetting, 67, 81
V-value of glass, 21, 101, 102

W

wavefront aberrations, 38, 41, 42, 120, 121, 123, 126, 127
wavefront, relationship to rays, 41,
weights, 31, 34

X

x-ray telescopes, 210

Z

Zeiss Biogon, 73, 84, 85, 89
Zeiss Ikon, 73,
Zeiss Hologon, 15, 16
Zeiss Sonnar, 73, 82, 83, 84
zoom lenses, 1, 59, 78, 108, 115
zoom relays, 188
zoom telescopes, 189

Michael Kidger was born in Birmingham, England, on July 6, 1937. He received scholarships at the age of 17 to Imperial College, London, where he graduated in 1958, and was awarded an MSc in Applied Optics in 1959. He spent a short time in industry with the optical firm Taylor, Taylor and Hobson of Leicester. In 1963 he joined the optical design team at Imperial College under Professor Charles Wynne. In 1966 he accompanied Wynne to work on bubble chamber optics at the Brookhaven Laboratories. In 1967 Kidger was appointed lecturer in the applied optics section of Imperial College—a post he held for 20 years. He retained a part-time teaching post at IC and left in 1987 after an association lasting 33 years. His PhD dissertation, "The Application of Electronic Computers to the Design of Optical Systems, Including Aspheric Lenses," was published in 1971. In 1982 he formed the company Kidger Optics Ltd. with his wife Tina. He was a regular participant and exhibitor at SPIE meetings and served on several SPIE committees, including the Scholarship Committee. He gave optical design courses worldwide up until his death. Michael Kidger died in Australia on February 2, 1998, at the age of 60.